Random Vibrations and Reliability

Proceedings of the IUTAM Symposium, held at Frankfurt/Oder (GDR) from October 31 to November 6, 1982

---

edited by Prof. Dr. sc. nat. Klaus Hennig
Institute of Mechanics
of the Academy of Sciences of the German Democratic Republic

Akademie-Verlag · Berlin 1983

Editor: Prof. Dr. sc. nat. Klaus Hennig

Die Beiträge dieses Bandes wurden vom Originalmanuskript der Autoren reproduziert.

Erschienen im Akademie-Verlag, DDR-1086 Berlin, Leipziger Straße 3-4

Lizenznummer: 202 • 100/416/83
Printed in the German Democratic Republic
Gesamtherstellung: VEB Kongreß- und Werbedruck, 9273 Oberlungwitz
Umschlag: Karl Salzbrunn
LSV 1125
Bestellnummer: 763 268 9 (6786)
DDR 68,-- M

PREFACE

Nonlinear random vibrations and reliability of mechanical systems represent fields in Mechanics to which recently a high amount of research effort is directed in many countries as well as in the German Democratic Republic. So it is quite natural that these topics were the main subjects of this IUTAM Symposium on Random Vibrations and Reliability, held in Frankfurt/Oder, G.D.R., from October 31 to November 6, 1982. It was organized by the Institute of Mechanics of the Academy of Sciences of the G.D.R.

It was a considerable part of the top ranking researchers in these fields, who, coming from Austria, China, Czechoslovakia, F.R.G., India, Italy, Japan, Poland, Sweden, U.K., U.S.A., U.S.S.R., Vietnam and the G.D.R., presented their most recent results and who had spirited and stimulating discussions on a very high level about results, the lines of further development in their own and the colleagues' research fields, on possible and already realized applications etc. The spirit of collegiality and the desire in getting help and helping to further develope this relatively new and exciting field of research made the atmosphere of the Symposium a warm and friendly one.

The lectures dealt with three main topics:

- Reliability of mechanical systems
  Results on reliability in connection with random, especially earthquake and wind excitaions, first-passage failure of randomly excited structures, reliability prediction, applications to the construction of structures in mechanical and civil engineering
- Identification of random systems
  Parameter estimation, structural identification, identifications to the dynamics of machines, vehicles and structures
- Random vibrations
  Forced, parametrically excited and self-excited random vibrations with various nonlinear influences, stochastic stability, narrow-band and wide-band excitations, applications of Markov's theory, comparison of different analytical and simulation methods, applications to mechanical and civil engineering

These Proceedings contain all papers read at the Symposium.

The Scientific Committee of the Symposium consisted of

K. Hennig (G.D.R.) and G. Schmidt (G.D.R.) - chairmen -
A. H-S. Ang (U.S.A.), B. L. Clarkson (U.K.), S. H. Crandall (U.S.A.), M. F. Dimentberg (U.S.S.R.), T. Nakamizo (Japan), L. Pust (Czechoslovakia), S. N. Simanov (U.S.S.R.), W. Wedig (F.R.G.)

The success of the Symposium would not have been possible without the nonabating activity of the local Organizing Committee under direction of H. Friedrich and his assisting co-workers B. Heimann and W. Kaiser and of many other colleagues from the Institute of Mechanics, which can not be mentioned here all by name. I wish to thank all of them for their hard work done in preparing and during the Symposium.

Further I would like to thank the sponsors of the IUTAM Symposium:
The International Union of Theoretical and Applied Mechanics;
The Academy of Sciences of the G.D.R.
as well as the Akademie-Verlag, Berlin, who made publication of these Proceedings possible.

K. Hennig

PARTICIPANTS

H. Aberspach (G.D.R.)
N. Ahlberendt (G.D.R.)
A. H-S. Ang (U.S.A.)
E. Backhaus (G.D.R.)
S. Barthel (G.D.R.)
H.-J. Beer (G.D.R.)
B. Bellach (G.D.R.)
O. Beyer (G.D.R.)
M. Bilý (Czechoslovakia)
F. Casciati (Italy)
F. L. Chernousko (U.S.S.R.)
S. H. Crandall (U.S.A.)
T. Dahlberg (Sweden)
M. F. Dimentberg (U.S.S.R.)
P. Donath (G.D.R.)
M. Engelke (G.D.R.)
L. Faravelli (Italy)
K. F. Fischer (G.D.R.)
L. Fischer (G.D.R.)
U. Fischer (G.D.R.)
U. Fordran (G.D.R.)
H. Friedrich (G.D.R.)
G. Haberland (G.D.R.)
H.-J. Handtke (G.D.R.)
E. Heene (G.D.R.)
B. Heimann (G.D.R.)
K. Hennig (G.D.R.)
F. Holzweißig (G.D.R.)
J. Hult (Sweden)
W. Kaiser (G.D.R.)
L. Knöfel (G.D.R.)
R. Koterazawa (Japan)
F. Kozin (U.S.A.)
V. Kropáč (Czechoslovakia)
C. Lange (G.D.R.)
Y. K. Lin (U.S.A.)
G. Lindemann (G.D.R.)
A. Lingener (G.D.R.)
H. Linke (G.D.R.)
B. Mädiger (G.D.R.)
W. D. Mark (U.S.A.)
G. Mauersberger (G,D.R.)
B. Michel (G.D.R.)
H. Mohr (G.D.R.)
P. C. Müller (F.R.G.)
J. Murzewski (Poland)
T. Nakamizo (Japan)
S. Narayanan (India)
M. Peschel (G.D.R.)
E. Platen (G.D.R.)
M. Pokorný (Czechoslovakia)
A. Renger (G.D.R.)
J. B. Roberts (U.K.)
W. Schiehlen (F.R.G.)
G. Schmidt (G.D.R.)
G. Schröder (G.D.R.)
G. I. Schueller (Austria)
R. Schulz (G.D.R.)
M. Schwaar (G.D.R.)
K. Sobczyk (Poland)
G. Spaethe (G.D.R.)
P-T. D. Spanos (U.S.A.)
L. Sperling (G.D.R.)
H. Stobbe (G.D.R.)
V. Tamuzs (U.S.S.R.)
S. A. Timashev (U.S.S.R.)
W. Tischer (G.D.R.)
M. Türschmann (G.D.R.)
R. Ueckerdt (G.D.R.)
Nguyên Van Dao (Vietnam)
E. H. Vanmarcke (U.S.A.)
B. Vater (G.D.R.)
W. Wedig (F.R.G.)
J. Wicher (Poland)
J. Wittenburg (F.R.G.)
Wei Qiu Zhu (China)
F. Ziegler (Austria)

CONTENTS

KEYNOTE ADDRESS

Jan Hult, IUTAM Secretary General

Herr Professor Dr Grote, Kollegen Hennig und Schmidt, meine Damen und Herren!

Im Namen von Professor Drucker, Präsident der IUTAM, heisse ich Sie alle willkommen zu dieser Tagung, zu diesem Symposium!

On behalf of the President of IUTAM, Professor Drucker, I wish you all welcome here!

Die Akademie der Wissenschaften, DDR, joined our Union as a member in 1973, and we are now going to take part in the first IUTAM Symposium to be arranged in the German Democratic Republic. Each such occasion, when a pin can be put in a new place on the IUTAM globe, is of great importance to our Union because it marks its international scope.

Our topic this week in Frankfurt/Oder will be Random Vibration and Reliability. My teacher and friend, Stephen Crandall, who is present here, reviewed this field of science in an article some 20 years ago. He stated it as the combination of probability theory and applied mechanics.

The first such undertaking was made in physics and resulted in the kinetic theory of gases and the now classical statistical mechanics. These disciplines, however, consider micro-models, whereas here we shall be discussing macro-models. The objects behind our studies may be vehicles, such as ships in rough waves, cars or trucks on rough roads, or airplanes in turbulent wind. But they may also be bridges under fluctuating loads or buildings subject to earthquake loading. Or they may be general systems such as are considered in control theory.

The starting point of our discipline was a paper by a very well known German scientist. In his analysis of Brownian motion in 1905 Albert Einstein's model was a mass particle with damping subject to a fluctuating force. The next significant study appeared not until 1931, when van Lear and Uhlenbeck analyzed random vibration of strings and beams. So we may, perhaps, state that our discipline is only 50 years of age. It is thus a young field, and it shows signs of great vitality.

This was shown clearly in the first IUTAM Symposium on Stochastic Problems in Dynamics, which was held in Southampton in 1976. Ten of you, who are here now, presented and discussed new results there, six years ago. In the mean time you have been active, each of you, in your own research at home. This brings to mind the expression "the invisible college".

The middle of the 17th century was a turbulent time in England. A civil war had shattered much of established ways of life. In Oxford, however, a small group of men used to meet to discuss scientific matters in an informal,unorganized way. Among them were Robert Boyle, the physicist, John Wallis and John Wilkins, mathematicians both, William Petty, the economist, Thomas Willis, doctor of medicine, and Christopher Wren, then active as an astronomer. Later Robert Hooke joined them, as an assistant to show physical experiments.

This group, which did not reveal its existence to authorities, came to be known among friends as "the invisible college". Later on, when order was restored in England, they appeared in public, and soon were transformed to the Royal Society.

What did these men do in their meetings? They presented new results, both theoretical and experimental, but above all they discussed. They asked questions, and they exchanged views informally and freely. I am convinced some of them must later in life have considered these years their best! May this symposium be such an occasion of discussion and exchange of views in a similar spirit!

Our symposium has been prepared by a Scientific Committee and a local Organizing Committee. Let me thank all of you in those two committees for all the good work you have done for IUTAM. I have had the privilege of following some parts of it closely, and I have become convinced that we shall have a fine week here in Frankfurt.

So now, let's get started and let's hope for some good vibration and some random events in the discussions. That will make our symposium a lively one, and I think that in the end it will also give the results of our scientific work more reliability.

SECTION R E L I A B I L I T Y

RELIABILITY OF NONLINEAR - HYSTERETIC STRUCTURAL SYSTEMS TO EARTHQUAKE EXCITATIONS

A. H-S. Ang[1] and Y. K. Wen[1]

A comprehensive model for predicting structural damage to earthquakes is described. The major components of the model include the characterization of the random ground motion of an earthquake, the structural modeling and response analysis considering the nonlinear-hysteretic behavior of the system, and the determination of the seismic hazard function when lifetime reliability is required. The requirements for defining seismic damage and the associated damage-producing potential of future earthquakes are delineated.

INTRODUCTION

When subjected to high intensity earthquakes, structures can generally be expected to suffer some degree of damage. If properly engineered, the resulting damage should not exceed some tolerable limit under an earthquake excitation of specified intensity, or over the life of the structure. In either case, of course, a method for calculating or estimating the degree of damage would be required; furthermore, for the purpose of determining the lifetime reliability, a method for predicting the damage potential of future earthquakes is needed.

In earthquake resistant structures, the failure or collapse would more likely be caused by the repeated oscillatory ground shaking rather than by a single high-amplitude motion. For this reason, the strength of a structure (e.g., the initial yield strength) by itself may not be the proper measure of structural capacity for earthquake resistance. Properly designed (e.g., with sufficient ductility), a structure may have considerable reserve capacity beyond that corresponding to the initial yielding. The dynamic response of structures in this range, of course, will be nonlinear and inelastic. Indeed, the resiliency of the structure is derived from its inelastic and hysteretic behavior.

Therefore, for the purpose of damage prediction, the nonlinear hysteretic behavior of structures must be included or taken into account. Furthermore, as earthquake ground motions may be represented as random processes, nonlinear hysteretic random vibration methods would be required for the response analysis.

MODEL FOR STRUCTURAL DAMAGE PREDICTION

A damage prediction model suitable for the purposes mentioned above is presented. Earthquake excitations and structural responses are modeled as random processes, and the structural and excitation parameters are described by random variables. The model essentially serves to synthesize various analytical elements for assessing the possible degrees of structural damage to a given earthquake, or to the lifetime maximum earthquake. For the latter purpose, assessment of the various levels of seismic hazard is also required. The principal elements of the damage prediction model may be described briefly as follows.

Ground Motion Characterization

The ground motions, particularly the strong motion phase, of a given earthquake may be modeled by a stationary Gaussian random process. In this form, it can be characterized by a power spectral density function, such as the Kanai-Tajimi spectrum

[1]Professor of Civil Engineering, University of Illinois, Urbana, IL 61801, USA.

$$S(\omega) = S_o \frac{1 + 4\beta_g^2(\omega/\omega_g)^2}{[1 + (\omega/\omega_g)^2]^2 + 4\beta_g(\omega/\omega_g)^2} \tag{1}$$

in which:

ω_g = the dominant frequency of the ground motion;

β_g = a band width parameter; and

S_o = a scale factor.

The scale factor, S_o, is related to the peak ground acceleration, a_{max}, through the following relation;

$$a_{max} = (PF)[\int S(\omega)dw]^{1/2} \tag{2}$$

where, PF is the "peak factor," for which average values have been suggested by Vanmarcke and Lai (1980). Since a structure may be weakened or damaged more by a number of stress reversals than by a single high stress excursion during an earthquake, the duration of a strong motion earthquake is, therefore, also important in addition to the amplitude and frequency content of the ground motions. The earthquake duration would be a function of the magnitude as well as distance, and empirical relationship for this function has been proposed (Kameda and Ohsawa, 1983).

Structural Model and Response Analysis

The responses of structures subjected to earthquake loadings are often in the inelastic range, especially when high levels of damage are possible. The restoring force of a structure, therefore, may become hysteretic and deteriorates in stiffness, strength or both, when the number of oscillations is large, as would be the case for long-duration earthquakes. Therefore, analytical models for predicting structural response and damage have to include the capability of describing the hysteresis and degradation of the restoring force. The model recently developed by Wen (1980) and Baber and Wen (1981) is appropriate for this purpose. The essence of the model may be described with a single-degree-of-freedom system (extension to multi-degree-of-freedom systems should be obvious). The single-degree-of-freedom equation of motion is

$$m\ddot{u} + c\dot{u} + g(u,t) = - ma \tag{3}$$

in which the restoring force is

$$g(u,t) = \alpha ku + (1-\alpha)kz \tag{4}$$

where:

k = the pre-yielding stiffness;

α = ratio of post-yielding stiffness to pre-yielding stiffness; and

$(1-\alpha)kz$ = the hysteretic part of the restoring force, in which z is described by the following nonlinear differential equation;

$$\dot{z} = \frac{1}{\eta}[A\dot{u} - \nu(\beta|\dot{u}||z|^{n-1}z - \gamma\dot{u}|z|^n)]. \tag{5}$$

Equation 5 assures that the restoring force is hysteretic (i.e., dependent on the time history of the motion u), with the hysteretic force parameters A, β, γ, n, ν and η governing the amplitude, shape of the hysteretic loop, and transition from elastic to inelastic ranges.

Degradation of the restoring force can be included by prescribing the parameters to be functions of the total hysteretic energy dissipated, see Eq. 7 below, which depends on the severity of the response as well as the number of oscillations. For

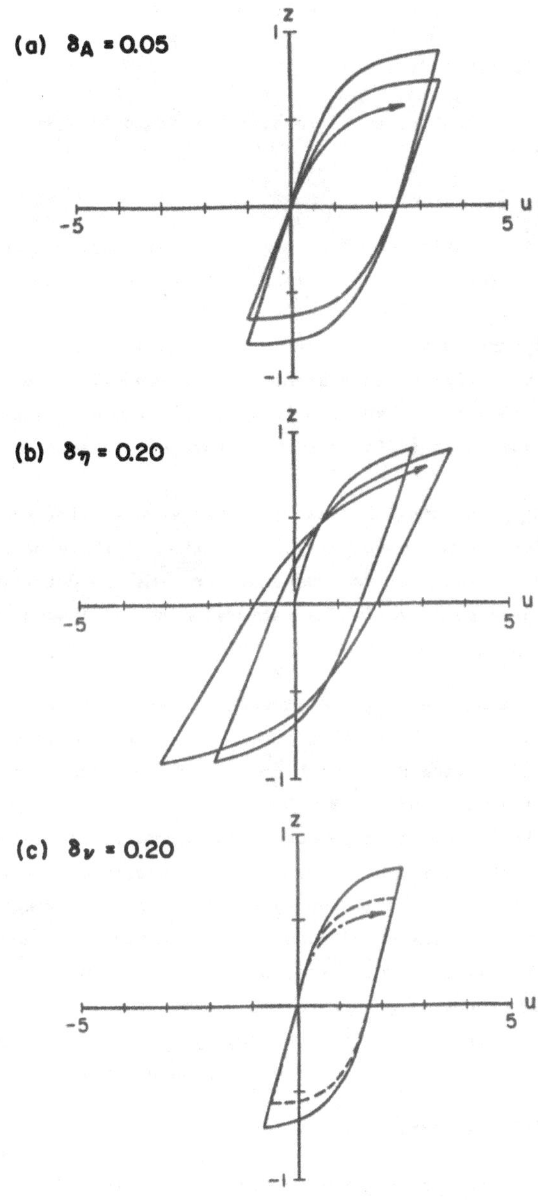

Figure 1 System Degradation Models

example,

$$A(t) = A_o - \delta_A \varepsilon_T(t) \tag{6}$$

gives both stiffness and strength degradation; δ_A is the deterioration rate, and ε_T is the total hysteretic energy dissipated, given by

$$\varepsilon_T(t) = (1-\alpha)k\int_0^t z(\tau)\dot{u}(\tau)d\tau. \tag{7}$$

Similarly, by allowing ν and η to be increasing functions of $\varepsilon_T(t)$; e.g.,

$$\nu = \nu_o + \delta_\nu \varepsilon_T(t), \tag{8a}$$

$$\eta = \eta_o + \delta_\eta \varepsilon_T(t), \tag{8b}$$

one obtains systems with only strength or stiffness degradation, where the parameters ν and η control the degradation rates. Examples of these degradation models are shown in Figure 1.

Depending on the structural configuration and characteristics, the structure as a whole may be modeled as a simple shear-beam, rigid column-flexible beam, or more general discrete hinge system (Baber and Wen, 1982). In the last model, plastic hinges can form in the beam and/or columns; in addition to the story displacements, rotation of the joints are also allowed.

Inelastic Response Analysis. The differential equation for the restoring force, Eq. 4, allows a simple close form equivalent linearization of the equation of motion (Wen, 1980). In essence, a solution for the response requires the determination of the covariance matrix [S] of the response variables satisfying the following matrix equation

$$\frac{d[S]}{dt} = [G][S] + [S][G]^t - [B] \tag{9}$$

in which [G] and [B] are matrices of the structural system and ground excitation parameters, respectively. The stationary solution of Eq. 9 for nondeteriorating systems is obtained iteratively by solving Eq. 9 with $\frac{dS}{dt} = 0$; whereas, the nonstationary solution requires numerical integration of Eq. 9.

The solution of Eq. 9 for the response variable statistics would include the covariance matrix of all the structural displacements and velocities. The power spectral density (in the case of stationary response) can be obtained through an eigenvalue analysis of the matrix [G]. The statistics of maximum response can be obtained by using currently available approximate procedures, such as those based on a Poisson upcrossing assumption or spectral moment (e.g., Davenport, 1964; Vanmarcke, 1975). A response quantity that is particularly useful for predicting potential structural damage is the total energy dissipation $\varepsilon_T(t)$ of Eq. 7. The mean value of $\varepsilon_T(t)$ is

$$E[\varepsilon_T(t)] = \int_0^t E[z(\tau)\dot{x}(\tau)]d\tau \tag{10}$$

in which $E(z\dot{x})$ is one element in the matrix [S]. Evaluation of the variance of $\varepsilon_T(t)$ requires the covariance matrix between two time instants given by

$$C(s,r) = \begin{cases} \Phi(s)\Phi^{-1}(r)C(r,r) & ; \quad \text{for } s > r \\ C(s,s)[\Phi^t(s)]^{-1}\Phi^t(r) & ; \quad \text{for } s \leq r \end{cases} \tag{11}$$

in which Φ is obtained from the matrix differential equation $\dot{\Phi} = G\Phi$, with initial condition $\Phi(0) = I$, the unit matrix, and $C(r,r) = S(r)$.

The mathematical accuracy of the method described above for all response statistics has been verified with extensive Monte Carlo simulation at all excitation levels (Baber and Wen, 1981). A comparison of the root-mean-square (rms) interstory displacements of a 4-story building modeled as a shear-beam is shown in Figure 2, which shows a representative degree of accuracy of the method of analysis.

Definition of Damage

Damage or failure of an inelastic structure subjected to earthquake excitation is difficult to define. In addition to the maximum response commonly used for describing the failure of linear systems, a structure may be weakened or damaged more by a large number of oscillations of the ground shaking than by a single high excursion. In this light, the energy dissipated by a structure, such as through hysteresis, is also important in defining damage or failure. Therefore, structural damage may be described in terms of the maximum response (e.g., the ductility ratio), the energy dissipated, or a combination of the two.

A definition of impending collapse as a function of the ductility ratio and total energy dissipation has been suggested by Banon, et al. (1981) for deterministic excitations. With the nonlinear-hysteretic random vibration method described above, the analytical evaluation of the statistics of these two important variables should permit also the definition of impending collapse and damage under random process excitations.

Lifetime Prediction

Aside from the damage to a given earthquake, the potential seismic damage over the life of a structure is often important for engineering planning and design. For this latter purpose, the damage prediction model must include the determination of all levels of potential seismic hazard over the life of a structure. Consistent with the definition of structural damage described above, the pertinent seismic hazard curve must be a function of the earthquake duration, in addition to its amplitude and frequency content. In this regard, the available seismic hazard evaluation models (e.g., Der-Kiureghian and Ang, 1977) may be used, with appropriate modification, to assess the hazard curves in terms of the maximum spectral intensity (i.e., the maximum ordinance of the appropriate ground motion PSD).

The duration will depend on the earthquake magnitude and distance; however, for the purpose of lifetime damage prediction, its relation with the seismic hazard ordinate (such as the maximum spectral intensity) is required. The necessary relationship will depend on the statistical correlation between the duration and the spectral intensity.

AN ILLUSTRATIVE EXAMPLE

For illustration, consider a three-story reinforced concrete building, with the structural frame shown in Figure 3. This particular building was damaged during the 1976 Tangshan earthquake in China, and subsequently collapsed during an aftershock in the same year. More detailed structural properties of the building may be found in Wei and Dai (1981).

Building Description

The building was modeled as a three degree-of-freedom shear-beam structure. The mass of each story was estimated assuming a live load of 500 kg/m^2 in addition to the dead weight of the structure. Its stiffness was computed by assuming that the concrete

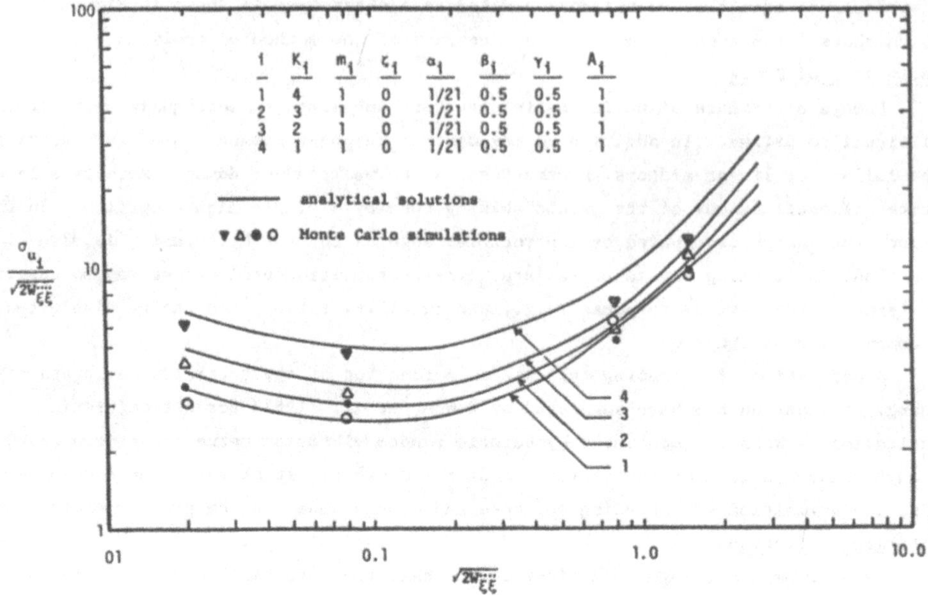

Figure 2 Stationary Response of 4 D.O.F. System, White Noise Excitation

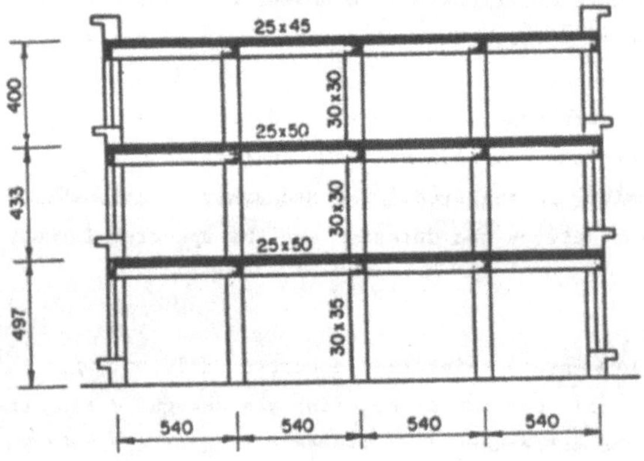

Figure 3 Section of Reinforced Concrete Building Frame

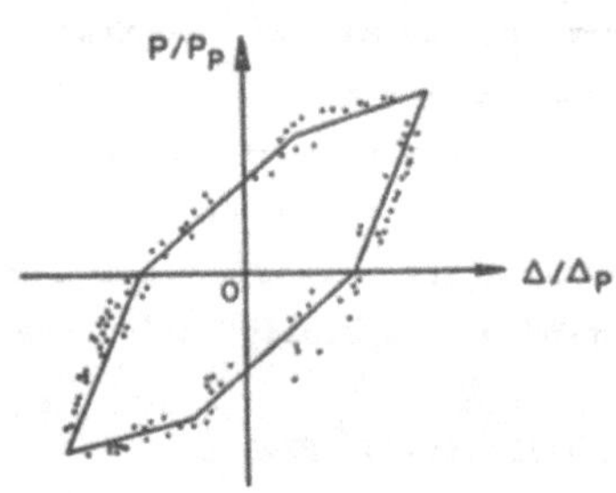

Figure 4 Hysteresis Loop Shape (Reproduced from Zhu and Zhang, 1981)

(with a compressive strength of 150 kg/cm^2) has an elastic modulus of 183 t/cm^2.

The hysteretic model parameters used in the analysis of the building, shown in Figure 3, were obtained by first estimating the maximum moment capacity and corresponding shear force and displacements of each story column. The maximum moment capacities for the column sections were estimated using an equivalent stress block as outlined in ACI (1977) and assuming that the concrete crushes at a strain of 0.003 whereas the steel yields at a strain of 0.002. The normalized hysteretic parameters were then chosen such that the strength of the hysteretic loop remains consistent with the maximum moment capacities of the columns, and the shape of the hysteresis approximates that reported in Zhu and Zhang (1981) and reproduced in Figure 4. The normalized hysteretic parameters and the mass and stiffness of each of the stories of the building in Figure 3 are summarized in Table 1.

The first and second mode natural frequencies of the building are 0.81 cps and 2.23 cps, respectively. A viscous damping of 2% of critical was assumed, whereas the post-yield stiffness was assumed to be zero.

The structure was subjected to a base excitation with peak accelerations ranging from 0.1 g to 0.5 g for a duration of 20 seconds. The excitation was described by the Kanai-Tajimi spectrum, Eq. 1, with a peak factor of PF = 2.5.

Calculational Results

The root-mean-square (rms) response of each story for 10 and 20 seconds are summarized in Table 2. It may be observed that at low levels of excitation, there is only a small increase in the rms response with time as the level of degradation is low. However, at high excitation levels, the rms response of the first story increases considerably with time, whereas there is even a slight decrease in the response of the higher stories. This is more clearly seen in Figure 5.

Shown in Figure 6 are the hysteretic energies dissipated in each of the three stories for an acceleration level of 0.3 g. For comparison, the hysteretic energies dissipated were also calculated for a system without degradations in strength and stiffness. It can be seen that the energies absorbed by the second and third stories level off with time; hence, beyond certain point, almost all energy imparted to the system by the earthquake has to be dissipated through the first story, resulting therefore in large rms response in the first story as indicated in Figure 5.

The effect of degradation is further examined in Figures 7 and 8, which show the hysteretic energy dissipation and degradation of the hysteretic parameter, A, at various excitation levels. For excitations with peak accelerations $\geq$ 0.4 g, the strength and stiffness of the reinforced concrete in the first story would have degraded to virtually zero after 15 seconds of excitation, indicating that the limiting energy dissipation capacity in the first story is approximately 213 t-cm, as shown in Figure 7. Any further energy input, therefore, must be absorbed by the upper stories; actually, beyond this stage the building will have collapsed.

Observations and Comments

It may be emphasized that the actual building was severely damaged in the second floor columns during the Tangshan earthquake of 1976; these were subsequently strengthened and the entire building collapsed during an aftershock due to failures of the first-floor columns. This sequence appears to be consistent with the results indicated in Figure 5

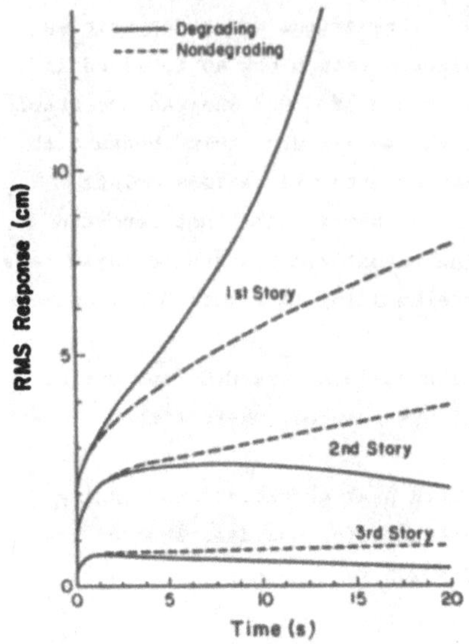

Figure 5 RMS Inter-story Displacement; $a_{max} = 0.3$ g

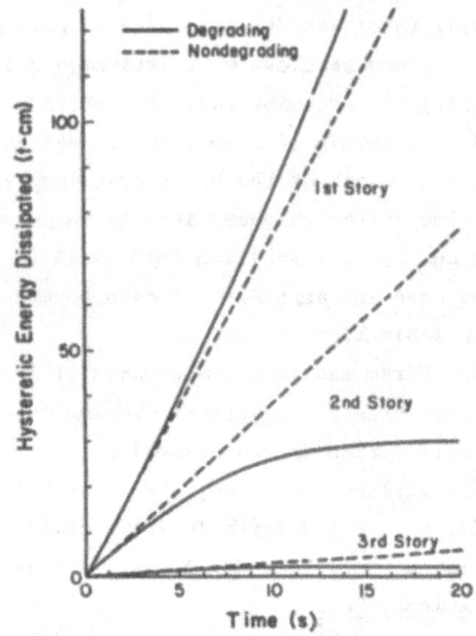

Figure 6 Hysteretic Energy Dissipation, $a_{max} = 0.3$ g

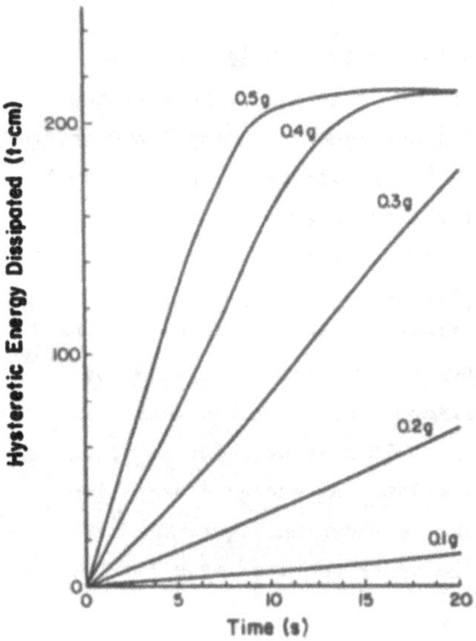

Figure 7 Hysteretic Energy Dissipation in 1st Story

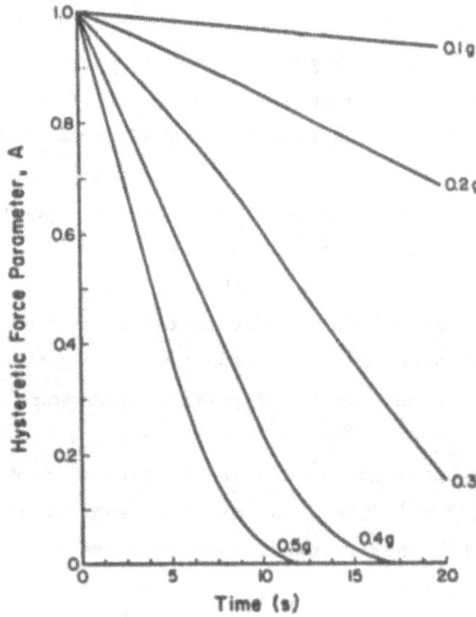

Figure 8 Degradation of Hysteretic Parameter A with Time

of the analytical results presented here, and serves as a partial validation of the model described herein.

CONCLUSIONS

Some degree of damage is usually unavoidable when structures are subjected to earthquakes; therefore, prediction of structural damage is necessary for proper earthquake resistant design. An analytical model for this purpose is proposed and described. A more complete indicator of structural damage requires a function of the hysteretic energy absorbed and the maximum structural distortion (or ductility ratio). Such a two-variable damage function needs further additional developments; however, for practical purposes, one-parameter damage function in terms of the hysteretic energy or the maximum distortion may be more useful. The studies and developments leading toward these objectives have been presented and illustrated herein.

ACKNOWLEDGMENTS

The studies reported herein are part of a research program at the University of Illinois supported by the Civil and Environmental Engineering Division of the National Science Foundation under Grant CEE 80-02584. The calculations for the illustrative example were performed by Y-H. Kwok, Research Assistant, and X-X. Tao of the Institute of Engineering Mechanics, Harbin, China, while visiting as a Scholar at the University of Illinois in 1982.

REFERENCES

ACI Standard 318-77, "Building Code Requirements for Reinforced Concrete," American Concrete Institute, 1977.

Abrams, D. P. and Sozen, M. A., "Experimental Study of Frame-Wall Interaction in Reinforced Concrete Structures Subjected to Strong Earthquake Motion," Civil Engineering Studies, Structural Research Series No. 460, University of Illinois, May 1979.

Baber, T. T. and Wen, Y. K., "Random Vibration of Hysteretic Degrading Systems," Journal of the Engineering Mechanics Division, ASCE, Dec. 1981.

Baber, T. T. and Wen, Y. K., "Stochastic Response of Multistory Yielding Frames," Earthquake Engineering and Structural Dynamics, 1982.

Banon, H., Biggs, J. M. and Irvine, H. M., "Seismic Damage in Reinforced Concrete Frames," Journal of the Structural Division, ASCE, Vol. 107, No. ST9, Sept. 1981.

Bartels, R. H. and Stewart, G. W., "Solution of the Matrix Equation AX + XB = C," Communications of the ACM, Vol. 15, No. 9, 1972.

Davenport, A. G., "The Distribution of Largest Value of a Random Function with Application to Gust Loading," Proceedings Institute of Civil Engineers, London, Vol. 28, 1964.

Der Kiureghian, A. and Ang, A. H-S., "A Fault-Rupture Model for Seismic Risk Analysis," Bull. of Seismological Soc. of America, Vol. 67, No. 4, August 1977, pp. 1173-1194.

Kameda, H. and Ohsawa, I., "Effect of Ground Motion Duration on Seismic Design for Civil Engineering Structures," Memoirs of Faculty of Engineering, Kyoto University, Japan, Vol. 45, Part 1, January 1983.

Vanmarcke, E. H., "On the Distribution of the First-Passage Time for Normal Stationary Random Processes," Journal of Applied Mechanics, Vol. 42, Mar. 1975.

Vanmarcke, E. H. and Lai, S-S. P., "Strong-Motion Duration and RMS Amplitude of Earth-Quake Records," Bull. of Seismological Soc. of America, Vol. 70, No. 4, August 1980.

Wei, Lian and Dai, Guoying, "A Survey of the Collapse of a Three-Story Reinforced Concrete Framed Structure in a Factory during the Tangshan Earthquakes," Earthquake Engineering & Engineering Vibration, China, Vol. 1, No. 1, June 1981.

Wen, Y. K., "Equivalent Linearization of Hysteretic Systems," Journal of Applied Mechanics, Transaction ASME, March 1980.

Zhu, Bolong and Zhang, Kunlian, "A Study of Restoring Force Characteristics of Reinforced Concrete Flexural Members with a Constant Axial Load," Earthquake Engineering & Engineering Vibration, China, Jan. 1981.

TABLE 1: STRUCTURAL PARAMETERS OF BUILDING

Story	Stiffness per Column (t/cm)	Mass (t)	Hysteretic Model Parameters A_o	β	γ	δ_A
1	1.92	829	1.00	0.16	0.09	0.0047
2	1.82	815	1.00	0.15	0.09	0.0042
3	2.32	426	1.00	0.33	0.18	0.0187

TABLE 2: R.M.S. INTERSTORY DISPLACEMENT RESPONSE (cm)

Story	Time (s)	Acceleration (g) 0.1	0.2	0.3	0.4	0.5
1	10	1.4	3.2	8.6	36.3	142.5
	20	1.5	4.7	37.8	330.2	657.2
2	10	1.1	1.9	2.6	2.7	3.0
	20	1.1	2.0	2.2	2.6	3.1
3	10	0.3	0.5	0.5	0.5	0.5
	20	0.3	0.5	0.4	0.4	0.5

ANALYSIS AND APPLICATIONS OF NON-STATIONARY RANDOM PROCESSES FOR RELIABILITY ESTIMATION

Matej BÍLÝ, Jozef ČAČKO*

SUMMARY

The paper deals with non-stationary processes, whose statistical characteristics (probability density and power spectral density) change in time. If the reason for non-stationarity can be separated, such processes are analyzed as stationary ones. Otherwise their statistical characteristics are obtained either in a classic form for a set of repeated realizations or in an evolutionary form, the sliding start of analysis creating a parameter. It is shown how they can be used for laboratory testing and the algorithms reproducing the time-dependent probability density of ordinates or power spectral density are derived for the on-line process simulation.

INTRODUCTION

The reliability estimation of complicated mechanical systems represents a rather complicated problem which in the up-to-date evolution of scientific engineering disciplines has undoubtedly gained a leading place. Because most structures are exposed to random environmental loads, investigations are being carried out all over the world with the aim to find correlations between the environmental load effects and resulting fatigue life, which becomes the limit state in more than 70 percents of service breakdowns.

Applications of random process theory to reliability (mainly fatigue) problems started about 50 years ago with the well known Gassner spectrum, representing a transformed random process to a set of sinusoidal cycles. Although this approach has been successful in many instances, it contains many drawbacks and apart from the fatigue area has no other use.

A further qualitative step was done when computers came to a general use and so the random process theory could be brought to its engineering applications. Systematic measurings and analyses of environmental processes have shown, however, that some processes exhibit time-dependent trends and other long time changes which cannot be neglected, because they often have an adverse effect on the structure examined.

*Institute of Materials and Machine Mechanics of the Slovak Academy of Sciences, Bratislava, Czechoslovakia

This has evoked an increasing attention paid to methods which could take into account such time-dependent properties, creating a nature of non-stationary processes, appearing in more than 50 percents of practical cases.

ANALYSIS OF NON-STATIONARY RANDOM PROCESSES

Supposing a long enough realization of a random process x(t) is available. If its non-stationarity properties are evident due to mean or variance changes in time, there is no need to prove it by a test of non-stationarity. Practically, however, this is a simple procedure for a numerical analysis [1] and so it is useful to find an objective justification for more complicated computations, compared with the stationary process approach. It is worth mentioning here that a standard test can discover the non-stationarity in mean or variance but not in the correlation function (although the non-stationary variance also reflects this property).

Having come to the opinion that the analysed process possesses clear non-stationarity properties, two possible ways of its analysis can be adopted (Fig. 1)[1]

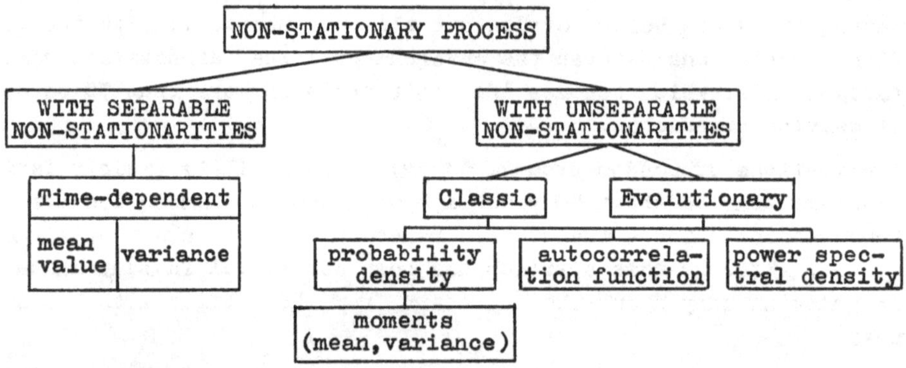

Fig. 1. Possible ways of analysis of non-stationary processes

[1] It should be stated that non-stationary processes represent such a new area of investigations that it is not surprising to find unidentical definitions and characteristics of their properties in the literature. One must be, therefore, careful when applying results derived by various authors.

The first way relies on the possibility to separate a reason for the process non-stationarity either by subtracting of a time-dependent component or by its filtering. This in fact means that the non-stationary process is supposed to have a form

$$x(t) = y(t) + A(t) \qquad \text{or} \qquad x(t) = B(t)\, y(t),$$

where $y(t)$ is a stationary process and $A(t)$, $B(t)$ are time-dependent deterministic functions, slowly changing relatively to $y(t)$. If $A(t)$ or $B(t)$ can be identified and separated from $x(t)$, the subsequent analysis deals with a stationary process $y(t)$ only.

Practical experience proves that such a separation is relatively easily carried out if the time-dependent mean value is concerned. But even in this case the whole procedure is not unique and opinions on the most suitable separable function can substantially differ. This is why this approach hardly gives adequate results without a wide experience, knowledge of a physical reason for non-stationarity and also without a quick possibility to compute and verify the effect of such separation. Frequently the functions $A(t)$ and $B(t)$ are not identified but simply neglected, which can, naturally, lead to various errors in the statistical characteristics to be estimated.

The second way of non-stationary random process analysis does not suppose any preliminary properties and the computed statistical characteristics implicitly contain time as a parameter (this means that they are not invariant). In this case one can either follow an assumed mathematical model of a physical phenomenon and try to identify its parameters (which is probably a more promising way than the next one), or estimate all possible statistical properties without any presumed model with the hope to use at least some of them afterwards.

Irrespective of this choice one should decide which sort of characteristics are to be selected in order to describe the non-stationarity properties. According to Fig. 1 the oldest classic characteristics are based on the statistical behaviour of a set of repeated realizations obtained under identical conditions (Fig. 2). Depending on properties of the k-dimensional distribution function $F_k(x_i, t_i, i=1,2,\ldots)$ (in the scope of the correlation theory, however, $k = 1,2$) one obtains various non-stationary processes with a time-dependent mean value

$$m(t_j) = \frac{1}{n} \sum_{i=1}^{n} x_i(t_j),$$

or/and variance

$$M^2(t_j) = \frac{1}{n-1} \sum_{i=1}^{n} \left[x_i(t_j) - m(t_j)\right]^2 ,$$

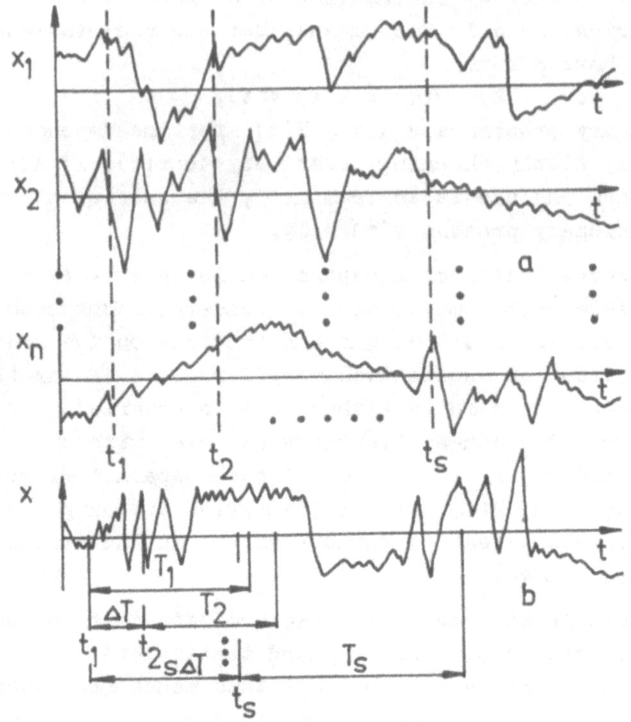

Fig. 2.
Scheme of classic (a) and evolutionary analysis (b)

or/and probability density function

$$f(x,t_j) = \lim_{\substack{\Delta x \to 0 \\ i=1,2,\ldots,n}} \frac{P[x < x_i(t_j) \leq x + \Delta x]}{\Delta x}$$

or/and autocorrelation function

$$R(t_j,\tau) = \frac{1}{n}\sum_{i=1}^{n} \bar{x}_i(t_j)\, \bar{x}_i(t_j + \tau),$$

or/and power spectral density

$$S(f,t_j) = 2\,\Delta t\,[R(t_j,0) + 2\sum_{p=1}^{s} R(t_j,\tau) \cos 2\pi f \Delta t]$$

or/and with some other parameters ($\bar{x}$ denotes a centred process).

This classic approach to non-stationary process analysis has, however, serious drawbacks from the point of view of its practical applications. The most serious one is a monitoring of a number of process realizations (at least ten) obtained under identical environmental conditions (the extreme is an earthquake process), computation of averaged characteristics over these realizations etc.For these and other reasons a new approach to the non-stationary random process analysis has been elaborated based on their evolutionary properties. A logical step from the classic statistical characteristics to the evolutionary ones resembles to the introduction of the regodic property of stationary processes with one principal difference, however: the statistical characteristics of an ergodic process do not depend on time, whereas the evolutionary statistical characteristics of one realization of a non-stationary process are functions of time (practically they depend on the starting point of sampling and analysis). In this way one gets non-stationary processes with a time-dependent evolutionary

mean value

$$m_e(t) = \frac{1}{N_e} \sum_{j=1}^{N_e} x(t_j),$$

or/and variance

$$M_e^2(t) = \frac{1}{N_e} \sum_{j=1}^{N_e} [x(t_j) - m_e(t)]^2$$

or/and probability density function

$$f_e(x,t) = \lim_{\substack{\Delta x \to 0 \\ s \leq t}} \frac{P[x < x(s) \leq x + \Delta x]}{\Delta x}$$

(P denotes the probability of occurrence of a realization x(t) in a given interval)

or/and autocorrelation function

$$B_e(t,\tau) = \frac{1}{N_e} \sum_{j=1}^{N_e} \bar{x}(t_j)\, \bar{x}(t_j + \tau)$$

or/and power spectral density

$$S_e(t,f) = \sum_{p=-k}^{k} B_e(t,p\Delta t)\, e^{-i2\pi f p \Delta t}\, \Delta t ,$$

where e = 1,2,... denotes the e-th analysed part of a process x(t) with the length T_e, $\tau = p\Delta t$, k - integer number, Δt - interval of sampling, N_e - number of the e-th part process ordinates (obviously $T_e = N_e \Delta t$), j = 0,1,2,... corresponds to the j-th discrete time interval.

It is to be noted, however, that these expressions contain the parametric time implicitly only. In other words, this means that the result of this analysis is either a sequence of numbers (means, variances) or a set of functions (probability density functions, autocorrelation functions, power spectral densities), from which one can try to determine an explicit function of time. As usually it is not easy and this is why the final result is presented in the form of discrete points or functions with parameters corresponding to the starting point of their computation. As an example Fig. 3 shows the probability density and power spectral density of a non-stationary process obtained for the discrete time intervals ΔT. If this interval is too coarse, one could further use the spline polynome approximation [4] for example, to get a finer description of these non-stationary functions. Such an approach is quicker (and at the same time accurate enough) than calculation of probability densities and power spectral densities for more intervals ΔT.

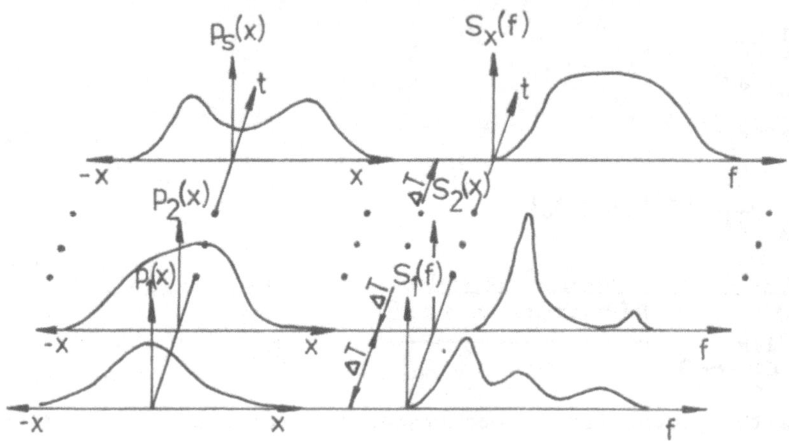

Fig. 3. Non-stationary fields of probability densities and power spectral densities obtained for discrete time intervals ΔT (sliding start of computation)

From the practical point of view the evolutionary statistical characteristics have certain advantages compared with the classical ones. The main of them consist in analysing one realization only without any need to repeat measuring.

One should understand, however, that different approaches yield different results and the evolutionary and classic characteristics are, in general, incomparable.

APPLICATIONS OF NON-STATIONARY RANDOM PROCESSES

Although the problem of reliability estimation of mechanical systems has become accute during the last twenty years, its important part, i.e. fatigue strength, has been dealt with for more than hundred years. Despite of immense information gained about this subject, a direct application of the random process theory to practical reliability problems is very seldom, even if only stationarity properties are supposed. Apart from a few theoretical proposals of fatigue damage accumulation hypotheses exploiting some stationary statistical characteristics, the use of random processes (stationary or non-stationary ones) for <u>calculation</u> of reliability characteristics is at its beginning (not considering the counting methods of transformation of random processes to blocks of sinusoidal cycles).

The recent advent of electrohydraulic loading machines equipped with digital computers and close loop control has caused, however, that the random process theory has taken a large step to its engineering application in the reliability and especially fatigue field, giving a real chance to verify <u>experimentally</u> the fatigue and dynamic properties of structures under conditions corresponding to environmental random (stationary or non-stationary) loading processes. This approach presupposes, however, that the monitored environmental process can be linked up to a driving signal of a loading system, or in other words, can be simulated in a convenient way. Because the computer controlled loading systems are used not only to generate ordinates of random processes and to realize them on the actuator but also to read and compute certain characteristics, to output data and to make decisions in <u>real time</u>, the simulation must be as quick as possible not to keep the machine waiting for the next command.

Although the ideal universal method of simulation should guarantee the identity of all parameters available (say, in the scope of the correlation theory), many theoretical, technical, economic circumstances as well as "the testing philosophy adopted" caused that various simulating algorithms have been developed. In what follows some results derived by the authors for non-stationary processes will be given.

Simulation of probability density of ordinates

The algorithm of simulation of a non-stationary random process x(t) respecting its ordinate probability density content depends on the form of this information. Now we shall suppose that the probability density function has an <u>analytical form</u> f(x,t). Then the simulation can be based on finding the relationship

$$x(t) = g\,[y(t),\, a(t),\, b(t), \ldots], \qquad (1)$$

where y(t) is a stationary random process and a(t), b(t),... are non-stationary random processes or deterministic functions, respectively (excluding constants and stationary functions, leading to simulation of stationary processes).

Let us show two examples of how Eq. (1) can be used, simulating the Gaussian and Weibull non-stationary probability densities $f_G(x,t)$ and $f_W(x,t)$, respectively.

The Gaussian probability density function has the form

$$f_G(x,t) = [(1/\sqrt{2\pi})M(t)] \exp [- (x - m(t))^2/2M^2(t)] \tag{2}$$

and the Weibull one

$$f_W(x,t) = A(t)[x - a(t)]^{m(t)} \exp\left\{ - \frac{A(t)[x - a(t)]^{m(t) + 1}}{m(t) + 1} \right\}, \tag{3}$$

where m(t), M(t), A(t) and a(t) are deterministic or non-stationary functions.

After the transformation in the first case

$$y(t) = [x(t) - m(t)] /M(t) \tag{4}$$

and in the second case

$$y(t) = \frac{A(t)}{m(t) + 1} [x(t) - a(t)]^{m(t) + 1} \tag{5}$$

one gets

$$f_G(y,t) = f_G(y) = (1/\sqrt{2\pi}) \exp (- y^2/2) \tag{6}$$

or

$$f_W(y,t) = f(y) = \exp (-y) . \tag{7}$$

Both processes y(t) are, therefore, stationary with a Gaussian normalized distribution (zero mean and unit variance) and exponential distribution, respectively. The corresponding simulating relationships are then obtained by using the inverse transformation of Eqs. (4) and (5), i.e.

$$x_{Gi} = M(t) \alpha_i + m(t) \tag{8}$$

or

$$x_{Wi} = a(t) + [\frac{m(t) + 1}{A(t)} \beta_i]^{[1/(m(t) + 1)]} , \tag{9}$$

where α_i and β_i are normally and exponentially distributed random numbers, resp.

If the parameters m(t), M(t), A(t) and a(t) are deterministic functions, they are simply computed in the discrete time intervals $t_i = i \Delta t$ and substituted to the simulating formulae (8) and (9). More complicated a situation occurs, however, when they represent non-stationary processes. In this case they have their own parameters (deterministic or also non-stationary), which can be further computed or simulated etc. In such a way one gets various non-stationary processes with complicated composed probability density functions.

In a similar manner one can get any non-stationary random process with a simple or composed distribution. Although the generation of random numbers with various probability density functions slows down the computation of process ordinates it is still fast enough for standard computers allowing simulation up to 50 cps.

When the time-dependent probability density function was obtained in the form of a two-parametric field (similar to Fig. 3), one simulates the non-stationary process part by part, supposing different probability density functions for each time interval ΔT. These partial stationary probability densities are simply simulated using, for example, transformation tables from the uniform to a required experimental probability density prepared and stored in the computer memory beforehand.

A finer procedure has been developed for the calculation of moments when one process simulation should be changed to the subsequent one not waiting until the time ΔT elapses [3].

Simulation of power spectral density

There are again two possibilities how to simulate a non-stationary random process with a given time-dependent power spectral density S(t,f). The first one supposes that S(t,f) has an analytical form and enables simulation of Gaussian processes only.

Let us suppose that the non-stationary random process y(t) is obtained as the output of the linear differential equation with the non-stationary input x(t) or in the convolutory form

$$y(t) = \int_0^\infty h(\xi)\, x(t-\xi)\, d\xi \, . \tag{10}$$

The corresponding autocorrelation function then becomes

$$R_y(t,\tau) = \left\langle \int_0^\infty h(\xi)\, x(t-\xi)\, d\xi \int_0^\infty h(\eta)\, x(t+\tau-\eta)\, d\eta \right\rangle , \tag{11}$$

where $h(\xi)$ is the unit impulse response function and the brackets $\langle . \rangle$ stand for the mean value operator.

Using the Wiener-Khinchin relationship one gets the expression for the power spectral density $S_y(t,f)$. If it is further supposed that a deterministic function $\psi(t,\xi)$ can be found, for which the following relationship is valid:

$$R_x(t-\xi,\vartheta) = \psi(t,\xi)\, R_x(t,\vartheta) \tag{12}$$

and denoting

$$k(t,\xi) = \psi(t,\xi)\, h(\xi) \, , \tag{13}$$

the expression for the power spectral density $S_y(t,f)$ can be expressed as

$$S_y(t,f) = H(f)\,K(t,f)\,S_x(t,f) = H(t,f)\,S_x(t,f), \qquad (14)$$

where H(f) is the system transfer function and

$$K(t,f) = \int_0^{\infty} k(t,\xi) \exp(\,i2\pi f\xi)\,d\xi. \qquad (15)$$

Providing the functions $S_y(t,f)$, $S_x(t,f)$ and $\psi(t,\xi)$ are known, the unknown unit impulse response function h(t) of the dynamic system can be determined. Because usually lim h(t) = 0 for $t \to \infty$, one obtains the final simulating formula

$$y_i = \sum_{j=0}^{k} c_j\, x_{i-j}\,. \qquad (16)$$

If the non-stationary power spectral density is again expressed in the form of a <u>two parametric</u> evolutionary <u>field</u> (Fig. 3), the simulation is carried out part by part. In such a way it is possible to simulate both the probability density and the power spectral density.

CONCLUSIONS

Applications of non-stationary random processes to reliability estimation of mechanical structures belongs nowadays to the most difficult problems of engineering sciences. They require a well trained working team equipped with a modern measuring and processing technique. It is natural that this type of a complicated approach will be used for the most expensive and critical components, conditioning to a large extent the service reliability and safety.

All methods mentioned hitherto have been programmed for the MTS fatigue machine with the PDP 8/e computer. The speed of generation of random numbers with required properties ranges between 50 and 700 numbers per second. This is fully acceptable for most practical tests.

REFERENCES

1. Bendat J., Piersol A. G.: Measurement and Analysis of Random Data. John Wiley, New York, 1967.
2. Bílý M., Bukoveczky J.: Digital simulation of environmental processes with respect to fatigue. Journal of Sound and Vibration <u>49</u>, 1977, 551-568.
3. Čačko J.: About the identification of unstable stochastic processes. Stroj. Čas. <u>32</u>, 1981, No. 1 (in Slovak).
4. Hiroshi A.: Smooth surface fitting, the SFCFIT subroutine. Communications of the ACM <u>17</u>,1974, No.1.

A SIMPLIFIED RELIABILITY APPROACH IN STOCHASTIC NON-LINEAR DYNAMICS

Fabio Casciati[(1)] and Lucia Faravelli[(1)]

The reliability of random non-linear mechanical systems subject to non-stationary stochastic excitations is investigated. Two different approaches are presented. The first procedure, whose operative aspects are the object of research still in progress, in only sketched; it is founded on the theory of non-linear regression. The second approach, simpler and hence less accurate than the first procedure, is discussed in some detail. This second method is founded on the assumption of a multiplicative model for the carrying capacity of the system, each factor of this multiplicative scheme being independent of each other, lognormal distributed and contributed by single sources of uncertainty. A numerical example of the simplified approach is provided with reference to a plane frame subject to seismic excitation.

1. Introduction

A stochastic non-stationary excitation of practical interest is characterized by short pulses of non-zero magnitude, each pulse being a realization of a non-stationary stochastic process and each random interarrival time being much longer than the random duration of pulses. A mechanical system subject to an excitation of this type can be analysed in the time domain by investigating its properties at the end of each pulse for given initial conditions |7|. Obviously, if the original initial conditions are restored at the end of each pulse by repairing the system, the analysis of the system becomes easier. In particular, the reliability of the system coincides with the probability of survival of the system subject to a single excitation pulse. An analytical solution, however, is not available either for this simple kind of reliability analysis problem. Classical random vibration theory cannot be applied and there is no alternative to simulation approaches |4|, especially when non-linear mechanical systems are analysed.

This paper considers the problem of assessing the reliability of a non-linear mechanical system during a single excitation pulse. The solutions proposed in the following do not avoid numerical experiments, but attempt to rationalize their planning and to obtain a complete exploitation of their results.

2. Governing relations for mechanical systems

Consider a discrete or discretized mechanical system subject to a single-parameter stochastic excitation W(t) of random duration d. At any instant of time t in (0,d), the system is governed by the incremental equilibrium equation |1|:

$$\underline{C}(t,\underline{\theta})\ \underline{\Delta u}(t) = \underline{\Delta q}(t,W,\underline{\theta}) \qquad (01)$$

(1) Department of Structural Mechanics - University of Pavia - Italy

where $\underline{\Delta u}(t)$ is the vector of the displacement increments in the time increment Δt and the matrix $\underline{C}$ and the vector $\underline{\Delta q}$ are the effective stiffness and load terms. For quasi-static loading, $\underline{C}$ and $\underline{\Delta q}$ are the stiffness matrix at time t and the increment of the load vector q(t) from t to (t+Δt) respectively. When dynamic effects are not negligible, $\underline{C}$ and $\underline{\Delta q}$ also include inertial and damping contributions according to a selected time-integration scheme. The vector $\underline{\Theta}$ (of entry Θ_i) denotes all the uncertain quantities of interest for the definition of the system. Each set of homogeneous quantities $\underline{\Theta}^j$ may be regarded as the product of a random variable Θ_{oj} by a stochastic vector $\underline{\Gamma}^j$

$$\underline{\Theta}^j = \Theta_{oj}\, \underline{\Gamma}^j \tag{02}$$

each Γ_i^j being independent of each other. A product of the type (02) can be introduced to describe the yielding properties of the material (j=1) and the limit values $\underline{R}^*$ in the local failure criterion (j=2). For other quantities, as masses, damping ratios, stiffnesses and non-yielding parameters of the material constitutive law (j=3,...,6 etc), the effect of the stochastic terms $\underline{\Gamma}^j$ may be negligible (i.e. $\underline{\Gamma}^j$=1) and, hence, the random variable Θ_{oj} may be sufficient for describing the relevant source of uncertainty. According to equation (02), a separate form may be adopted also for the load function W(t):

$$W(t) = Y\, W_s(t) \tag{03}$$

where Y is a random intensity parameter and $W_s(t)$ is a non-stationary stochastic function. For sake of simplicity, each realization $w_s(t)$ of $W_s(t)$ is thought, in the following, to be a segment of length d of a stationary random process modulated by a suitable non-constant intensity function. It follows that the process $W_s(t)$ is defined by the duration d, the intensity function and the random parameters of the power spectral density function relevant to the associated stationary process.

The failure of the system is defined by a critical inequality |1| that may be expressed either in kinematic terms

$$u_i(t) \geq u_i^* \tag{04}$$

or in terms of local rupture

$$R_i(t,\, u(0,t)) \geq R_i^* \tag{05}$$

where u_i^* and R_i^* are some prescribed limit values. In the sequel, u_i^* will be thought to be deterministic, whereas R_i^* is thought to be random with properties derived by experimental data. In equation (05), u(0,t) denotes the displacement history in the interval (0,t) and R_i is a suitable combination of damage indicators to be compared with the limit value R_i^*. The first ultimate limit state (equation (04)) will be denoted as "progressive failure" while the achievement of the limit state associated with equation (05) will be called "brittle failure". In the latter case, in fact, the analysis of the system stops as soon as first local rupture occurs, while the analysis of the system degraded by previous local ruptures may be required before limit state (04) is reached. If the left hand-sides of equations (04) and (05) are non-monotonic functions of the time, the system survives if it survives at every time. In view of reliability calculations, the time axis can be discretized into a collection of critical instants, which can be regarded as element assembled in series: their failure, in fact, is actually sufficient for the system to fail |13|. A more attractive fashion, however, may be the formulation of critical inequalities in terms of quantities that are monotonic functions of time, as

$$u_i^o(t) = (\max u_i(t),\ \text{in } (0,t)) \geq u_i^*, \tag{04'}$$

$$R_i^o(t) = (\max R_i(t, u(0,t)), in\ (0,t)) \geq R_i^*. \qquad (05')$$

Thus the system survives if it survives at t=d, where reliability has to be assessed. Given a specific value $\underline{\theta}$ of the vector $\underline{\Theta}$ and a realization $w_s(t)$ of $W_s(t)$, all sources of uncertainty are removed and a deterministic indicator $K_Y(y)$, function of the intensity, may be introduced for each ultimate limit state |9|: $K_Y^P(y)$ is 0 for all intensities y such that the system does not enter progressive failure and 1 for all y for which the system does it; $K_Y^B(y)$ is 0 for all intensities such that local rupture does not occur and 1 for all y that involve at least one local rupture. These two indicators are not necessarily monotonic functions of y. As an example, consider the plane frame of Fig. 1 subject to the ground motion $w(t) = y\ w_s(t)$, where $w_s(t)$ is the well-known El Centro 1940 N-S accelerogram normalized by setting the peak acceleration to the gravity acceleration. Since the higher nodes on the left-hand side of the frame reach displacements greater than the prescribed limit for y=0.5, one obtains $K_Y^P(0.5)=1$. By constrast, for y=0.64, two first-floor beam elements fail instead of the third floor external column and no displacement exceeds the prescribed limit: therefore $K_Y^P(0.64)=0$. Similar considerations on the location of most damaged zones hold also for brittle failure and lead one to analogous conclusions on the non-monotonic nature of the function $K_Y^B(y)$. |3|.

If $(\underline{\Theta}|y,w)$ denotes the random vector $\underline{\Theta}$ conditional on Y=y and $W_s(t)=w_s(t)$, the probability of failure of the system is given by

$$P_f(y|w_s) = \underset{\Theta|y,w_s}{E} [K_Y(y)]. \qquad (06)$$

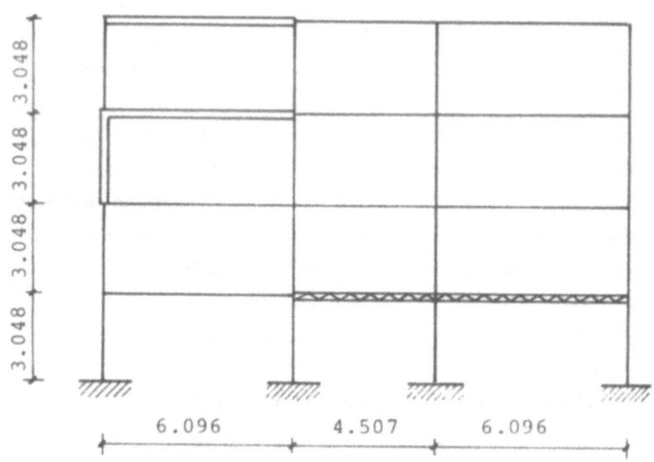

Figure 1 - Plane frame under investigation. The elements ═══ fail when the structure is excited by the El Centro accelerogram with peak acceleration scaled up to 0.5g; the elements ⩘⩘ fail when the peak acceleration is scaled up to 0.64 g. Lengths are expressed in meters.

Function (06) is called the fragility curve of the system given the excitation $w_s(t)$. The unconditional probability of failure, i.e. the unconditional fragility curve, is

$$P_f(y) = \underset{w_s|y}{E} \left[P_f(y|w_s) \right]. \tag{07}$$

The fragility curves $P_f(y|w_s)$ and $P_f(y)$ are assumed in the sequel to be monotonic functions of the intensity y |9|. Therefore, $P_f(y|w_s)$ and $P_f(y)$ can be regarded as the cumulative distribution functions $F_{Y_R|w_s}$ and F_{Y_R} of the auxiliary variables $Y_R|w_s$ and Y_R, which have a clear meaning of resistances if and only if also $K_Y(y)$ is a monotonic function.

In principle fragility curve (07) can be found by repeated simulation of $(W_s|y)$ and $(\underline{\Theta}|y,w_s)$, followed by deterministic system analysis. Two procedures can be pursued. For every y, a sample value of $K_Y(y)$ is obtained by deterministic analysis of every system defined by simulated values of W_s and $\underline{\Theta}$: the average of the sample values calculated for the different systems simulated provides an estimation of the probability of failure $P_f(y)$. Alternatively, for every simulated sample $(w_s,\underline{\theta})$, one can search by deterministic analysis the minimum value y_R of y that causes failure. Then, fragility curve (07) is obtained by estimating the cumulative distribution function of Y_R from the sample of values y_R obtained for the different systems simulated. This approach, correct when $K_Y(y)$ is a monotonic function of y, gives results on the safety side when $K_Y(y)$ is not monotonic. Finally, convolution |1| of the unconditional fragility curve with the probability density function of the seismic intensity parameter Y provides the actual probability of failure of the system.

3. Fragility models

The computation of fragility curves by simulation presents several inconsistencies. Accuracy of the result requires a large number of numerical time-history analyses, but the relevant computational effort is either unbearable or unjustified by the rôle of fragility curves in reliability assessment. Further, the results of a simulation process are meaningful "per se", but cannot be employed in order to forecast the response of a system whose data are perturbed. Besides, these results do not permit one to separate the uncertain quantities having significant influence on reliability from the quantities whose uncertainty does not alter the safety of the system.

Hybrid approaches making use of analytical fragility models relating Y_R with the basic uncertain quantities $\underline{\Theta}$ and W_s may be more convenient for reliability analysis purposes |4|. The parameters of the fragility model adopted have to be determined by numerical investigation.

3.1 Regression model

For the mechanical system under investigation, the basic uncertain quantities $\underline{\Theta}$ and W(t) have been partitioned according to equations (02) and (03). Each variable of the vector $\underline{\Theta}_o$ has a fixed influence on the value of Y_R, and a functional dependence of Y_R on $\underline{\Theta}_o$ can be introduced. Conversely, the effects on Y_R of the vectors $\underline{\Gamma}^j$ and of the function $W_s(t)$

are random: therefore, they are modelled as random variables. The fragility model can therefore be assumed to have the form |15|:

$$Y_R(\underline{\Theta},W_s) = Y_R(\underline{\Theta}_o,\underline{\Gamma}^j,W_s) = g(\underline{\Theta}_o) + \sum_j(\varepsilon^j|\underline{\Theta}_o = \overline{\underline{\theta}}_o) + (\varepsilon_s|\Theta_o = \overline{\theta}_o) + \varepsilon_o. \qquad (08)$$

In equation (08) $g(\underline{\Theta}_o)$ expresses the functional dependence of Y_R on $\underline{\Theta}_o$ and is frequently assumed to have a polynomial form of low order in $\underline{\Theta}_o$ on the basis of the Taylor series idea. The coefficients of this polynominal form can be derived by usual parametric methods. Moreover, ε_o is a random term for lack of fit and ε^j and ε_s are the random effects on Y_R of the vector $\underline{\Gamma}^j$ and of the random function $W_s(t)$ respectively. The quantities ε^j and ε_s are assumed to be mutually independent in equation (08): however, in order to allow for possible dependence, the sum

$$\zeta = \sum_j (\varepsilon^j|\underline{\Theta}_o=\overline{\underline{\theta}}_o) + (\varepsilon^s|\underline{\Theta}_o=\overline{\underline{\theta}}_o) + \varepsilon_o \qquad (09)$$

can be directly estimated instead of the single terms. This calculation is possible via total simulation of the quantities $\underline{\Gamma}^j$ and $W_s(t)$, followed by deterministic analysis of the system with fixed values $\overline{\underline{\theta}}_o$ of $\underline{\Theta}_o$. This deterministic analysis leads to the calculation of the associated failure intensity y_R. Then, from the sample of values y_R relevant to the different systems simulated and from the knowledge of the term $g(\overline{\underline{\theta}}_o)$, the distribution function of $\zeta = y_R-g(\overline{\underline{\theta}}_o)$ can be estimated.

A basic step in the derivation of model (08) is the design of the numerical experiments, i.e. the selection of a set of observation points $(\underline{\theta},w_s)$: this aspect is the subject of research still in progress by the authors and associates. However, it is evident that the number of experiments required may be very high and application of model (08) is justified only for very significant systems, whose response is highly influenced by the quantities $\underline{\Theta}_o$ and their uncertainty.

For less important systems or for systems mainly governed by the stochastic terms $\underline{\Gamma}^j$ and W_s, a simplified approach to the problem of assessing system reliability is discussed in the sequel. Obviously, a large variety of mixed models in also possible |9|.

3.2 Random model

If a model of type (08) is supposed to hold for the natural logarithm of Y_R, and the separate effect of each source of uncertainties ($\underline{\Theta}_o$, $\underline{\Gamma}^j$ and W_s) is regarded as a random quantity ε, the terms $g(\underline{\Theta}_o)$ and ε_o disappear from the expression of $\ln Y_R$ and one can write

$$Y_R = \exp(\varepsilon_1+\varepsilon_2+\varepsilon_3+\varepsilon^1+\ldots+\varepsilon_s) = \exp(\sum_k \varepsilon_k) = Y_{R_o}\Pi_k Y_k' \qquad (10)$$

where Y_{R_o} is a suitable deterministic scale factor and Y_K' the contribution to Y_R of the k-th source of uncertainty. The product $Y_{R_o}.Y_k'$ is the auxiliary variable characteristic of the system in case the k-th quantity is the only uncertainty and the remaining components are fixed to suitable reference values. In equation (10) the ε_k, and hence the Y_k', are regarded as mutually independent variables. The additional assumption that the variables Y_k' in equation (10) have lognormal distribution greatly simplifies the fragility analysis. The model in fact is completely defined by the medians $\breve{m}_k$ of Y_k' and the variances $\sigma^2_{L_k}$ of $\ln Y'_k$ (the logarithmic variances of Y'_k), since Y_R has itself lognormal distribution with median

$$\breve{m} = Y_{R_0} \Pi_k \breve{m}_k \tag{11}$$

and logarithmic variance

$$\sigma^2_L = \sum_k \sigma^2_{L_k} . \tag{12}$$

In the case that the Y'_k are assumed to be lognormal[(1)] random variables, model (10) is denoted as *random model*. It has been recently developed |11||9| and applied to complicated mechanical and structural systems.

Model (10) is less accurate than model (08) because of the additional assumptions introduced to make model (10)operative . However, this simplified approach presents some advantages. It permits one to separate the contributions to the uncertainty of Y_R, of the uncertainty on the basic design variables and hence to neglect meaningless terms. Moreover, application of model (10) requires the estimation of the central values of k random variables, and few numerical simulations (20 to 50) are generally sufficient to give central values with satisfying accuracy. Naturally if k is high, the required computational effort may become unbearable; in this case one can consider the single Y'_k to be contributed by a set of random variables instead of by a single source of uncertainty, as the authors have shown in recent papers |5||6|.

A further advantage of the random fragility model is the possibility of introducing Bayesian estimations of median and logarithimic standard deviation of some of the right-hand-side factors, as illustrated by the numerical example developed in the following.

4. A numerical example

A numerical application of model (10) has been developed for the reinforced concrete plane frame of Figure 1 designed to resist seismic excitation. The same frame had been used in previous studies |2||5||6||8|. Three sets of random parameters are considered to contribute the uncertainty of Y_R. The first set, associated with the lognormal random variable Y'_1, contains the parameters necessary to describe the ground motion $W_s(t)$. The randomness of the duration of the accelerogram (according to the definition given in |14|) and the randomness of the parameters of the modified Kanai-Tajini spectral density function have been defined according to Ref. |12|. The second set, associated with Y'_2, contains the equicorrelated random vector of the resistances $\underline{R}^*$ at the section selected as critical. These sections are the ends of the finite elements by which the frame has been discretized. The damage parameter R_i to be compared with R^*_i in equation (05') takes into account low-cycle fatigue effects and is defined by the relation |5|

$$R_i(t) = \left[(1.10\ D_i(t)^{0.38})^2 + (67.27\ \xi_i^{0.3625}\ \phi_{ui}(t)^{0.4138})^2\right]^{1/2} \tag{13}$$

In equation (13) D_i is the ratio between the energy dissipated by inelastic rotation at

---

(1) Shifted lognormal distributions have also been proposed in order to allow for the experimental result that no failure is generally recorded for very low values of intensity.

the i-th critical section and the maximum elastic energy stored in the relevant member in antisymmetric bending. The symbol ϕ_{ui} denotes the inelastic rotation at the i-th section and ξ_i is the corresponding dimensionless neutral axis depth in the ultimate state: $\xi=x/b$, with x = distance of the centroid of the cracked section from compressed boundary; b = distance of steel reinforcement from compressed boundary. The associated resistances $\underline{R}^*$ are assumed to have lognormal distributions with coefficient of correlation $\rho=0.5$, mean 12.1 and variance 11.2 |5|| 2|. The third set of parameters, associated with Y_3', contains the uncertain quantities that define the behaviour of the mechanical system: masses, stiffnesses, damping ratios, hardening ratios and yielding moments. The randomness of these quantities has been defined according to Refs. |5| and |10|.

In Ref. |5|, the reliability of the frame of Figure 1 during a seismic event has been analysed according to model (10) (i.e. following the scheme of Figure 2) for the ultimate limit state defined by brittle failure (equation (05')). The results are summarized in Table I in terms of the average spectral pseudovelocity

$$Y = \frac{1}{\ln\gamma_\omega} \int_{\ln\omega_s-\ln\gamma_\omega}^{\ln\omega_s} S_V(\omega_s,\beta_s)\,(d\ln\omega) \tag{14}$$

that has been adopted as intensity of the seismic excitation. In equation (14) $S_V(\omega_s,\beta_s)$ is the pseudovelocity response spectrum ordinate for (ω_s,β_s), ω_s being the first natural frequency of the elastic system and β_s being the associated damping ratio. The averaging interval parameter γ_ω has been taken equal to 2.

Table I - Median and logarithmic standard deviation of Y_{RB}, Y_{1B}', Y_{2B}' and Y_{3B}' obtained in Ref. |5| for the limit state defined by brittle failure: $Y_{RO} = 32.4$ cm/s

Parameter	Median	Logarithmic standard deviation
Y_{RB}/Y_{RO}	0.60	0.386
Y_{1B}'	1.00	0.122
Y_{2B}'	0.65	0.341
Y_{3B}'	0.92	0.134

In this paper the reliability of the same frame is analysed for the ultimate limit state defined by progressive failure (equation (04')). In order to proceed dynamic analysis after first rupture, the two models of the degraded system drawn in Figure 3 have been considered. These two models have been used in Ref. |6| to implement a general-purpose code for dynamic analysis of non-linear systems. In the first model (Figure 3a), if an element reaches local failure (equation (05')), the element is substituted by a truss element. In the second model (Figure 3b), beam elements failed are substituted by truss elements, while column elements failed are removed. The first model being too conservative, the second scheme has been adopted in this paper for developing reliability calculations. Obviously, when the first element failing is a column, brittle and progressive failure generally coincide.

The calculation of the relevant fragility curves has been tackled both via total

Figure 2 - Application of model (10) to the structural problem illustrated in Figure 1

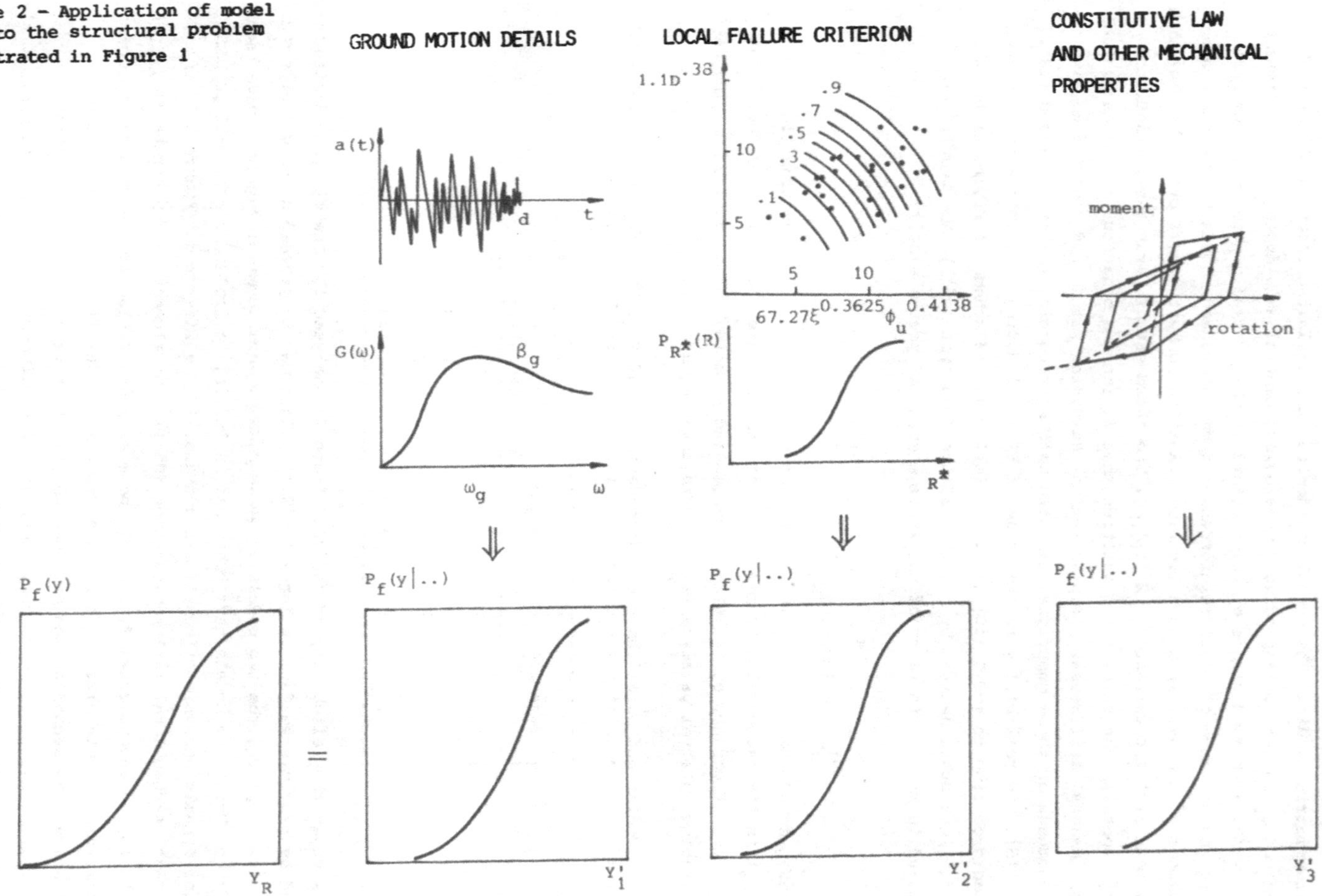

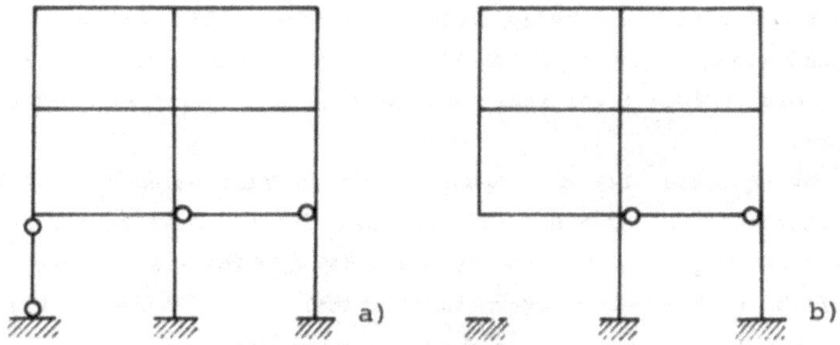

Figure 3 - Alternative models for proceeding dynamic analysis after an element fails: a) beams and columns failed are substituted by truss elements; b) beams failed are substituted by truss elements and columns failed are removed.

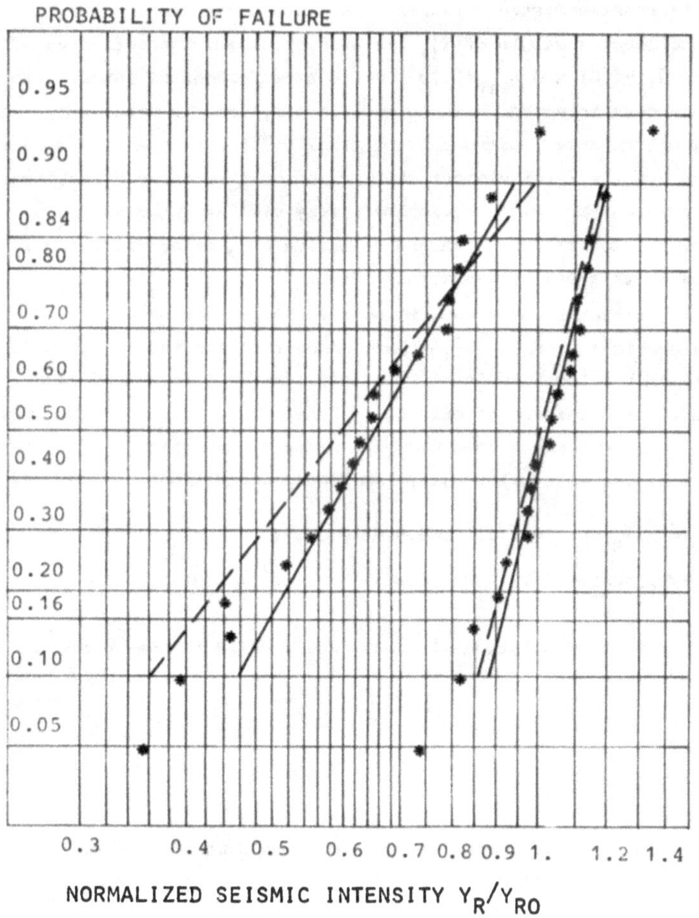

Figure 4 - Fragility curves via global simulation and cumulative distribution functions of Y_1': solid lines are relevant to progressive failure; dashed lines to brittle failure.

simulation of the uncertain quantities and by application of model (10). Twenty structural problems have been simulated by artificially yielding twenty accelerograms, twenty yielding moment vectors, twenty local resistance vectors and twenty sets of the other mechanical parameters.

For the intensity defined by equation (14) and normalized by the same value Y_{RO} =32.4 cm/s used in Ref.|5| , total simulation provided as estimates of the median $\breve{m}_P$ and of the logarithmic standard deviation σ_{LP} of Y_{RP} the values 0.649 and 0.264 respectively. By graphical interpolation of the sample points drawn in lognormal paper (Figure 4) the value 0.27 of σ_{LP} has been estimated. Since in the case of brittle failure total simulation has been proved to underestimate the actual value of σ_{LB}, further information appears to be suitable. This information may be obtained by a large increase of the sample size (say 100 instead of 20) or by the adoption of random model (10). Fixed the other quantities to reference values, a reference structure is defined: the minimum y causing the failure of this reference structure has been calculated for the 20 simulated ground motions and hence a sample of Y'_{1P} has been obtained. The estimates of its parameters are: median $\breve{m}_{1P}$=1.03 and σ_{L1P}=0.137, while after graphical interpolation a value $\sigma_{L1P}=\sigma_{L1B}$=0.122 has been selected. For the reference system, in fact, only two cases present the first local rupture located at a critical section relevant to a beam element; generally columns are the first element failing and brittle and progressive failure coincide. It follows that the distributions are very similar , the median $\breve{m}_{1P}$ being a little higher, and the logarithmic standard deviation σ_{L1P} being fictitiously increased by the two cases where a beam element fails first.

With reference to Y'_2 and Y'_3, there is no reason to modify the properties of the third contribution to Y_R, while previous studies |2||6| have shown that the median $\breve{m}_2$ of Y'_2 gains nearly ten per cent and the logarithmic standard deviation σ_{L2} is halved when progressive failure is considered instead of brittle failure. The result on σ_{L2} holds if first rupture frequently occurs at beams elements. In the present case, this is not true and the decrease in the value of σ_{L2} will be lower, say 40%. Then, from equations (11) and (12)

$$\breve{m}_{1P}\cdot\breve{m}_{2P}\cdot\breve{m}_{3P} = 1.03(1.1\text{x}0.65)0.92 = 0.677,$$

$$\sigma^2_{L1P}+\sigma^2_{L2P}+\sigma^2_{L3P} = 1.122^2+(0.6\text{x}0.341)^2+0.134^2 = 0.273$$

which confirm the results of total simulation without excessive additional computational effort.

AKNOWLEDGMENTS

This research was supported by grants from the Italian Research Council (C.N.R.) and the Italian Ministry of Public Instruction (MPI).

REFERENCES

|1| Augusti G., Baratta A. and Casciati F.: Probabilistic Methods in Structural Engineering, Chapman & Hall, 1983

|2| Banon H. and Veneziano D.: Seismic Safety of Reinforced Concrete Members and Structures, Earthquake Engineering and Structural Dynamics, 10(2), 1982, pp.179-194

|3| Belotti F. and Bissolati E.: Comportamento di strutture intelaiate in cemento armato in zona sismica (in Italian), SM Thesis, Dep. of Struct. Mechanics, University of Pavia, 1982

|4| Casciati F.: Probabilistic Analysis of Inelastic Structures, Nuclear Engineering and Design, 71, 1982

|5| Casciati F. and Faravelli L.: Seismic Risk for Structure Frames, Proc. of 7th Intern. Symposium on Earthquake Engineering, Roorkee (India), 1982

|6| Casciati F. and Faravelli L.: Progressive Failure for Seismic Reliability Analysis, to be published

|7| Casciati F., Faravelli L. and Gobetti A.: Reliability Analysis of Frames with Geometric Non-Linearity subject to Stochastic Ground Accelerations, Proc. of 3rd Int. Conf. on Application of Statistics and Probability in Soil & Structural Engineering (ICASP), Sydney, 1979, pp. 756-764

|8| Casciati F., Faravelli L. and Gobetti A.: Fragility Curves and Uncertainty of the Seismic Action, Proc. of Int. Conf.on Soil Dynamics and Earthquake Engineering, Southampton, 1982, pp. 619-631

|9| Casciati F., Faravelli L. and Veneziano D.: The Seismic Fragility of R.C. Frames, Proc. of 7th ECEE, Athens, Vol. 4, pp. 143-152

|10| Hasselman T.K. and Simonian S.S.: Structural Uncertainty in Seismic Risk Analysis, NUREG/CR-1560, UCRL-15218, Lawrence Livermore Laboratory, USA, 1980

|11| Kennedy R.P., Cornell C.A., Campbell R.L., Kaplan S. and Perla H.F.: Probabilistic Seismic Safety Study of an Existing Nuclear Power Plant, Nuclear Engineering and Design, 59(2), 1980, pp. 315-338

|12| Lai, S.P.: Overall Safety Assessment of Multistorey Steel Buildings subject to Earthquake Loads, Ph. D. Thesis, M.I.T., Cambridge, USA, 1980

|13| Pinto P.E. and Nuti C.: Seismic Reliability of R.C. Structures, EUROMECH 155, Lyngby, Denmark, 1982; to appear in DIALOG 5-82

|14| Vanmarke E.H.: Representation of Earthquake Ground Motions: Scaled Accelerograms and Equivalent Response Spectra, Report No. 14, State-of-the-Art for Assessing Earthquake Hazard in the United States, U.S. Army Engineering Waterway Experiment Station, Vicksburg, USA, 1979

|15| Veneziano D.: Reliability Analysis of Complex Structural and Geotechnical Facilities, EUROMEC 155, Lyngby, Denmark, 1982; to appear in DIALOG 5-82

POLAR CRANES FOR NUCLEAR POWER STATIONS IN EARTHQUAKE AREAS

U. Fischer[1], M. Engelke[2]

The construction of nuclear power stations in earthquake areas necessitates special consideration of the design of the polar crane in the containment shell. Possible models for the dynamic behaviour of the crane during periods of seismic activity and various calculation methods and their results are compared. The simple 2-mass-model and the well-known response spectrum method have proved to give an informative overview without posing many problems.

Introduction

When nuclear power stations are to be built in areas where there is a danger of seismic activity, in addition to the normal stringent safety requirements, the dynamic stresses caused by the earthquake must also be taken into consideration. The polar bridge crane (Fig 1) located in the reactor room of a nuclear power station is of particular importance as far as safety requirements are concerned. The crane moves on a circular track, the foundation of which is formed by the reactor building itself. This crane must still be in perfect working condition immediately after an earthquake because, should it be necessary to shut down the nuclear power station, the crane has an essential role to play.

As far as vibrations are concerned, the crane can be seen as a vibrator with support excitation. The usual and most effective method for the determination of the crane vibration caused by earthquakes is the use of the relative displacement x relative to the displacement at the support u. In this way, assuming linear behaviour, the differential equation of motion can be expressed in the form

$$\underline{A}\,\ddot{\underline{x}} + \underline{B}\,\dot{\underline{x}} + \underline{C}\,\underline{x} = -\underline{A}\,\ddot{\underline{u}} \qquad (1)$$

In Eq. 1

$\underline{x}$ = the column vector of relative displacement
$\underline{u}$ = the column vector of support displacement
$\underline{A}$ = the mass matrix
$\underline{B}$ = the damping matrix
$\underline{C}$ = the stiffness matrix

---

1) Technische Hochschule Otto von Guericke Magdeburg, 3010 Magdeburg, PSF 124
2) VEB Schwermaschinenbaukombinat Ernst Thälmann Magdeburg, 3011 Magdeburg

The acceleration of the crane track which is characterised by the elements of vector $\ddot{\underline{u}}$ is itself a "response" by the building to the movement of the ground caused by the earthquake. It is assumed to be a unified value throughout the entire crane track. Usually it is possible to consider the vertical and horizontal movements of the crane both separately, along and at right angles to the bridge axis. The functions $\ddot{\underline{u}}$ cannot of course be regarded as fixed for possible individual earthquakes. They are regarded as realisations of stochastic processes for which the choice of calculation process to be used determines whether the realisations have to be defined as functions of time or whether the statistical characteristics are sufficient.

1. Choice of model

1.1 Vertical load

Fig. 2 shows one possible representation of the polar crane. The discretisation of the crane girder by the use of finite beam elements to establish a differential equation system according to Eq. (1) causes relatively few problems. Actual calculations have been made for $11 \leqq n \leqq 19$.

The model not only makes it possible to calculate the dynamic forces, but also allows the simultaneous determination of the load at various cross-sections of the crane girder.

However, it has been shown that for informative examinations for designing the line of polar cranes this type of system has too many parameters which have to be varied. Therefore, examinations were also made to see how far the system could be simplified without too great a loss of informative content. Fig. 3 shows a simplified representation of the polar crane.

The crane bridge can be regarded as a slim girder with a constant cross section, and the masses of the trolley and the load can be seen as point masses. The distance between the trolley and the edge of the girder is variable, as is the spring rigidity of the cables which is dependent on the height of the load. The auxiliary trolley with auxiliary hoisting gear is not shown.

A further simplification is the reduction of the system to a two mass oscillator (Fig. 4). To do this it is necessary to reduce the mass of the crane bridge at the load point. If this is done by approximation using the static curve of bending of a girder on two supports with a constant cross-section under the effect of concentrated force, for parameters c_1 and m_1 the relationships

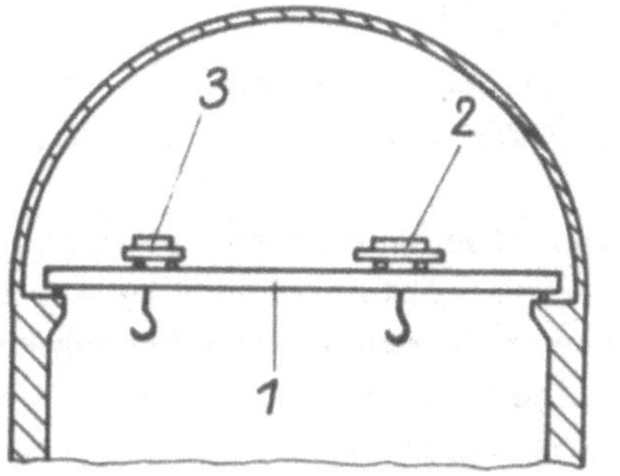

Fig. 1: Typical layout of a polar crane
1 - polar crane bridge
2 - main trolley
3 - supporting trolley

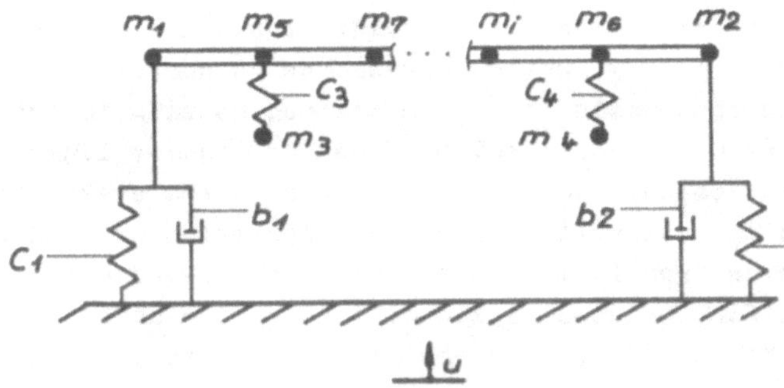

Fig. 2: n-mass model (n>6) of a polar crane

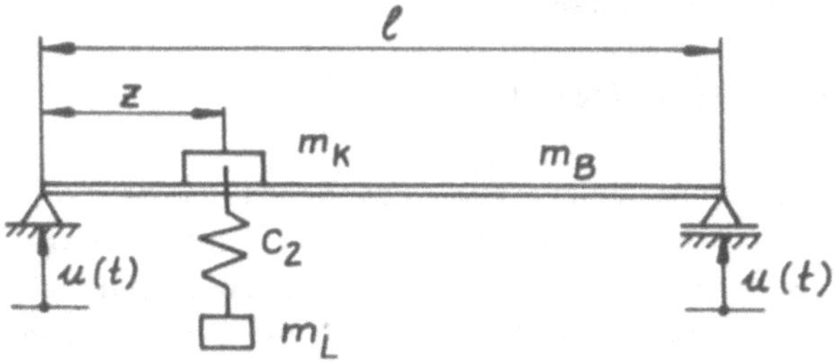

Fig. 3: 2-mass-crane model

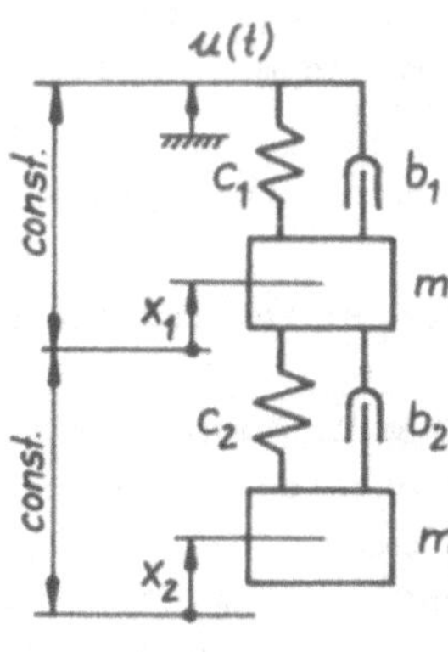

Fig. 4: 2-mass model for vertical excitation

$$c_1 = \gamma(z) \cdot \frac{EI}{\ell^3} \tag{2}$$

$$m_1 = m_K + \mu(z)\, m_B \tag{3}$$

are obtained. The factors γ and μ are shown in Fig. 5, cf /2/, p. 74.

By using integration in the time domain the chosen coordinates make it possible to allow for inconsistencies caused by slackening of the cable. To estimate the loss in accuracy due to the chosen discretisation process the eigenfrequency of the system with $c_1 = b_1 = m_K = m_L = 0$ (free beams) was compared with the exact solution, and the error was found to be 0.42 % when $z = \ell/2$ and just 30 % when $z = \ell/20$. Taking into consideration that this is a particularly unfavourable case, it can be seen that this degree of accuracy is completely satisfactory.
The damping values b_1 and b_2 for the model as seen in Fig. 4 have not yet been determined. If these measurements are not known it is necessary to base calculations on references concerning the damping degree ϑ of the system. For this reason, the Soviet regulations /3/ allow a logarithmic decrement of 6 %, which corresponds to approximately $\vartheta = 0.01$ to be used as a basis for calculations. Holzweißig and Dresig (/4/, p. 49) give $\vartheta = 0.014$ for steel bridges. As long as linear models are adequate for the calculation of the reponse of the system to seismic excitation, it is advisable to limit consideration of principal damping by means of a transformation to principal co-ordinates of the non-damped system. For the application of integration in the time domain, e.g. for sectional linear systems, it is necessary to determine the constants b_1 and b_2. As the damping effect can basically be attributed to the material damping, it is advisable to assume b_1 and b_2 to be proportional to c_1 and c_2. The factor of proportionality should then be assumed to be

$$b_1/c_1 = b_2/c_2 = 2\vartheta/\omega_1 \tag{4}$$

where ω_1 is the eigenfrequency of the basic vibration.

1.2 Horizontal load perpendicular to the bridge axis

On consideration of the fact that the horizontal components of earthquake excitation can be up to twice as big as the vertical components, and that the bridge rigidity is generally lower, it is obvious that the guarantee of safety in the horizontal direction is very necessary. The model as seen in Fig. 4 is also useable for this case, if the value $m_L \cdot g/l_s$ with a free cable length of l_s is set for c_2. In most cases, however, the partial frequency

$$1/2\pi \cdot \sqrt{g/\ell_2} \tag{5}$$

is so low that the hunting of the load is not likely to cause any resonance and m_L need not be taken into consideration. The result is the model of a 1-mass-oscillator with support excitation.

Due to the reasons given above, the stresses caused by the horizontal movements prove to be relatively high. The stress is reduced considerably if the wheels are allowed to glide along the rails, or if the brakes on the bogie truck are adjusted so that only limited horizontal forces can be transmitted. In this case, the model of an oscillating chain also proves to be unsuitable. Instead, it is recommendable to select a model with several degrees of freedom, which can be derived from the discretisation of the beam. As the ease with which the wheels move along the rails varies frequently, a step-by-step numerical integration of the equations of motion is recommended.

1.3 Horizontal stress along the bridge axis

Due to the high eigenfrequency of the lengthwise vibrations of the bridge compared with the excitement frequency, the bridge girder can be treated as a rigid body. The acceleration forces to be transmitted from the wheels of the bogie truck remain lower if the forces from the trolley chassis transmitted toward the bridge axis are restricted by sliding.

2. Measurement of the earthquake excitation

The stresses caused by seismic activity which have to be considered when dimensioning polar cranes for nuclear power stations can be calculated from the strength of the earthquakes to be expected in the particular area, and the reactions they cause in the reactor building. The following possible given data and representation of earthquake excitation for the position of a crane were examined:

1. Accelerations of the crane track as functions of time
2. Response spectrum of the crane track acceleration

If, due to the chance characteristics of the earthquake, it is not possible to predict the duration of the earth vibrations, calculations must be based on recorded times of other earthquakes (so-called dimensioning earthquakes) which are characteristic for the area concerned. Such records can also be used in conjunction with numerical integration of the differential equations, to simulate the movement and stress of the crane. In order to achieve sufficient accuracy in the evaluation of these results, several realisations of comparable earthquakes which are probable for the location should be chosen. Here the calculation of artificial realisations is recommendable. Often $\ddot{u}$ is reached by filtering a stationary Gaussian white noise $\dot{w}(t)$ through an intensity function $I(t)$ and an impulse response h (cf /7/ to /9/):

$$\ddot{u}(t) = \int_0^t h(t-\tau)\, I(\tau)\, \dot{w}(\tau)\, d\tau \tag{6}$$

where according to /8/

$$\left. \begin{aligned} I(t) &= e^{-\alpha t} + e^{-\beta t} \\ h(t) &= \frac{e^{-\delta\omega t} \sin\left(\omega t \sqrt{1-\vartheta^2}\right)}{\omega t \sqrt{1-\vartheta^2}} \end{aligned} \right\} \tag{7}$$

with corresponding constants α, β, ω, ϑ to be chosen.

The most commonly used method is that of the response spectrum. A response spectrum means the eigenfrequency-dependent maximum responses (acceleration, speed and displacement) of a linear 1-mass-oscillator with a certain damping for a given earthquake. The response spectrum can be calculated from acceleration-time functions as a smoothed envelope curve over the eigenfrequency of a 1-mass-oscillator. The response spectrum shown in Fig. 6 is unsmoothed due to the realisation for different damping at the level of the crane track. If the response spectrum is not given for a specific location, estimates based on the mean measuring values can be taken from the literature. From the curves given by KOS /5/, for example, the following functions can be approximately formulated:

$$\left. \begin{aligned} \ddot{y}_h &= \ddot{y}_{mh} \left[1 - 2{,}25 \left(\lg \frac{f}{4\,\mathrm{Hz}}\right)^2\right]; \quad 1\,\mathrm{Hz} < f < 16\,\mathrm{Hz} \\ \ddot{y}_v &= \ddot{y}_{mv} \left[1 - 2{,}5 \left(\lg \frac{f}{7\,\mathrm{Hz}}\right)^2\right]; \quad 2\,\mathrm{Hz} < f < 20\,\mathrm{Hz} \\ \ddot{y}_{mh} &= \frac{0{,}108\,\alpha \cdot 10^{(i-4)/3} + 0{,}02813 \cdot 2^{i-4}\, h/m}{\sqrt{1 + 200\,\vartheta}}\, g; \quad 0 \leqq \vartheta \leqq 0{,}1 \\ \ddot{y}_{mv} &= \inf\left(0{,}32\,\alpha \cdot 10^{(i-4)/3}\, g;\ 0.45\, \ddot{y}_{mh}\right) \end{aligned} \right\} \tag{8}$$

α = 0.45 - 2.24 for very good to bad foundations.

Here

$\ddot{y}_h$, $\ddot{y}_v$ = absolute maximum acceleration horizontally and vertically during the earthquake

f, = eigenfrequency and degree of damping of the 1-mass-oscillator

g = the acceleration due to gravity

h = height of the crane track above ground

i = intensity of the earthquake according to the "modified Marcelli scale" (MMI)

The fact that even for damping-free cases only finite values are obtained for the "response acceleration" is due to the consideration of the finite duration of the earthquake. The formulae given here are in

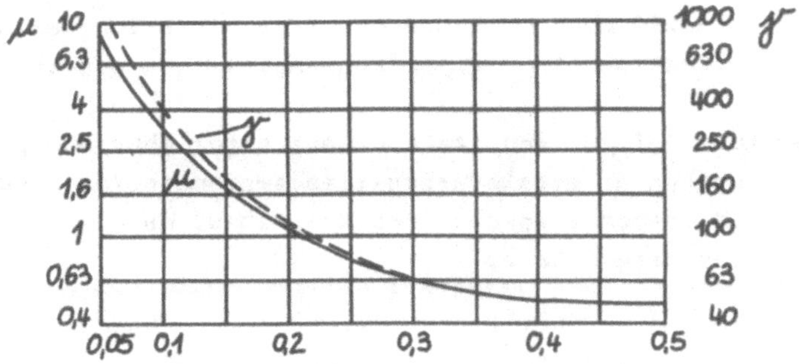

Fig. 5: Coefficients of mass-discretisation

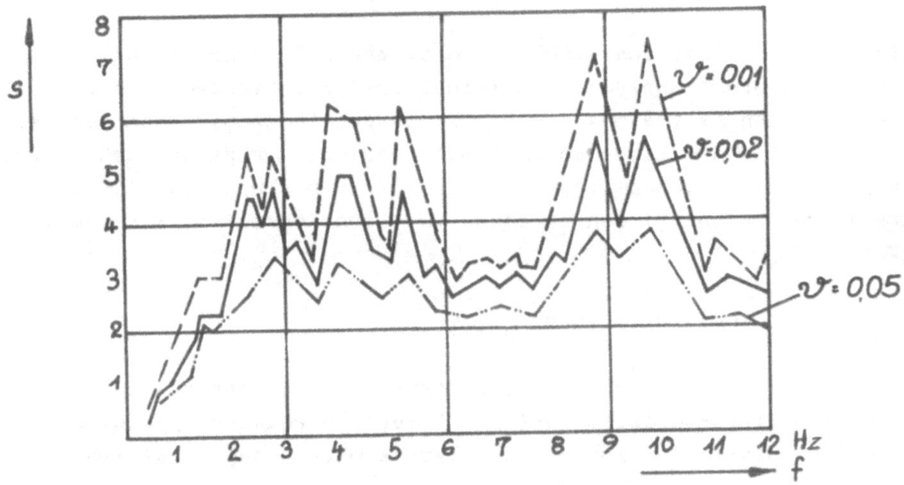

Fig. 6: Response spectra

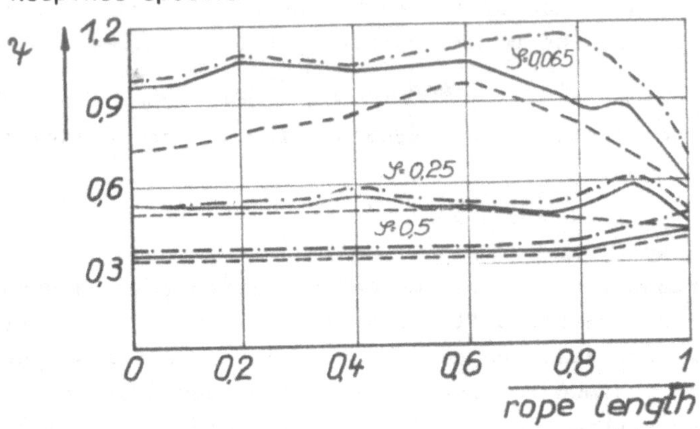

Fig. 7: Dynamical factors on vertical earthquake excitation

________ by the time history method

-------- by the response spectra method I

-.-.-.-. by the response spectra method II

any case only useable when at least one free vibration falls within the given frequency domain. Other eigenfrequencies need not then be considered.

In the literature (cf /6/ and its sources) algorithms and programmes for the construction of acceleration-time sequences for simulated earthquakes from response spectra are described. We do not intend to treat them in any detail here.

3. Methods of calculation

3.1 Integration of the equations of motion in the time domain

This process allows the simultation of the behaviour of the crane without any great analytical difficulties. Its disadvantage is that sure deduction of the maximum stress is only possible when several realisations of the excitation (actually measured or artificially produced) are applied. Non-linear features (cable slackness, lifting, sliding and catching) can be taken into consideration relatively easily.

3.2 Response spectrum method

The response spectrum method is restricted to linear systems. For this method the differential equation (1) must be reduced approximately to the differential equation of oscillators with a degree of freedom of

$$\ddot{x}_k + 2\vartheta_k \omega_{k0} \dot{x}_k + \omega_{k0}^2 x_o = -\ddot{u} \,. \tag{9}$$

This is done by expansion according to the modal vectors $\underline{z}_k$ with the eigenfrequencies ω_k. The column vector $\underline{z}_k$ must be normed so that

$$\underline{z}_k^T \underline{A}\, \underline{z}_k = 1 \qquad \text{and} \qquad \underline{z}_k^T \underline{C}\, \underline{z}_k = \omega_k^2 \,. \tag{10}$$

As $\underline{x} = \alpha_k \underline{z}_k$ and is the largest acceleration of the 1-mass-oscillator $\ddot{y}$ ($\ddot{y}_h$ or $\ddot{y}_v$ as equation (8)),

$$\ddot{\alpha}_k = \underline{z}_k^T \underline{f}\, \ddot{y} \tag{11}$$

can be calculated by comparison.

As the deflections for slightly damped oscillators are generally considerably larger than the deflection of the support, the fact that the co-ordinates x_j are relative co-ordinates is usually disregarded. The thus determined maximum accelerations (the elements of the vector $\ddot{\alpha}_k \underline{z}_k$) of the kth modal vector, the maximum displacements (the elements of the vector $\ddot{\alpha}_k \underline{z}_k / \omega_k^2$) or the stresses derived therefrom can be summed up as follows:

$$\sigma = \sqrt{\sum_{k=1}^{n} \sigma_k^2} \tag{12}$$

Here σ_K stands for the maximum value of the kth modal vector. Obviously this summarizing rule is very arbitrary and so this method is marked by its simplicity of application. It is usual only to take into consideration the eigenfrequencies which offer the largest contribution.

4. Results

The following are results which were achieved in comparative examinations of a polar crane manufactured by VEB Schwermaschinenbau– Kombinat Ernst Thälmann, Magdeburg, GDR. The crane has a support width of 43 m and is designed for a maximum load of 320 t.
Using the model according to Fig. 4 for the vertical load and the mass and spring values calculated by means of the static line of bending for the bridge, the first two eigenfrequencies of the crane with rigid supports can be estimated with sufficient accuracy. Dependent on the position of the trolley (in the area between the centre position and a point 1/15 of the width of the support from the end of the bridge) and the length of the cable, these lie between 1.8 and 12 Hz and thus in the possible resonance zone of earthquakes. The third eigenfrequency with an oscillation node in the bridge which is not reached in this model is greater than 13 Hz. Due to the asymmetry of the mode of oscillation this is only slightly excited.
Examinations of the vertical vibrations of the polar crane during earthquake excitation of Grade MSK VII with a 2-mass model produce results which are given in Fig. 7. In order to produce comparable values the response spectrum as Fig. 6 and the realisation of an earthquake in the form of an acceleration function in time were used as a basis for the method used. The Figure shows the quotient ψ, which represents the additional dynamic load, relative to the static load in the cables, depending on the length of the cable and the position of the crane trolley. The response spectrum and time history methods produce results which deviate only slightly from each other. Non-linearities due to lifting of the wheels or slacking of the cables can only be reached by the time history method and can give load values which are both lower and higher than those given by linear calculation.
The more accurate model with the number of degrees of freedom between 11 and 19 which verifies the dynamic load factors obtained with the basic model makes possible a more accurate calculation of the additional dynamic load in the individual girder cross-sections.

Literature

/1/ Kos, M.: Seismische Beanspruchung der Krane
Konstruktion 32 (1980) 3, 105 - 109

/2/ Fischer, U., Stephan, W.: Mechanische Schwingungen
Leipzig, VEB Fachbuchverlag, 1981

/3/ Zeitweilige Normen für die Projektierung von Kernenergieanlagen in Erdbebengebieten
WSN-15-78 MINERGO, Moskau 1979 (russ.)

/4/ Holzweißig, F., Dresig, H.: Lehrbuch der Maschinendynamik,
Leipzig, VEB Fachbuchverlag 1979

/5/ Kos, M.: Seismische Berechnung von Laufkranen
fördern und heben 30 (1980) S. 458 - 463

/6/ Klement, H.-D.: Berechnung spektrumkompatibler Erdbebenverläufe.
Diss. TH Darmstadt 1979

/7/ Jennings, P., Housner, G. W., Trai, N.C.:
Simulated earthequake motions. Techn. Report
Earthquake Eng. Res. Lab., Cal. Techn. April 1968

/8/ Levy, R., Kozin, F., Moorman, R. B. B.:
Random processes for earthquake simulation,
Journal of the Eng. Mech. Div., A SCE 97 (1971),
EM 2, 495 - 517

/9/ Shinozuka, M., Sato, Y.: Simulation of nonstationary random processes, Journal of the Eng. Mech. Div. ASCE 93 (1967), EM 1,11

RELIABILITY OF MECHANICAL SYSTEMS SUBJECT TO STOCHASTIC EXCITATIONS

Klaus Hennig[1)]

1. Introduction

Statements on reliability of randomly excited structures raise considerable difficulties if, according to reality, physical and geometric nonlinearities have to be taken into consideration. For solving such problems one can offer either the direct procedure by means of realizations using measuring values of random loads or, in the case of given statistical load parameters, simulation methods followed by determination of the wanted statistical characteristics. This procedure seems to be favourable especially because then it is possible to use well-known numerical methods for corresponding deterministic problems. Outlines of such methods are followed by some relations dealing with the determination of first passage probabilities. An example of a randomly excited cylindrical shell is considered. Finally problems of fatigue failure within the framework of a general fatigue damage concept are treated.

2. Randomly excited forced continua vibrations

For the purpose of calculation of continua vibration problems it is often useful to tackle the corresponding initial boundary value problem not by approximation of the local derivatives but to take advantage of the possibility to formulate the problem as a variational problem. Hereafter, this way will be sketched where the principle of virtual displacements

$$\delta J = \int_R \varrho(x)\, \delta\bar{u}^T(x,t)\, \bar{u}_{,tt}(x,t)\, dR + \int_R \delta\bar{\varepsilon}^T(x,t)\, \bar{\sigma}(x,t)\, dR - \int_R \delta\bar{u}^T(x,t)\, \bar{\bar{q}}(x,t)\, dR - \int_{S_p} \delta\bar{u}(x,t)\, \bar{\bar{p}}(x,t)\, dS + \delta W_G(t) = 0 \qquad (1)$$

is the starting point for approximate calculation of forced randomly excited nonlinear continua vibration problems.

In (1) $R \subset R^3$ means the domain in Euclidean space R^3 occupied by the body before deformation, $\bar{R} = R \cup S$ is its closure, $S = S_u \cap S_p$ means the piecewise smoothly proposed boundary, where S_u is the part with given displacements $\bar{\bar{u}}$ and S_p is the part with given stresses $\bar{\bar{p}}$; x is an abbreviation for the coordinates of the points $P \in R$; $\bar{u}(x,t)$ is the displacement column vector, $\bar{u}^T{}_{,tt}$ the second time derivative of the transposed

1) Professor Dr.-Ing. sc., Academy of Sciences of the G.D.R., Institute of Mechanics, Rudower Chaussee 5, DDR-1199 Berlin-Adlershof

vector $\bar{u}$; $\bar{\varepsilon}$ denotes the strain vector and $\bar{\sigma}$ the stress vector, $\bar{\bar{q}}$ is the impressed force vector corresponding to R, ϱ the density of the undeformed body, and $\delta W_G(t)$ is the virtual work of polygenetic forces (hence not to derive from scalar functions).

We write the strain-displacement relations in the form

$$\bar{\varepsilon}(x,t) = \underline{D}(x)\,\bar{u}(x,t) + \bar{b}(\bar{u}(x,t)) \tag{2}$$

where $\underline{D}$ is a linear matrix differential operator and $\bar{b}(\bar{u})$ is a vector function operator concerning the nonlinear part of the strain vector. Considering these elastic bodies we assume the stress-strain relations to have the form

$$\bar{\sigma}(x,t) = \bar{F}(\bar{\varepsilon}(x,t)). \tag{3}$$

Now, according to Kantorovitch's method [8] we put into (2) the (N-dimensional) semi-discretization

$$\bar{u}_N(x,t) = \underline{F}(x)\,\bar{p}(t) \tag{4}$$

where $\underline{F}(x)$ is the matrix of permissible coordinate functions (cf. [1,2]). Considering

$$\begin{aligned}\bar{\varepsilon}(x,t) &= \underline{D}(x)\,\underline{F}(x)\bar{p}(t) + \bar{b}(\underline{F}(x)\bar{p}(t)) := \underline{B}(x)\bar{p}(t) + \bar{e}(x,\bar{p}(t))\\ &:= \bar{\varepsilon}(x,\bar{p}(t)),\end{aligned} \tag{5}$$

$$\bar{\sigma}(x,t) = \bar{F}(\underline{B}(x)\bar{p}(t) + \bar{e}(x,\bar{p}(t))) := \bar{F}(\bar{\varepsilon}(x,\bar{p}(t))) \tag{6}$$

from (1) it follows the nonlinear system of ordinary differential equations

$$\underline{M}\,\ddot{\bar{p}} + \bar{R}_G(\bar{p},\dot{\bar{p}}) + \bar{K}(\bar{p}) = \bar{R}(t) \tag{7}$$

where

$$\begin{aligned}\underline{M} &= \left\{\int_R \varrho(x)\,\underline{F}^T(x)\,\underline{F}(x)\,dR\right\},\\ \bar{K}(\bar{p}) &= \int_R \left(\underline{B}(x) + \left\{\frac{\partial \bar{e}(x,\bar{p}(t))}{\partial \bar{p}}\right\}\right)^T \bar{F}(\bar{\varepsilon}(x,\bar{p}(t)))\,dR\\ \bar{R}_G(t) &= \int_R \underline{F}^T(x)\,\bar{\bar{q}}(x,t)\,dR + \int_{S_p} \underline{F}^T(x)\,\bar{\bar{p}}(x,t)\,dS.\end{aligned} \tag{8}$$

Here we have assumed $\delta W_G(t)$ being in the form $\delta\bar{p}^T\bar{R}_G(\bar{p},\dot{\bar{p}})$. The meaning of $\dot{\bar{p}}$ is $\frac{d\bar{p}(t)}{dt}$ etc, and $\bar{0}$ is the N-dimensional zero vector.

For determination of approximate solutions of (7) we apply the Newton-Raphson method (cf. [3]).

After some calculations using the abbreviations

$$\underline{D}_1(\bar{p},\dot{\bar{p}}) = \left\{\frac{\partial \bar{R}_G(\bar{p},\dot{\bar{p}})}{\partial \dot{\bar{p}}}\right\}, \quad \underline{D}_2(\bar{p},\dot{\bar{p}}) = \left\{\frac{\partial \bar{R}_G(\bar{p},\dot{\bar{p}})}{\partial \bar{p}}\right\}, \underline{E}(x,\bar{p}) = \left\{\frac{\partial \bar{F}(\bar{\varepsilon}(x,\bar{p}))}{\partial \bar{\varepsilon}}\right\},$$

$$\underline{K}(\bar{p}) = \left\{ \int_R \left(\underline{B}(x) + \left\{\frac{\partial \bar{e}(x,\bar{p})}{\partial \bar{p}}\right\}\right)^T \underline{E}(x,\bar{p}) \left(\underline{B}(x) + \left\{\frac{\partial \bar{e}(x,\bar{p})}{\partial \bar{p}}\right\}\right) dR \right.$$

$$\left. + \int_R \frac{\partial}{\partial \bar{p}} \left(\left\{\frac{\partial \bar{e}(x,\bar{p})}{\partial \bar{p}}\right\}^T \right) \bar{F}(\bar{\varepsilon}(x,\bar{p}))\ dR \right\}, \tag{9}$$

$$-\bar{S}(\bar{p}) = \int_R \left(\underline{B}(x) + \left\{\frac{\partial \bar{e}(x,\bar{p})}{\partial \bar{p}}\right\}\right)^T \left[\bar{F}(\bar{\varepsilon}(x,\bar{p})) - \underline{E}(x,\bar{p})\bar{\varepsilon}(x,\bar{p}) \right.$$

$$\left. + \underline{E}(x,\bar{p})\left(\bar{e}(x,\bar{p}) - \left\{\frac{\partial \bar{e}(x,\bar{p})}{\partial \bar{p}}\right\}\bar{p}\right)\right] dR$$

$$- \left\{ \int_R \frac{\partial}{\partial \bar{p}} \left(\left\{\frac{\partial \bar{e}(x,\bar{p})}{\partial \bar{p}}\right\}^T \right) \bar{F}(\bar{\varepsilon}(x,\bar{p}))\ dR \right\} \bar{p} - \underline{D}_1(\bar{p},\dot{\bar{p}})\dot{\bar{p}} - \underline{D}_2(\bar{p},\dot{\bar{p}})\bar{p} + \bar{R}_G(\bar{p},\dot{\bar{p}})$$

we get the sequence of linear ordinary differential equation systems

$$\underline{M}\ddot{\bar{p}}_{n+1} + \underline{D}_1(\bar{p}_n,\dot{\bar{p}}_n)\dot{\bar{p}}_{n+1} + \left[\underline{D}_2(\bar{p}_n,\dot{\bar{p}}_n) + \underline{K}(\bar{p}_n)\right]\bar{p}_{n+1} = \bar{S}(\bar{p}_n) + \bar{R}(t), \tag{10}$$

$n = 0,1,2,\ldots,$ for $\bar{p}_{n+1}(t)$.

It should be remarked that for strain-stress relations being not transformable by inversion into the form (3) the vector

$$\bar{S}(\bar{p}) = \tag{11}$$

$$\int_R \left(\underline{B}(x) + \left\{\frac{\partial \bar{e}(x,\bar{p})}{\partial \bar{p}}\right\}\right)^T \underline{E}(x,\bar{p}) \left[\bar{g}(\bar{\sigma}(x,\bar{p})) - \bar{\varepsilon}(x,\bar{p}) - \bar{e}(x,\bar{p}) + \left\{\frac{\partial \bar{e}(x,\bar{p})}{\partial \bar{p}}\right\}\right]\bar{p}\ dR$$

$$+ \left\{ \int_R \frac{\partial}{\partial \bar{p}} \left(\left\{\frac{\partial \bar{e}(x,\bar{p})}{\partial \bar{p}}\right\}^T \right) \bar{\sigma}(x,\bar{p})\ dR \right\} \bar{p} + \underline{D}_1(\bar{p},\dot{\bar{p}})\dot{\bar{p}} + \underline{D}_2(\bar{p},\dot{\bar{p}})\bar{p} - \bar{R}_G(\bar{p},\dot{\bar{p}})$$

has to be used instead of (9), where $\bar{\sigma}(x,\bar{p})$ is defined by $\bar{\sigma}(x,\bar{p}) := \underline{E}(x,\bar{p})\bar{e}(x,\bar{p})$ and $\bar{\varepsilon}$ by (5).

If in the framework of correlation theory there are statistical characteristics of excitation, then it is possible to find out the corresponding characteristics of the linear vibration response by means of the well-known methods in the time domain as well as in the frequency domain. It is easy to see that this procedure is not applicable in (7) or (10), respectively (with the exception of the linear special case), cf. [4,5].

In many practical problems, however, advantage can be taken of the fact, that excitations often are given "realization like" in form of measurement values. Then it is appropriate immediately to integrate (7) or (10) numerically using those measurement values [12] and then to determine the statistical parameters needed for characterization of the random vibration response and for estimation of the reliability of the considered structure (cf. [6,7,8]). For application of this procedure under given statistical characteristics of loads the use of additional simulation methods is necessary (cf. [9]). On principle this procedure seems to be favourable, because thus well-known numerical methods for corresponding deterministic calculation of structures can be applied; further, in general here we have not so strong restrictions with respect to nonlinearities in comparison with other approximate methods for random initial boundary value problems.

3. Failure probabilities

Let $Z(t)$ be a random process decisively for the failure on the critical point of the structure in the time interval $[0,T)$. We denote the probability for the random process $Z(t)$ to exceed the threshold z_0 at least once during the time interval $[0,T)$ - in practice often connected with failure of structure - by $P_1(z_0;0,T)$. Then $P_0(z_0;0,T) = 1 - P_1(z_0;0,T)$ is the probability not to exceed the threshold z_0 in the time interval $[0,T)$. This probability (cf. [10,11]) is given by

$$P_{0N}(z_0;0,T) = \exp \sum_{s=1}^{N} (-1)^s \underbrace{\int \cdots \int}_{s}_{0\leq\tau_1<\ldots<\tau_s\leq T} k_s(\tau_1,\ldots,\tau_s)\, d\tau_1\ldots d\tau_s$$

for $N \to \infty$. Since it is difficult to determine the higher cumulant functions k_s, $s > 2$, and the evaluation of the repeated integral is troublesome, too, it often is convenient to restrict consideration to the first approximations $N = 1$ or $N = 2$ which are sufficient for practical purposes in the most cases. If the threshold z_0 is sufficiently high (e.g. $z_0 \geq 6\sigma_Z$), the difference between first and second approximation is disappearing; thus for many problems the first approximation $N = 1$ is sufficient. This first approximation is identical with the relation we get assuming Poisson distributed threshold crossing probabilities.

As application at first we consider a unilateral clamped circular cylindrical shell of constant thickness h under stationary random loadings and linear elastic, homogeneous isotropic constitutive relations characterized by Young's modulus E and Poisson's ratio ν (cf. [13]); r is the radius of the medium surface and l the length of the shell.

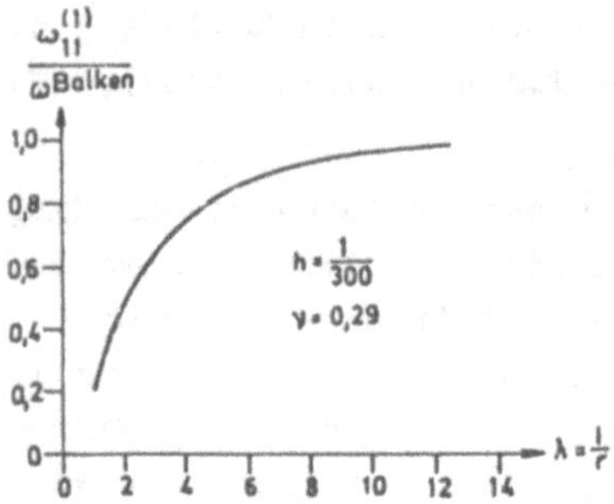

Figure 1

Figure 1 shows the rate of the first eigenfrequency ω_{11} of the shell to the eigenfrequency of a clamped beam with the same cross section as the cylindrical shell in dependence upon $\lambda = l/r$ for $h/r = 300^{-1}$ and $\nu = 0.29$. Hence it appears that only from about $\lambda \approx 10$ the shell model is replaceable by a beam model.

In the following consideration of forced vibrations we suppose the excitation acting only in direction perpendicular to the medium surface. Thus the load vector reduces from $p(x,0,t) = \{p_x(x,0,t), p_o(x,0,t), p_Z(x,0,t)\}^T$ to $p(x,0,t) = \{0,0,p_Z(x,0,t)\}^T$. Here we have chosen $p_Z(x,0,t)$ in the form $p_Z(x,0,t) = \sum_{m=0}^{M} q(x,t) \cdot b_m \cos m\Theta$, where $q(x,t) = \frac{1}{2} c_w \varrho_L \overset{w}{V}{}_x^2(t)$ is the dy-

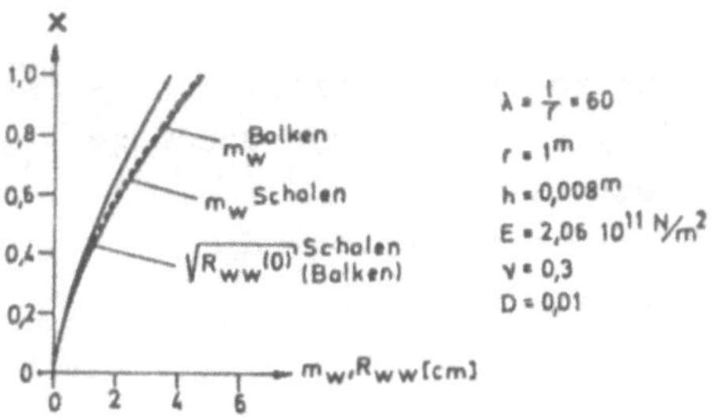

Figure 2

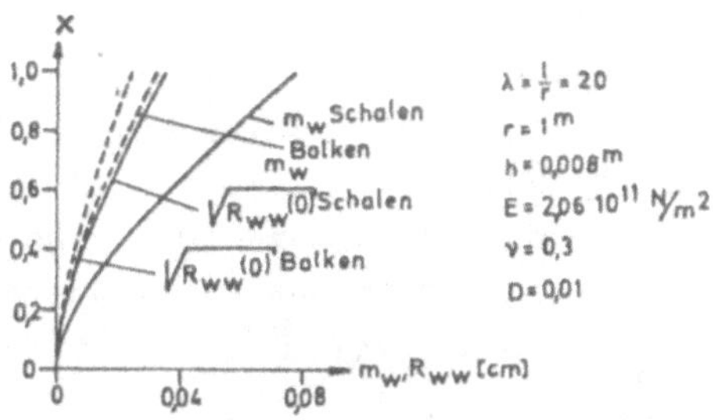

Figure 3

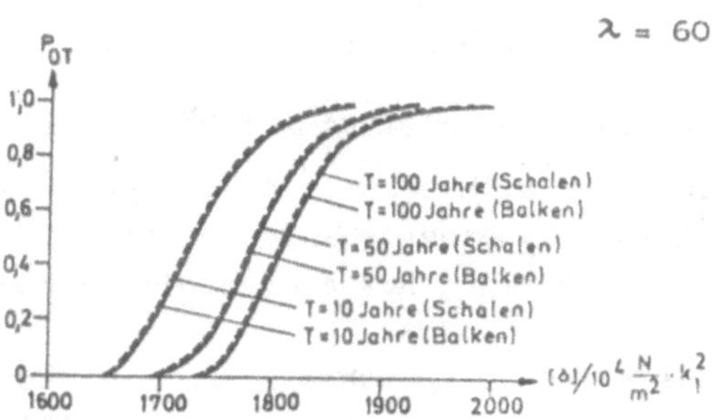

Figure 4

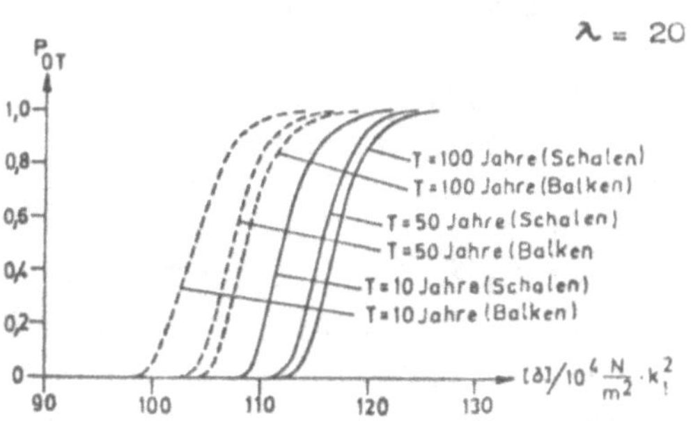

Figure 5

namic pressure of wind and c_w a special aerodynamic parameter; the Fourier coefficients b_m describe the wind distribution in circumferential direction Θ while $\overset{w}{V}_x(t)$ is the temporary wind velocity in height x.

Using Davenport's wind model [14] in Figs. 2 and 3 there are shown the mean values and variances of deflection for $\Theta = 0$ for steel chimneys with lengths $l = 60$ m or $l = 20$ m, respectively, versus the length of the shell; all this is calculated for the parameters $c_w = 0.4$, $b_o = -0.12$, $b_1 = 0.29$, $b_2 = 0.42$, $b_3 = 0.38$, $b_4 = -0.11$, $b_5 = 0.08$, $\bar{V}_{400,200} = 45$ m/s, respectively. Here $\bar{V}_{400,200}$ means the amount of medium wind velocity in the gradient height of 400 m, which occurs on an average only in 200 years in a low built town. From the as well in Figs. 2 and 3 presented results of beam theory one can see substantial differences to shell theory occurring already for $\lambda \leq 20$. Figs. 4 and 5 show the probabilities for nonexceeding admissible stresses by the maximum stresses (cf. [13]).

4. Fatigue damage accumulation

For classification of given damage hypotheses and for formation of new ones one can offer the concept of a general damage measure under following assumptions:

A) Damage measure of a structure element due to cyclic stress can be characterized by a scalar quatity D normed by D = 0 for the undamaged state and by D = 1 in the moment of fracture.

B) The growth of damage dD in the n-th cycle depends on the damage D so far read and on the amplitude S(n), i.e.

$$\frac{dD}{dn} = F(D,S(n)) \tag{12}$$

where $F(D,S) > 0$ as $S > S_d$, S_d means the endurance.

Under these assumptions it can be shown:

a) Different functions F in (12) can yield for any loading conditions S(n) the same lifetime L (e.g. $\frac{dD}{dn} = N^{-1}(S)$ where $D(L) = 1$ and $\frac{dD^*}{dn} = N^{-1}(S)\, D^{*(1-\varkappa^{-1})}$ where $\varkappa$ is a positive constant and $D^*(L) = 1$). The equivalence classes of equations (12) with respect to the property mentioned above represents one damage hypothesis in each case. The special form of the fictitious function D being not accessible by experiments remains unknown.

b) For the purpose of identification of a damage hypothesis the results of all two-step tests related to a fixed pre-load stage S_1 are sufficient. In additional two-step tests or in k-step tests with $k > 2$ we may check the presumptions A), B) (here a k-step test means a preloading in k-1 stress-steps and determination of the residual vinration cycle number in the k-th step).

c) Independence on succession and Miner's hypothesis are equivalent. They are represented by all kinetic damage equations of product form

$$\frac{dD}{dn} = F_1(D)\, F_2(S(n)) \;. \tag{13}$$

Thus as "k-step formula" we always get $\sum_{i=1}^{k} n_i\, N^{-1}(S_i) = 1$ from (13) and, taking into account the Wöhler's curve $S^m N$ = const. and $S(n) = S_i$ where $\underline{n_{i-1}} < n < \underline{n_i}$; $n_i = \underline{n_i} - \underline{n_{i-1}}$, $i = 1,\ldots,k$, comparison with experiments shows that the influence of load history on lifetime is in good agreement with the damage hypothesis

$$F(D,S(n)) = -\left\{D + \left[\left(\frac{S}{S*}\right)^{-m} - 1\right]^{-1}\right\} m\left(\frac{S}{S*}\right)^{m} N^{*} \ln \frac{S}{S*} \tag{14}$$

where $P(S^*,N^*)$ is a selected point of Wöhler's curve.

Under random loads in form of a zero-mean narrow-band process $\overset{\circ}{Z}(t)$ - with existing second t-derivative $\ddot{Z}(t)$ - and under consideration of point A) above formula (12) we get the random initial boundary problem

$$\frac{dD}{dt} = -\ddot{Z}(t)\delta(\dot{Z}(t))\, \underline{1}(\overset{\circ}{Z}(t)-S_d)\, F(D,\overset{\circ}{Z}(t)) \;; \qquad D(0) = 0$$

with norm condition $D(L) = 1$; (.) is the Dirac function, $\underline{1}(.)$ the Heavidide function, and the random variable L means the lifetime. As can be seen from this it is only possible to establish an analytical formula for L in the special case where $F(D,\overset{\circ}{Z}(t)) = F_1(D)F_2(\overset{\circ}{Z}(t))$, i.e. on the basis of Miner's hypothesis.

With $L = \overset{\circ}{L} + m_L$ and under the assumption $\sigma_L = (E[\overset{\circ}{L}{}^2])^{1/2} \ll E[L] = m_L$ for stationary Gaussian narrow-band processes $\overset{\circ}{Z}(t)$ with zero mean value we obtain Miles' formula for the estimation of the medium lifetime m_L (cf. [4]).

Especially during the period of stable crack growth there are some touches between the " K-concept" and the kinetic damage equations (12). Thus starting from crack growth relations of the form

$$\frac{da}{dn} = F_1(a)\, F_2(S(n)) \tag{15}$$

where $a(n_o) = a_o$ as $S = \text{const.}$ one gets for instance $a = \phi((n-n_o)\tilde{F}_2(S)+\phi^{-1}(a_o))$ and from this in the case of a k-step loading $(n_1,S_1,\ldots,n_kS_k)$ the damage accumulation equation

$$\sum_{i=1}^{k} \frac{\tilde{F}_2(S_i)(n_i-n_o)}{\phi^{-1}\{\phi[F_2(S_i)(n_i-n_o)+\phi^{-1}(a_o)] + \alpha[g(n_k-n_o)-g(n_i-n_o)]\} - \phi^{-1}(a_o)} = 1\,. \tag{16}$$

Here we have assumed the initial crack length to be independent on the load amplitude and also the critical crack length a_{kr} to be a monotonic increasing function in dependence on the vibration cycle number; this function is assumed to have the form $a_{kr} = a_c + \alpha g(n-n_o)$, $a_c = \text{const.}$

If $\alpha = 0$, formula (16) becomes Miner's hypothesis. Consideration of (13) shows, that this results easily can be drawn from (15) if one puts $D = a/a_{kr}$, $a_o \leq a \leq a_{kr} = \text{const.}$

Concluding these considerations it should be remarked that the kinetic damage equations (15) permit lifetime calculations only for loads greater than S_d.

References

1 Michlin, S.G.: Variationsmethoden der mathematischen Physik, Akademie-Verlag, Berlin 1962

2 Strang, G.; Fix, G.I.: An Analysis of the Finite Element Method, Prentice Hall, Inc., Englewood Cliffs, N.J. 1973

3 Oden, J.T.: Finite Elements of Nonlinear Continua, McGraw-Hill Book Company, New York 1972

4 Heinrich, W.; Hennig, K.: Zufallsschwingungen mechanischer Systeme, Akademie Verlag, Berlin 1977

5 Renger, A.; Gupta, K.N.: Ein programmiertes und allgemeingültiges Berechnungsverfahren zur Bestimmung des dynamischen Verhaltens von linearen Schwingungssystemen mit vielen Freiheitsgraden bei stationärer Zufallserregung, Wiss. Z. Techn. Hochsch. Magdeburg 15(1971),589-601

6 Friedrich, H.: Zur Berechnung von Versagenswahrscheinlichkeiten bei nichtlinearen Schwingungssystemen, Proc. VIII Internat. Conf. Nonlin. Oscillations, Prag 1978, pp.257-262

7 Hennig, K.: Zur Berechnung von nichtlinearen fremderregten Zufallsschwingungen fester Kontinua, Proc. VIII Internat. Conf. Nonlin. Oscillations, Prag 1978, pp.315-324

8 Hennig, K.; Friedrich, H.; Knöfel, L.: Berechnung von fremderregten Zufallsschwingungen fester Kontinua, Maschinenbautechn. 26(1977), 11, 496-504

9 Friedrich, H.: Analytische und numerische Untersuchungen unter nichtlinearen, stochastisch fremderregten Schwingungssystemen, Diss. B, Techn. Hochsch. Magdeburg 1976

10 Friedrich, H.; Hennig, K.: Zur Berechnung von Überschreitenswahrscheinlichkeiten stochastisch beanspruchter mechanischer Systeme, Preprint P-42/79, ZIMM der AdW der DDR, Berlin 1979

11 Lin, Y.K.: First-excursion failure of randomly excited structures, AIAA J. 8(1970),4,720-725

12 Weber, M.: Klassifizierung der Verfahren zur Leitintegration dynamischer Systeme mit Betrachtungen über Stabilität und Verfahrensfehler, Report R-Mech-01/81, Inst. f. Mech d. AdW d. DDR, Berlin 1981

13 Nguyen-xuan Hung: Zur Untersuchung von deterministischen und stochastischen Schalenschwingungen, Diss. B, AdW der DDR, Berlin 1981

14 Vronay, D.F.; Smith, B.L.: Free vibration of circular cylindrical shells of finite length, AIAA J. 8(1970),3

15 Davenport, A.G.: The application of statistical concepts to the wind loading of structures, Proc. Inst. Civil. Engg. 19(1961),449-472

16 Lindholm, U.S.; Kana, D.; Abramson, H.N.: Breathing vibrations of a circular shell with an internal liquid, J. Aeronaut. Sci. 29(1962), 1052-1059

17 Wen-hwa Chu: Breathing vibrations of partially filled cylindrical tanks - linear theory, J. Appl. Mech., Trans ASME 1963

18 Corcos, G.M.: The structure of the turbulent pressure field in boundary layer flows, J. Fluid Mech. 18(1964),353-358

19 Chen, S-s.; Wambsgans, M.W.: Parallel-flow induced vibrations of fuel rod, Nucl. Engin. Des. 18(1972),253-258.

FATIGUE CRACK PROPAGATION AND RELIABILITY OF STRUCTURES UNDER RANDOM EXCITATIONS

Ryoichi Koterazawa[1)]

Summary Fatigue crack propagation studies were carried out under variable stress conditions where very small number of overstress was applied intermittently between very large number of cycles of understress below the threshold stress. Crack propagation rates were accelerated significantly (more than one hundred times) by understressing below the threshold. The acceleration was more pronounced in low strength materials. A method is presented to predict crack propagation under random excitations by a linear summation of the basic data obtained by two-level variable stress tests. Remarks were made on the reliability of structures under random excitations.

1. Introduction

It is well known that the fatigue strength of machine and structure components in actual services is sometimes considerably lower than the strength obtained by laboratory fatigue tests. One of the reason of this discrepancy has been shown to lie in the fact that most of the cyclic stresses applied to the components in service are understress below the fatigue limit, very few cycles of overstress being applied intermittently. The Miner's law (linear cumulative damage law) is far from valid in such cases (Nakamura and Horikawa, 1970).

The author has been conducting experimental studies on this problem regarding crack propagation stage of fatigue process and obtained results which show that significant acceleration (more than one hundred times) occurs in cases where a very small number of cycles (e.g. 2 cycles) of overstress are applied intermittently between very large numbers of cycles (of the order of 10^6 cycles) of understress below the threshold stress intensity ΔK_{th} (Koterazawa 1976, 1978, 1981, Koterazawa et al. 1976, 1978, 1979, 1982). These results indicate that ΔK_{th}, which has been thought to be an important design criterion in fracture mechanics, loses its physical meaning as the threshold for fatigue crack propagation under such variable stress conditions and this acceleration can be a vital factor that has to be considered in design of such components.

In this report, experimental results on the acceleration of crack propagation will be summarized and a method to predict the crack propagation under random excitations taking account of the acceleration will be presented. Remarks on the reliability of structures under such conditions will also be given.

2. Crack propagation under two-level variable stresses (basic experiments)

The test materials were a low carbon steel (0.1%C), a medium carbon steel (0.4%C), high strength steels (1Cr-1/4Mo steels, 0.4%C, heat treated to different strength levels) and aluminum alloys (5052-O, 2017-T3, 7N01-T6). Most of the tests were made with specimens of circumferentially notched round bar type (Fig. 1a) by rotating-bending fatigue machines. Some tests were made with single edge notched plate specimens in in-plane bending (Fig. 1b) by a plane-bending fatigue machine.

---

1) Institute of Engineering Mechanics, University of Tsukuba, Sakura, Niihari, Ibaraki, Japan

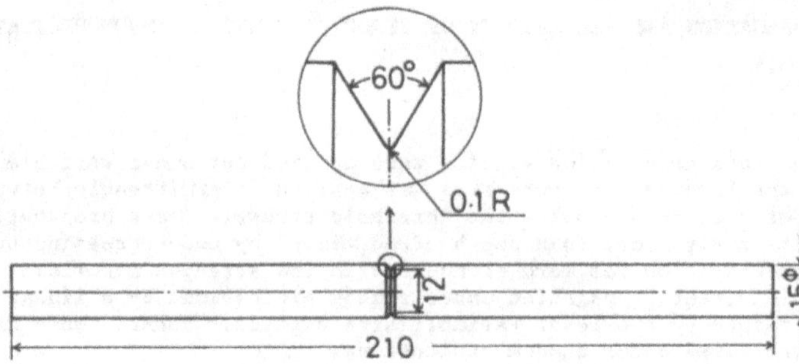

(a) round bar specimen for rotating bending test.

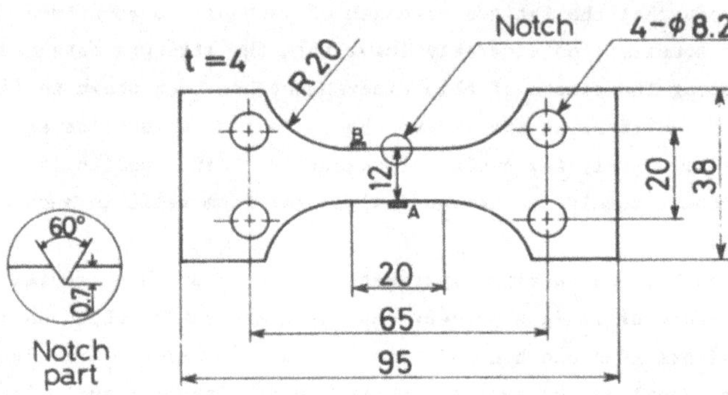

(b) plate specimen for in-plane bending test.

Figure 1. Test specimens.

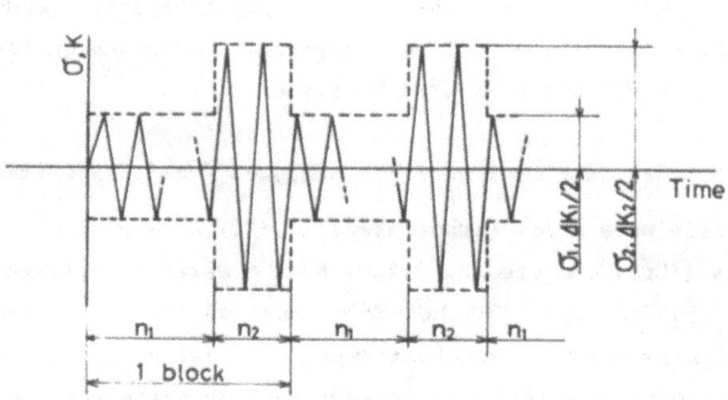

Figure 2. Loading program in two-level variable stress tests.

The specimens were, as a rule, precracked by alternating stress and tested under variable amplitude stress as shown in Fig. 2. The tests were started with understressing below ΔK_{th} followed by intermittent overstressing above ΔK_{th} as shown in the figure[1]. Finally, the tested specimens were fractured by a steady alternating stress or by an impulsive stress in liquid nitrogen. The total crack extension during periodic overstressing tests was measured on the fracture surface.

Some representative results with the round bar specimens are presented in Fig. 3. The acceleration ratio R_{ac} given in this figure is the actual crack extension under the periodic overstressing divided by a fictitious one which is obtained under the following assumptions : the crack propagation rate per cycle at any moment during a periodic overstress test is equal to that under steady cyclic stress of magnitude the same as the stress at that moment. The R_{ac} values for the cases of different understress ΔK_1 with the same overstress ΔK_2 are shown in the figure[2]. The occurrence of significant acceleration, more than one hundred in R_{ac} value, is clearly shown in this figure. The R_{ac} varies depending on materials strength, being larger in materials of lower tensile strength σ_B as indicated in the figure. That is, the acceleration is more pronounced in materials of lower strength. However, it seems that there is a lower limit in the value of understress below which R_{ac} is not large and that the limiting value does not depend appreciably on the materials, i.e. data curves for different materials tend to converge as the understress ΔK_1 decreases.

Test results with the plate specimens which show influence of overstress values[3] on the acceleration are shown in Fig. 4. There is a maximum in the acceleration at a certain value of the overstress ΔK_2. It is also shown in this figure that acceleration under repeated stress (R=0) is considerably less than that under alternating stress (R=-1). This would be due to well-known retardation effect by tensile overstressing.

An important point of this experiment lies in the large number of understress cycles in a program block n_1. In order to see the necessary number of n_1 to get significant acceleration, tests were done with different n_1 with other conditions being the same. The results are plotted in Fig. 5. In the range of smaller n_1, R_{ac} increases with n_1, but it is roughly constant in the range above $n_1 \sim 10^5$. In other words, number of $n_1 \sim 10^5$ is sufficient to give significant acceleration such as observed in this study. This is probably due to the recovery of dislocation structures by the large number of cycling of understress (Koterazawa, 1981).

In most of the overstressing tests conducted, the number of overstress cycles in a program block n_2 was 2. Such a small number was chosen because, as is well known, the first stress cycle of overstressing is much more effective for crack extension than subsequent cycles, so that smaller n_2 would give greater average crack extension per cycle during overstressing. As expected, the effect of n_2 is such as shown in Fig. 6, where the acceleration decreases inversely with the increase of n_2.

1) The overstresses were applied manually while the understress were applied by motor drive.
2) The magnitude of the overstress ΔK_2 was chosen such that the crack propagation rate under a steady stress of the same magnitude is around 0.1 µm/c.
3) Crack propagation rates under steady ΔK_2 are taken in abscissa in place of ΔK_2 values.

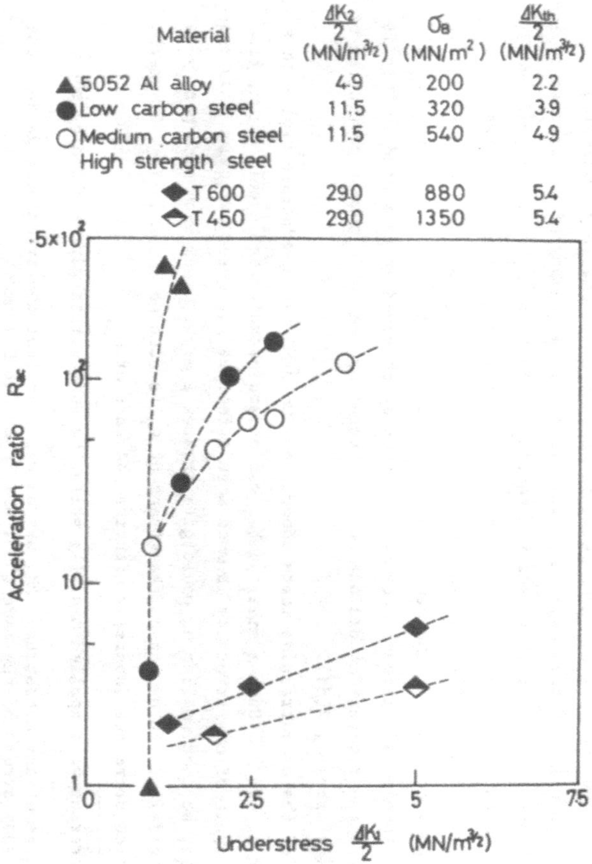

Figure 3. Acceleration of crack propagation by periodic overstressing ($n_1 \sim 10^6$, $n_2=2$, T600, T450: tempered at 600, 450 °C, respectively).

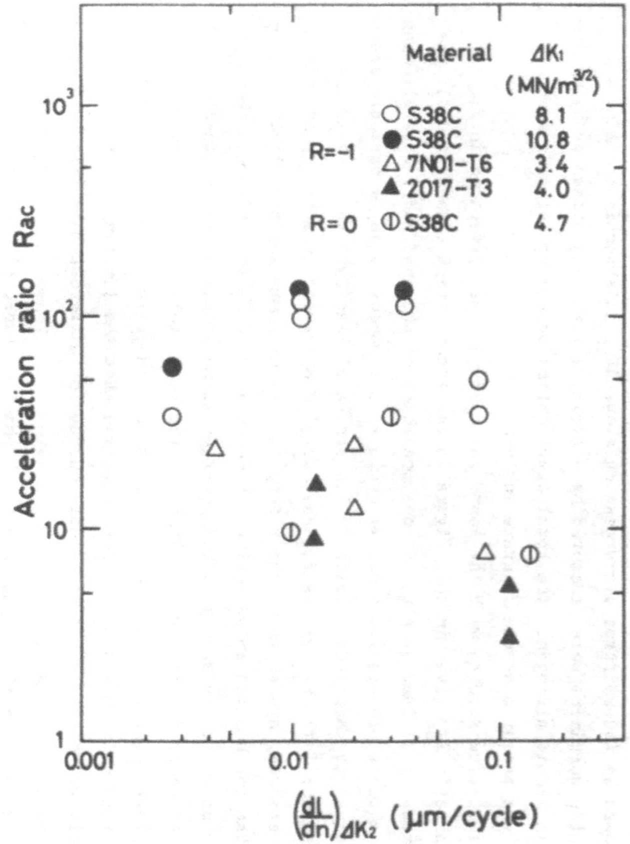

Figure 4. Effect of overstress value on acceleration of crack propagation ($n_1=2.2\times10^5$, $n_2=2$, $R=K_{min}/K_{max}$, S38C: medium carbon steel).

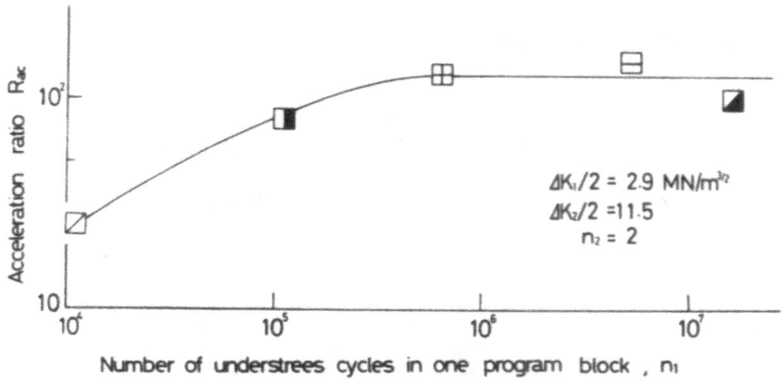

Figure 5. Effect of the number of understress cycles on acceleration of crack propagation (low carbon steel, rotating-bending).

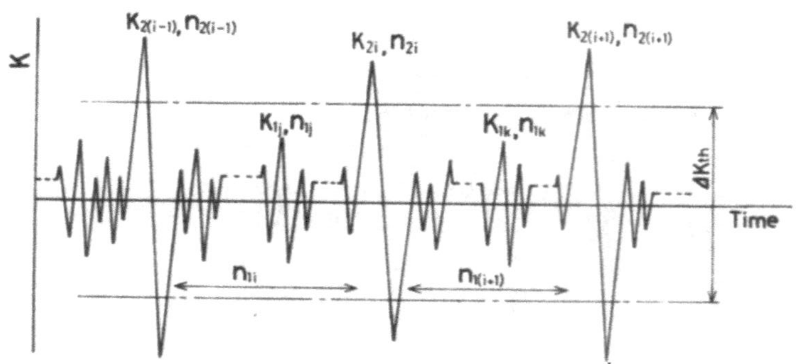

Figure 7. A schematic illustration of random stresses.

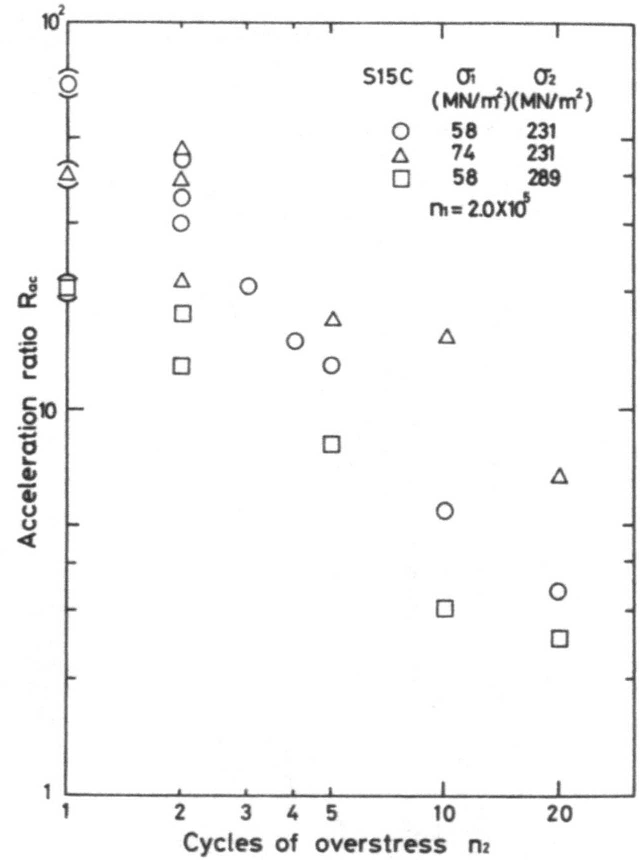

Figure 6. Effect of the number of overstress cycles on acceleration of crack propagation (low carbon steel, rotating-bending, K $MN/m^{3/2}$-0.05 σ MPa).

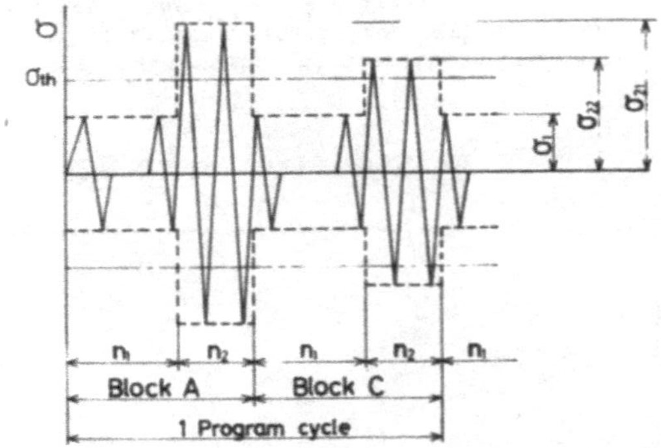

(a) Loading program I

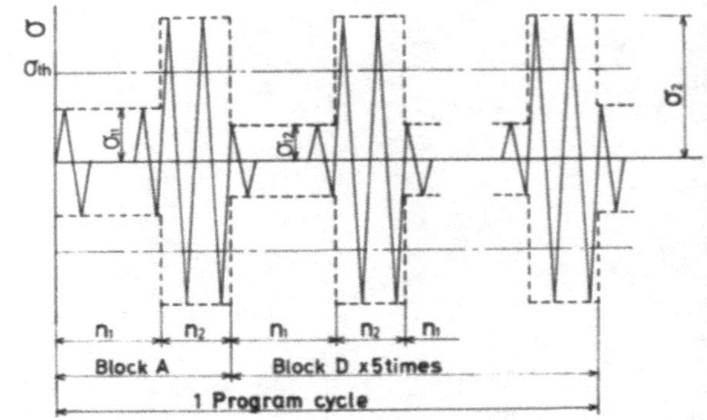

(b) Loading program II

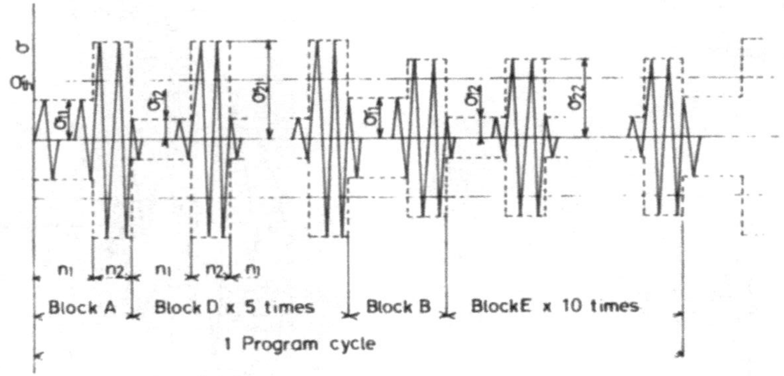

(c) Loading program III

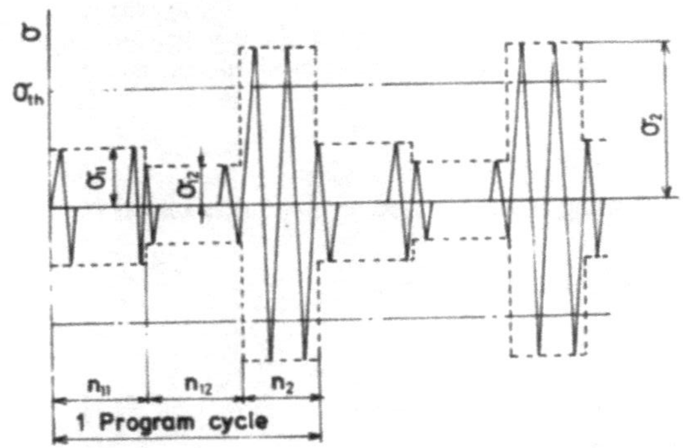

(d) Loading program IV

Figure 8. Loading programs in multi-level variable stress tests.

3. Crack propagation under multi-level variable stresses (random excitations)

A tentative method will be presented in this section to predict crack propagation under random stresses, in which most of the stresses are below ΔK_{th} and very few cycles of overstress is applied intermittently, on the basis of basic data of two-level variable stress tests.

As an example, let us consider a variable alternating stress such as shown in Fig. 7. Denote the total number of stress cycles in an understress block immediately before the ith overstress K_{2i} (number of cycles n_{2i}) as n_{1i} and the number of cycles of understress K_{1j} in that block as n_{1j}; denote those immediately after the K_{2i} as $n_{1(i+1)}$ and K_{1k}, n_{1k}. Assuming that acceleration ratios R_{ac} for arbitrary K_2, n_2, K_1, n_1 have been obtained by the two-level tests such as described in the preceding section, we will make a linear summation calculation to get the acceleration for the multi-level case such as illustrated in Fig. 7. In order to do this, we assume:

(1) The acceleration due to a block of understress is given by a linear summation of the effects of each understress in that block, i.e. $\sum_j (n_{1j}/n_{1i}) R_{ac}$.

(2) The acceleration due to an overstress K_{2i} is controlled by the understress blocks immediately before and after the K_{2i} and is given by the average of the effects of the two. Then, we have

$$R^*_{ac} = \frac{\sum_i (1/2)\{\sum_j (n_{1j}/n_{1i}) R_{ac}(K_{2i}, K_{1j}, n_{1i}) + \sum_k (n_{1k}/n_{1(i+1)}) R_{ac}(K_{2i}, K_{1k}, n_{1(i+1)})\}(dl/dn)_{K_{2i}} \cdot n_{2i}}{\sum_i \{(dl/dn)_{K_{2i}} \cdot n_{2i}\}} \quad (1)$$

In order to check the validity of this equation, multi-level variable stress tests shown in Fig. 8 were carried out with round bar specimens by means of rotating-bending machines. These tests were conducted in the same way as the two-level stress tests described in the preceding section. The results were compared with the predicted ones by Eq. (1) in Table 1. In this table, the ratio of experimental results to predicted ones R_{ac}/R^*_{ac} are from 0.7 to 1.3, which is well within the experimental error of this kind of fatigue tests. This indicates prospect of the prediction method proposed here.

Although the above example is for the case of alternating stress, the procedure is essentially the same also in the cases where the mean stress is included.

4. Remarks on the reliability of structures under random excitations

As discussed in the preceding sections, the linear cumulative damage laws combined with steady stress test data are far from valid in such cases as illustrated here which are expected to appear sometimes in actual services. Any reliability theory based on this type of law and data is unreliable in such cases. It should be based on variable stress tests such as illustrated in the previous sections. However, there are very few data of this sort available mainly because of the time consuming nature of tests of this kind. The number of understress cycles n_1 is extremely large as compared with the one usually employed in conventional variable stress fatigue tests, and this results in very large total test time necessary for this kind of experiments. As an important point regarding the reliability, it is practically impossible to get a sufficient number of test data for reliable statistical analysis, especially when one concerns with very low probability of failure such as in the case of aircraft structures, for example, where extremely large a number of test data is required. Therefore, mathematical theories of reliability based on statistics presently available are definitely useless for such cases. Formal application of such theories may lead to fatal accidents with appreciable probability. What is important for this kind of problem is not to make formal statistical analysis but to know physical reasons of failure by experiments or by failure analyses.

Table 1. Results of multi-level variable stress tests (low carbon steel, K $MN/m^{3/2}$-0.05 σ MPa).

<table>
<tr><th rowspan="2">Program No.</th><th>σ_{11}</th><th>σ_{12}</th><th>σ_{21}</th><th>σ_{22}</th><th rowspan="2">n_{11}</th><th rowspan="2">n_{12}</th><th rowspan="2">n_2</th><th rowspan="2">R_{ac}</th><th rowspan="2">R^*_{ac}</th><th rowspan="2">R_{ac}/R^*_{ac}</th></tr>
<tr><th colspan="4">(MPa)</th></tr>
<tr><td rowspan="2">I</td><td colspan="2" rowspan="2">58</td><td rowspan="2">232</td><td rowspan="2">139</td><td colspan="2" rowspan="6">$2.2x10^5$</td><td rowspan="11">2</td><td>249</td><td rowspan="2">307</td><td>0.81</td></tr>
<tr><td>260</td><td>0.85</td></tr>
<tr><td rowspan="2">II</td><td rowspan="2">58</td><td rowspan="2">29</td><td colspan="2" rowspan="2">232</td><td>42</td><td rowspan="2">65</td><td>0.65</td></tr>
<tr><td>55</td><td>0.85</td></tr>
<tr><td rowspan="2">III</td><td rowspan="2">58</td><td rowspan="2">29</td><td rowspan="2">232</td><td rowspan="2">174</td><td>103</td><td rowspan="2">83</td><td>1.24</td></tr>
<tr><td>98</td><td>1.18</td></tr>
<tr><td rowspan="5">IV</td><td rowspan="2">58</td><td rowspan="2">29</td><td colspan="2" rowspan="2">232</td><td colspan="2" rowspan="3">$1.1x10^5$</td><td>167</td><td rowspan="2">131</td><td>1.27</td></tr>
<tr><td>128</td><td>0.98</td></tr>
<tr><td>29</td><td>58</td><td colspan="2">232</td><td>139</td><td>131</td><td>1.06</td></tr>
<tr><td>29</td><td>58</td><td colspan="2">232</td><td>$2x10^5$</td><td>$2x10^4$</td><td>43</td><td rowspan="2">49</td><td>0.88</td></tr>
<tr><td>58</td><td>29</td><td colspan="2">232</td><td>$2x10^4$</td><td>$2x10^5$</td><td>73</td><td>1.49</td></tr>
</table>

As another important point regarding this problem, it should be pointed out that strength or lifetime of structures under complex conditions in actual services can not be deduced on the basis of conventional so-called fundamental tests conducted under simple conditions, such as the steady stress fatigue tests commonly employed. Tests should simulate essential points of actual service conditions including interactions between simple elementary cases like the ones presented in the preceding sections. It is a usual way of investigation to decompose complex actual service conditions into several simple cases and to study each case separately and then synthesize their results to get the behavior under complex conditions. This, seemingly, is being done in most of investigations without proper consideration on its validity. However, this can be done only if the complex system is linear or close to linear and does not have appreciable interactions between simple cases, and not in the system such as the one under consideration which is significantly non-linear and has significant interactions.

It should also to noted here that conventional variable stress fatigue tests are not useful because the number of understress cycles in these tests is usually comparable with the number of overstress cycles, or at most not far from it, so that accumulated damage or crack growth under understress cycles are small, or at most not large, as compared with the one under overstress, the effect being trivial and within the scatter of test data. Note that a factor of two is common in the scatter of fatigue life data , so that any meaningful effect should exceed this factor. Therefore, in order to obtain meaningful data, the number of understress cycles should be very large as shown in this study. However, most of the fatigue data under variable stress conditions presently available deals with trivial effects within this scatter, because it takes very much time to do the tests that yield significant effects as described above.

5. Conclusion

Fatigue crack propagation studies were carried out under variable amplitude stress conditions where a very small number of overstress was applied intermittently between a very large number of cycles of understress below the threshold stress intensity ΔK_{th}. The results are summarized as follows.

(1) Crack propagation rates were accelerated significantly (more than one hundred times) by understressing below ΔK_{th}.

(2) Acceleration was more pronounced in low strength materials. However, the lower limit of understress value below which no significant acceleration occurs did not change appreciably with strength of the material.

(3) Acceleration of crack propagation under multi-level variable stresses could be predicted with reasonably good accuracy by a proposed method based on a linear summation of the basic data obtained by two-level variable stress tests.

(4) On the basis of the test results presented here, remarks were made on the reliability of structures under random excitations and it was pointed out that the conventional reliability theories based on mathematical statistics would be useless for such cases as discussed in this report.

Very important would be the fact pointed out in this study that very low stresses below the threshold, which may easily be ignored in practical design, can have significant effects on fatigue crack propagation. Although the experimental data presented here are rather limited, the author believes that essential points were elucidated in this study. Systematic investigations along the same line are desired in order to provide sufficient data for reliable design that takes account of the acceleration effects pointed out here.

References

Koterazawa, R., 'Fatigue crack propagation under periodic overstressing', Int. J. Fracture, Vol. 12 (1976) p. 169.

Koterazawa, R., 'Fatigue crack propagation and discussion on design criterion under periodic overstressing', J. Soc. of Materials Science, Japan, Vol. 27 (1978) p.871.

Koterazawa, R., 'Fatigue crack propagation under periodic overstressing', Materials, Experimentation and Design in Fatigue, Proc. Fatigue '81, Westbury House 1981, p.159.

Koterazawa, R. and Shimo, D., 'Fatigue crack propagation and cumulative damage under periodic overstressing', J. Soc. of Materials Science, Japan, Vol. 25 (1976) p. 875.

Koterazawa, R., Shimo, D. and Kondo, J., 'Fatigue crack propagation and fractography under periodic overstressing', J. Soc. of Materials Science, Japan, Vol. 27 (1978) p. 54.

Koterazawa, R., Watanabe, H. and Takara, Y., 'Fatigue crack propagation under periodic overstressing with special attention to crack closure and dislocation structure', J. Soc. of Materials Science, Japan, Vol. 28 (1979) p. 868.

Koterazawa, R., Morisaki, I. and Yamada, T., 'Fatigue crack propagation under periodic overstressing and prediction of propagation rate', J. Soc. of Materials Science, Japan, Vol. 31 (1982) p. 685.

Nakamura, H. and Horikawa, T., 'Experiments on the variable stress pattern that gives minimum value of $\Sigma(n_i/N_i)$', Trans. JSME, Vol. 36 (1970) p. 1059.

ON THE RELIABILITY OF FLEXIBLE STRUCTURES UNDER NON-NORMAL LOADING PROCESSES

H.-J. Melzer[1] and G.I. Schuëller[2]

The dynamic analysis of structures under random loading is quite frequently carried out in the frequency range utilizing the classical power spectral method. As a consequence for statistical interpretation of the results the assumption of Gaussian properties of the process has to be made which might not be realistic in a number of practical cases. In this paper an expansion of the classical power spectral method is introduced which allows the treatment of non-Gaussian processes. For this purpose higher statistical moments are used to obtain analytical relations between input and output processes. The method is exemplified by analysis of structures under wind loading.

1. Introduction

Tall, slender structures, for example T.V. towers, high rise buildings, offshore platforms etc. are subjected to dynamic loadings due to wind, earthquake accelerations and waterwaves. These loads are random in space and time and therefore have to be modeled by stochastic processes. For analysis of such structures a number of methods, such as the Time History Method (THM), the Fourier Transformation (FT) and the "Stochastic Methods" (Stochastic Differential Equations, Covariance Analysis and Power Spectral Method (PSM)), are available. An obvious shortcoming of the THM is the fact, that for statistical interpretation of the results a large number of structural analyses has to be carried out - each for a particular realization of the input process. In general, this requires considerable computational efforts. To avoid this problem, stochastic methods are frequently utilized. However, in order to apply these methods a number of assumptions has to be made. The classical power spectral method (PSM), requires for example the assumption of linear system properties with small viscous damping. For distributional purposes - in addition - the assumption of normality of the input process is also required. According to the theory of stochastic processes only this type of process is fully defined by its mean value and its covariance matrix. Furthermore, it is the only process, which is invariant against linear operations. In other words Gaussian input in a linear system results in a Gaussian output. In practical application the assumption of Gaussian properties of loading process proves in many cases to be a very strong restriction.

1) Dr.-Ing., MBB, Ottobrunn, F.R. Germany

2) o.Prof.Dr.-Ing. habil., Universität Innsbruck, Austria

As shown in Fig. 1, the probability density functions of earthquake and wind loading do not necessarily reveal normal properties. With respect to its strong influence on reliability estimation of structures this fact is of considerable importance. In the following, an extension of the classical PSM is introduced which allows the treatment of non-normal properties in the frequency range. For this purpose higher statistical moments are utilized. Assuming ergodicity of the input and linear structural properties, the stochastic process of the loading as well as the response may be represented by its first one dimensional distribution. In fact non-Gaussian distributions can be represented by series - for example by Edgeworth- or Gram-Charlier Series etc. - approximation for which higher statistical moments are utilized. Moreover, extreme values may also be determined by also using the information of higher moments.

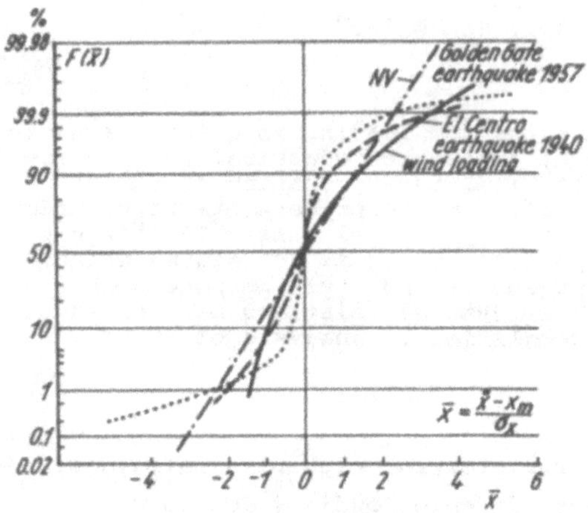

Fig. 1: Probability Density Functions, $F(\bar{x})$, of Observed Wind Pressure and Earthquake Acceleration Records Plotted on Normal Probability Paper

2. Method of Analysis

2.1 Aspects of Higher Statistical Moments (HSM)

As HSM are treated extensively in the literature (see for example [3,4,6,9]) only some basic principles are given here in order to enhance the understanding of further developments. The Central Statistical momoments of a random process $X(t) = X_t$ are defined as

$$\mu_\nu = E[(x_t - x_m)^\nu] = \int_{-\infty}^{\infty} (x_t - x_m)^\nu \, f(x)dx, \quad \nu = 1,2,\ldots, \tag{1}$$

where x_m is the mean value and ν indicates the order of the respective moment. The first two moments represent the central tendency (mean value) and the dispersion (standard deviation; $\mu_2 = Var(x_t) = \sigma_x^2$) of the process respectively. The third moment is a measure of its skewness or asymmetry and is defined by a non-dimensional coefficient $\gamma_1 = \pm \mu_3/\sigma_x^3$. The fourth moment defines the peakedness or excess of the process ($\gamma_2 = [\mu_4/\sigma_x^4]-3$). As γ_1 and γ_2 are zero for the Gaussian distribution, they may be used as indicator for deviations from the normal assumption [5].

2.2 Estimation of Probability Distribution Function (PDF) using HSM

The Edgeworth- as well as the Gram-Charlier (Type A) Series are based on a formally identical structure for the PDF:

$$f(\bar{x}) = \phi(\bar{x}) + \frac{C_3}{3!}\ \phi^{(3)}\ (\bar{x}) + \frac{C_4}{4!}\ \phi^{(4)}\ (\bar{x}) + \ldots \tag{2}$$

and for the cumulative distribution function (CDF)

$$F(\bar{x}) = \Phi(\bar{x}) + \frac{C_3}{3!}\ \phi^{(2)}(\bar{x}) + \frac{C_4}{4!}\ \phi^{(3)}(\bar{x}) + \ \ldots \tag{3}$$

where $\bar{x}_t = (x_t - x_m)/\sigma_x$. The coefficients C_K which contain the HSM information, however, differ beginning with the 6th order [3,4]. The functions $\phi(\bar{x})$ and $\Phi(\bar{x})$ represent the normal distribution; $\phi^{(n)}(\bar{x})$ are the derivatives of the normal distribution which may be obtained by Hermite polynomials $H_n(\bar{x})$:

$$\phi^{(n)}(\bar{x}) = (-1)^n\ H_n(\bar{x})\phi(\bar{x}). \tag{4}$$

The HSM of a random process may be calculated by estimating its correlation function from observed sample time histories. Assuming a stationary process X_t with zero mean value the second order correlation of a continuous process is

$$R_{x,2}\ (\tau_1) = \lim_{T\to\infty} \frac{1}{T} \int_0^T x(t)x(t+\tau_1)dt. \tag{5}$$

The variance of the process is then $\mu_{x,2} = R_{x,2}(0)$. Expanding equ (5) the higher order correlation functions $R_{x,3}$, $R_{x,4}$ etc and consequently the HSM $\mu_{x,3}$, $\mu_{x,4}$ etc may be obtained [3,4,7]. The meaning of higher order correlation becomes even more obvious when considering higher order spectra. These spectra may be obtained by performing multidimensional Fourier-Transformation (FT). The wellknown Wiener-Chintchin relation provides the power spectrum (2nd order). The expressions up to the 4th order are given in the following:

$$G_{x,2}(f_1) = \int_{-\infty}^{\infty} e^{-i2\pi f_1 \tau_1}\ R_{x,2}(\tau_1)d\tau_1, \tag{6}$$

Bi-spectrum (3rd order)

$$G_{x,3}(f_1,f_2) = \iint_{-\infty}^{\infty} e^{-i2\pi(f_1\tau_1\ +\ f_2\tau_2)}.\ R_{x,3}(\tau_1,\tau_2)d\tau_1 d\tau_2, \tag{7}$$

Tri-Spectrum (4th order)

$$G_{x,4}(f_1,f_2,f_3) = \iiint_{-\infty}^{\infty} e^{-i2\pi(f_1\tau_1+f_2\tau_2+f_3\tau_3)} R_{x,4}(\tau_1,\tau_2,\tau_3)\cdot d\tau_1 d\tau_2 d\tau_3. \tag{8}$$

Based on this information the HSM may be obtained:

$$\mu_{x,2} = \int_{-\infty}^{\infty} G_{x,2}(f_1)df_1, \tag{9}$$

$$\mu_{x,3} = \iint_{-\infty}^{\infty} G_{x,3}(f_1,f_2)df_1df_2, \tag{10}$$

$$\mu_{x,4} = \iiint_{-\infty}^{\infty} G_{x,4}(f_1,f_2,f_3)\cdot df_1df_2df_3. \tag{11}$$

From this one can see clearly that the spectra are representations of HSM in the frequency domain. The numerical evaluation of the integrals shown above is limited. So far the numerical evaluation up to the 3rd order (bi-spectrum) seems to be feasable [2,8].

2.3 Structural Response

2.3.1 The Transfer Function

The statistical parameters of the input process will be filtered by the structural system under consideration. The latter may be represented by a complex frequency response matrix $\hat{H}(f)$ which is by definition the FT of the response of a vibrational system under impulse loading $\delta(t)$. It is also refered to as the transfer function. For a multi degree of freedom (MDOF) system uncoupling of the system of equations of motions may be accomplished by using modal analysis. Therefore, the complex frequencies of the various modes - which are identical with the frequencies of the single degree of freedom (SDOF) systems - may be easily obtained:

$$\hat{H}(f) = \hat{H}_n(f) = \frac{1/k_n}{1-(\frac{f}{f_n})^2 + i2\xi_n \frac{f}{f_n}} \tag{12}$$

where $H_n(f)$ is the complex frequency response function of the nth mode, $f_n = \sqrt{k}/m$ (k = stiffness, m = mass) its natural frequency and $\xi = d/d_{crit}$ the damping ratio where the critical damping $d_{crit} = 2\sqrt{km}$.

2.3.2 SDOF Systems

Utilizing equs(6), (7), (8) and (12) the higher order spectra of the response, $G_{y\nu}(\cdot)$ for a SDOF system may be obtained by the following relations:

$$G_{y,2}(f) = G_{x,2}(f)\ H^*(f)\ H(f), \tag{13}$$

$$G_{y,3}(f_1,f_2) = G_{x,3}(f_1,f_2)H^*(f_1+f_2)\cdot H(f_1)H(f_2), \tag{14}$$

$$G_{y,4}(f_1,f_2,f_3) = G_{x,4}(f_1,f_2,f_3)\cdot H^*(f_1+f_2+f_3)H(f_1)H(f_2)H(f_3),$$

etc. (15)

where equs(13), (14), (15) represent the Power-, Bi- and Tri-Spectra of the response respectively. $H^*(f)$ is the conjugate of the complex frequency response. The HSM of the response can then be obtained in analogy to equs(9), (10) and (11).

2.3.3 "White Noise" Input

Equs (13), (14), (15) may now be simplified by assuming "white noise" properties of the stochastic process of excitation, X_t. For this case, the input spectra $G_{x,\nu}(\cdot)$ are constant values which lead to the following general expression for the HSM:

$$\mu_{y,n+1} = \mu_{x,n+1} \underset{1.....n}{\iint....\iint} H^*(f_1+....+f_n)H(f_1)...H(f_n)\cdot df_1...df_n. \quad (16)$$

In particular for the variance, skewness and peakedness:

$$\mu_{y,2} = \mu_{x,2} \int |H(f)|^2 \, df, \quad (17)$$

$$\mu_{y,3} = \mu_{x,3} \iint H^*(f_1+f_2)\cdot H(f_1)H(f_2)df_1df_2, \quad (18)$$

$$\mu_{y,4} = \mu_{x,4} \iiint H^*(f_1+f_2+f_3)\cdot H(f_1)H(f_2)H(f_3)\cdot df_1df_2df_3. \quad (19)$$

If the function $H(f)$ is defined by an analytical expression - for example in case of a SDOF-System or a modal complex frequency response function (see equ (12)) - the integrals may be solved analytically. In other words, the HSM of the response $\mu_{y,\nu}$ can be calculated from the HSM of the excitation $\mu_{x,\nu}$ directly, i.e. without numerical integration. Solutions for the variance (equ(17)) may be obtained by application of the method of residues (see for example [6]) and by utilizing Parseval's equation which provides a relation between the quadrants of the response function of the impulse $h(t)$ and the complex frequency response function $\hat{H}(t)$:

$$\int_0^T |h(t)|^2 \, dt = \int_0^T |\hat{H}(t)|^2 df \quad (20)$$

in which

$$h(t) = \frac{1}{\alpha_D m} e^{-\xi f_n t} \sin \alpha_D t \quad (21)$$

where the parameter $\alpha_D = f_n\sqrt{1-\xi^2}$ takes the effect of damping on the natural frequency into account.

Utilizing the above relations one obtains for the relation between the variance of the excitation and the response of a SDOF system

$$\mu_{y,2} = \mu_{x,2} \frac{(1-\xi^2)}{4(m\alpha_D)^2\alpha_D} \left[e^{-2\xi f_n T} \sin\alpha_D T \left(\frac{-2\xi}{\sqrt{1-\xi^2}} \sin\alpha_D T - 2\cos\alpha_D T\right) - \right.$$

$$\left. - \frac{\sqrt{1-\xi^2}}{\xi} e^{-2\xi f_n T} + \frac{\sqrt{1-\xi^2}}{\xi} \right] . \qquad (22)$$

For $T \to \infty$ i.e. for an infinitely long time interval equ (22) reduces to

$$\mu_{y,2} = \mu_{x,2} \frac{1}{4mkf_n\xi} . \qquad (23)$$

Additional relations between excitation and response for HSM may be obtained by continued application of the method of residues to the multidimensional integrals of the frequency response functions.

2.3.4 MDOF-Systems

For MDOF-Systems generally a number of load functions $x_k(t)$ contribute to the response y_i (see Fig. 2). The properties of a structure with respect to the transfer of a load at point k and the response at point i are represented by one member of the frequency response matrix $\hat{H}_{ik}(f)$. For calculation of the higher order spectra of the response y_i, all contributions of the various loading functions $x_k(t)$ have to be taken into account by superposition. For the Power- and the Bi-spectrum respectively this yields:

$$G_{y_{i,2}}(f) = \sum_k \sum_l \hat{H}^*_{ik}(f) H_{il}(f) G_{x_{kl,2}}(f), \qquad (24)$$

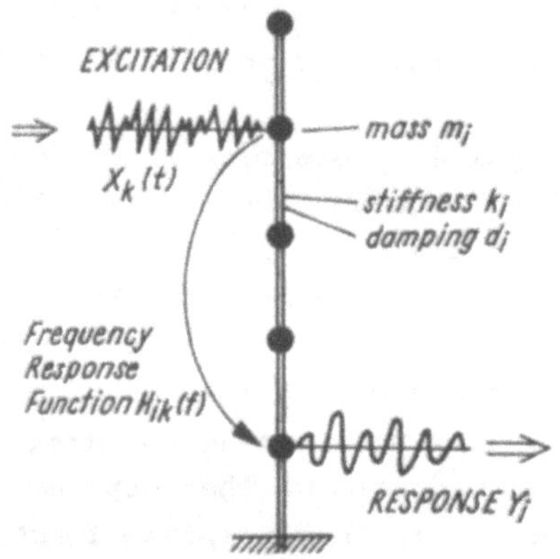

Fig. 2: Schematic Sketch of MDOF-System

$$G_{y_{i,3}}(f_1,f_2) = \sum_k \sum_l \sum_m \hat{H}^*_{ik}(f_1+f_2) \hat{H}_{il}(f_1) \hat{H}_{im}(f_2) G_{x_{klm,3}}(f_1,f_2) .$$

The summation has to be carried out over all loads k,l,m = 1,2,...K. Again, the statistical moments of the response are obtained by integration.

Alternatively, the relation between the statistical moments of the excitation and response for MDOF-systems may be established by modal decoupling of the differential equations of motion. It is well known that the modal contributions to the response $y_\nu(t)$ of a particular point r of a structure is

$$y_r(t) = \sum_{n=1}^{N} \phi_{rn} a_n(t) \tag{25}$$

where ϕ_{rn} are the eigenvectors and $a_n(t)$ the generalized coordinates. Utilizing the equation above one obtains for the 2nd and 3rd moments of the response, i.e. the variance and the skewness:

$$\mu_{y,2} = \frac{1}{T} \sum_{n_1=1}^{N} \sum_{n_2=1}^{N} \phi_{rn_1} \phi_{rn_2} \int_o^T a_{n_1}(t)\, a_{n_2}(t)dt, \tag{26}$$

$$\mu_{y,3} = \frac{1}{T} \sum_{n_1=1}^{N} \sum_{n_2=1}^{N} \sum_{n_3=1}^{N} \phi_{rn_1} \phi_{rn_2} \phi_{rn_3} \int_o^T a_{n_1}(t) a_{n_2}(t) a_{n_3}(t)dt. \tag{27}$$

For both relations, the time integral represents modal moments as well as cross moments.

2.4 Reliability Considerations

Knowing the statistical properties of the underlying load process and the structural response it is important-particularly for reliability based analysis and design purposes - to obtain information concerning extreme values of the response in connection with the number of exceedances of a critical threshold level Λ. This depends on the joint probability distribution of the process Y_t and its derivatives $\dot{Y}_t$, $\ddot{Y}_t$, i.e. $f_{y,\dot{y},\ddot{y}}\,(y = \Lambda\ ,\ \dot{y} = o,\ \ddot{y},t)$ [6]. The solution for the assumption of stationarity and normality of the process is well known [1]:

$$\eta_m = \gamma\sigma_y = [\sqrt{2\ \ln\nu T} + \frac{0{,}5772}{\sqrt{2\ \ln\nu T}}]\sigma_y \tag{28}$$

where η_m is the mean of the extreme values given a particular peak factor γ and the variance of the response σ_y. T is the time interval under consideration and $\nu^2 = \lambda_{2,y}/\lambda_{o,y}$ a ratio between spectral moments $\lambda_{n,y} = \int_{\Omega}^{\infty} f^n\, G_y(f)df$ where f is the structural frequency and $G_y(f)$ the PSD of Y_t.

For non-Gaussian processes the relation for the peak factor is to be expanded. The following general expression has been suggested [10]:

$$\eta^*_m = gM_{2o}^{1/2} = [\sqrt{2\ \ln\nu^* T} + \frac{0{,}5772+\ln(1+\phi)}{\sqrt{2\ \ln\nu^* T}}]\ M_{2o}^{1/2} \tag{29}$$

in which

$$\nu^* = [\frac{M_{02}}{M_{20}}]^{1/2} = \frac{E[Y^o{}_t, \dot{Y}_t^2]}{E[Y_t^2, \dot{Y}_t{}^o]} \tag{30}$$

and

$$\phi = \sum_{j=1}^{4} a_j \, (\sqrt{2 \ln \nu^* T})^j \tag{31}$$

where $a_j = F(M_{pq})$ is a function of the joint moment $M_{pq} = E[Y_t^p, \dot{Y}_t^q]$. Due to the additional term ln $(1+\phi)$ in equ (29) the non-Gaussian gust factor g may be 20-25 % larger than γ.

3. Numerical Examples

3.1 Simplified Calculation of HSM of the Response

Suppose a wind load is applied to a SDOF system (equ(12)). The statistical moments m_ν, μ_ν, - up to the 5th order - as determined from the time history of the response are shown in Fig 3. In comparison to simulation results, simplified calculations for the 2nd and 3rd moments respectively - utilizing equs (17) and (18) - are also shown. For this evaluation only part of the spectrum of excitation around the natural frequency of the complex frequency response function is used for integration rather than the entire range - as defined by equs (13),(14)- i.e. spectrum. From Fig 3 it can be seen that the accuracy of the approximation yields satisfactory results.

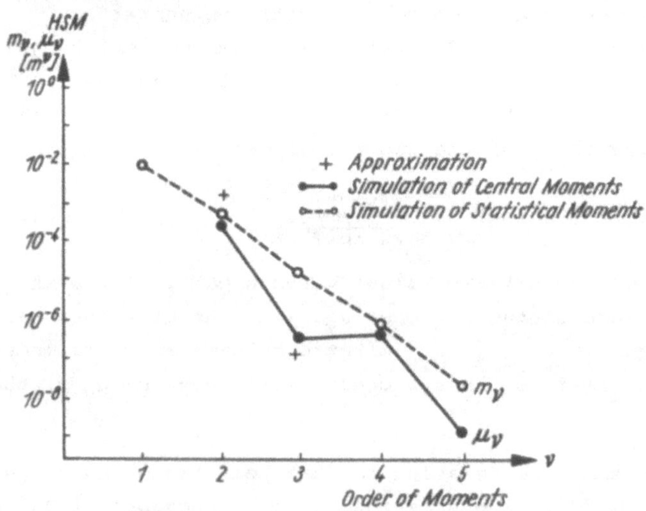

Fig. 3: Approximative Solution for HSM (μ_{y2}, μ_{y3}) of a SDOF System. Comparison with Simulation Results

3.2 Transformation of HSM

The change of the statistical properties of the response as compared with those of the excitation is advantageously represented in a so called β_1,β_2-Diagram, where β_1 and β_2 are coefficients of asymmetry (skewness) and peakedness (excess), i.e. $\beta_1 = \mu_3^2/\mu_2^3 = \gamma_1^2$ and $\beta_2 = \mu_4/\mu_2^2 = \gamma_2+3$. In Fig 4 one can observe the transformation of the coefficients β_1 and β_2 of a typical wind loading in the coefficients of the output, i.e. response. The point (0,3) is the so called normal point and represents the Gaussian distribution. The results show, that there is a tendendy of the SDOF System to transform even a strongly non-normal loading towards a normal response. In fact the skewness disappears but the peakedness does not. By invoking the central limit theorem it is quite frequently stated, that any system transforms a Non-Gaussian input in a normal output. The example shows, that this must not necessarily be true in all cases.

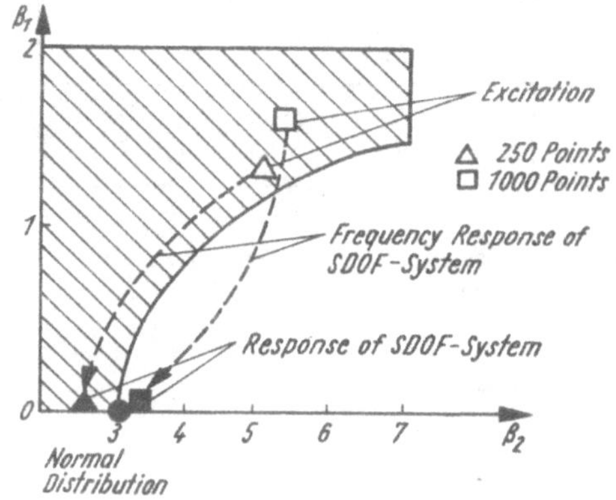

Fig 4.: β_1,β_2 Diagram for SDOF-System Under Wind Loading. Change of HSM Input/Output

3.3 Extrema of Non-Gaussian PDF

If a response of a system is assumed to be normal, the 3σ or 3.5σ fractiles of important maximum response parameters are generally used for design purposes. These fractiles refer to non-exceedance probablities of 99.865% on 99.977% respectively (see Fig 5). However, as pointed out in the previous sections, this assumption does not hold in a number of cases. The lower point of the antenna (restraint) of the Munich TV-Tower, for example, reveals for the horizontal deflection considerable deviations from the Gaussian distribution (Table 1). A repeated fit of various distributions to the time history of the response showed good values for both,

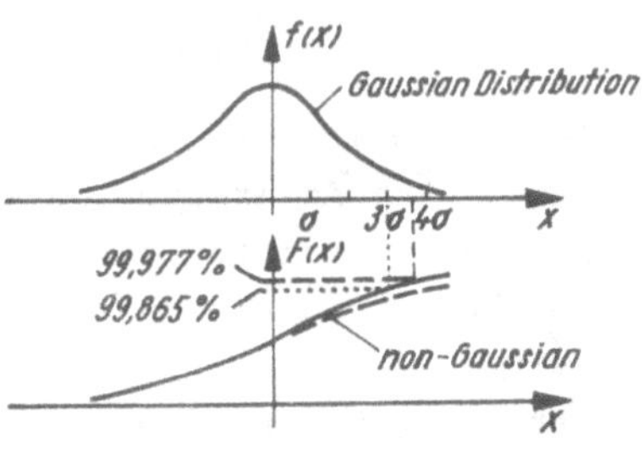

Fig 5.: Fractiles and Exceedance, Probabilities of Normal Distribution

the Weibull distribution and the Gram-Charlier series. Now extremal responses using Non-Gaussian distributions can be calculated. Table 2 shows in a comparison the change in fractile values for given non-exceedance probabilities P. It can be seen from the results, that Non-Gaussian distributions may yield up to 25% larger extremal response values as one would obtain adhering to the Gaussian assumption.

Deflection of Lower Antenna Point of Munich-TV-Tower	Standard Deviation $\sigma = \sqrt{\mu_2}$	Skewness $\gamma_1 = \mu_3/\sigma_3$	Peakedness $\gamma_2 = (\mu_4/\sigma^4)-3$
Actual Value	0.094 m	0.123	0.36
Value for Normal Distribution	0.094 m	0	0

Table 1: Coefficients of Skewness and Peakedness for Deflection at Lower Antenna Point of Munich-TV-Tower.

Distribution	Extrema for a particular non-exceedance probability P	
	P_1=99.865%	P_2=99.977%
Normal	3	3.5
Gram-Charlier	3.55	4.4
Weibull	3.25	3.775

Table 2: Non-Gaussian Fractiles

4. Conclusions

Considering the classical PSM which is presently used quite frequently to analyse dynamic systems under stochastic loading, the determination of HSM of input and output provides a considerable amount of additonal information, which is of great importance for design. It should therefor not be neglected in practical application. HSM can be calculated either from time histories directly or from higher order correlations or higher order spectra. The latter may be interpreted as frequency distributions of higher moments. This is in analogy to the power spectrum as frequency distribution of the variance. Therefore, the existance of higher order spectra is a measure of the Non-Gaussian properties of a process of for that matter non-linear structural behavior. HSM are to be used for series approximation of non-normal distributions. Its first term is respresented by the normal distribution and therefore contains the information obtained from the classical PSM. A more realistic description of the exceedance probability of certain extremal design thresholds results in a more reliable as well as economical design of structures under stocastic dynamic loading.

Acknowledgement

This research has been partially supported by the Deutsche Forschungsgemeinschaft under contract No.SFB96/B1o, which is gratefully acknowledged by the authors.

References

[1] Davenport, A.G.: Note on the Distribution of the largest value of a random function with application to gust loading, Proc. Inst. Civil Engrs., Vol 28, pp 187-196, 1964.

[2] Gabr, M.M. and Subba Rao, T.: A Note on the Estimation of the Bispectral Density Function of a Stationary Time Series, The Univ. of Manchester, Inst. of Science and Technology, Dept. of Mathematics, Techn. Rep. No. 123, Sept. 1979.

[3] Johnson, N.L., Kotz, S.: Distributions in Statistics, 1. Discrete Distributions (1969), 2. Continuous Univariate Distributions - 1 (1970), 3. Continuous Univariate Distributions -2 (1970), 4. Continuous Multivariate Distributions (1972), Wiley-Interscience Publ. John Wiley&Sons, New York, London.

[4] Kendall, M.G., Stuart, A.: The advanced theory of statistics, Vol.I Distribution Theory, 4. Aufl. 1977, Macmillan Publishing, New York, (vormals Griffin).

[5] Kotulski, Z. and Sobczyk, K.: Linear Systems and Normality, J. Statistical Physics, Vol. 24, No. 2, 1981.

[6] Lin, Y.K.: Probabilistic Theory of Structural Dynamics, McGraw-Hill, New York, 1967.

[7] Melzer, H.-J.: Tragwerksschwingungen unter Zufallslast mit nicht-Gaußischer Wahrscheinlichkeitsverteilung, Berichte zur Zuverlässigkeitstheorie der Bauwerke, LKI, Heft 54, 1981.

[8] Melzer, H.-J., Schuëller, G.I.: The Application of Bispectral Analysis for the Treatment of Non-Gaussian Wind Loading, Beiträge zur Anwendung der Aerolastik im Bauwesen, Heft 14, Innsbruck/Lausanne, 1981.

[9] Schuëller, G.I.: Einführung in die Sicherheit und Zuverlässigkeit von Tragwerken, Verlag Wilh. Ernst&Sohn, Berlin, 1981.

[10] Soize, C.: Gust Loading Factors with Nonlinear Pressure Terms, J. of the Struct. Div., ASCE, Vol. 104, No ST6, Paper 13844, June 1978, pp 991-1oo7.

TOPICS IN MICROMECHANICS OF DEFECTS WITH RESPECT TO RELIABILITY OF MECHANICAL SYSTEMS

BERND MICHEL [1)2)]

The author deals with the micromechanics approach to defect mechanisms in solids. Special attention was given to the current research work done in this field within The Institute of Mechanics, Academy of Sciences, Berlin.
Defect interactions are known to be responsible for real material behaviour of different kinds. It shows a decisive influence on strength of materials etc. It is related to several elementary processes connected with crack formation, crack propagation, crack branching, crack arrest etc. Last but not least it also has to be taken into account for the study of various defect accumulation phenomena such as formation of ensembles of microcracks, voids, dislocations, point defects, etc. The latter are known to be important for fatigue failure and various kinds of damage mechanisms in solids and structures. The stochastic approach to reliability, finally, should also require more detailed information on defect accumulation for establishing advanced models for damage and failure prediction.

The following problems were dealt with in detail:

- Mechanical interaction of defects near surfaces, interfaces, cracks, boundary layers etc. Much attention was given to the elastic interaction energy terms entering into the basic concepts of strength and fracture.
- Mechanical aspects of diffusion of point defects under the influence of internal stress fields; defect interaction was included. Accumulation of defects at near-surface and near-interface obstacles was considered.
- Problems of coupling between deformation and temperature fields. As an example a coupled field approach to crack tip temperature rise was dealt with.
- Evaluation of energy release rate and generalized conservation integrals (path-independent integrals) in fracture mechanics and micromechanics by means of theoretical and experimental methods, such as holographic interferometry.
- A hypothesis was presented on the existence of crack electron states, based on a combined model which includes continuum mechanics and quantum mechanics aspects as well.

---

1) Prof.Dr.sc.nat.Bernd Michel, Academy of Sciences of the GDR, Rudower Chaussee 5, DDR-1199 Berlin

2) Abridged version of the lecture given at the IUTAM Symposium in Frankfurt/Oder

1) Mechanical interaction of defects near surfaces and interfaces

Mechanical interaction of defects in solids is strongly influenced by external and internal surfaces (such as crack flanks), interfaces, surface layers etc. Analytical and numerical methods have been applied to get more detailed information on the interaction energy terms connected with the defects. Point defect- point defect interaction was considered, void-void interaction, interaction between point defects and near-surface inclusions and dislocations. The generalized "COTTRELL" interaction energy in surface-near regions of solids has been obtained. The influence of the so-called "modulus effect" (differences in the elastic constants of matrix and inclusions) was discussed. Layer thickness and other geometric and materials parameters have been investigated applying the powerfull technique of finite elements (FEM) to these problems in micromechanis, too. /1-6/.

2) Mechanical aspects of diffusion of radiation-induced defects in solids

The problems of diffusion of defects in solids are known to be among the most interesting in solid state physics. The overwhelming majority of reactions between point defects in solids has to be considered as diffusion-controlled /7/.

Stress-assisted diffusion of radiation-induced defects has been shown to be important in this connection, too. It is immediately related to reliability of nuclear reactors. The author has proposed a mechanical- thermal field model which is based on continuum thermodynamics of an elastic solid with internal stress fields. A more general diffusion equation for the defects could be derived /8-1o/. It includes the widely used diffusion models of COTTRELL /11/, AIFANTIS /12/, CAHN /13/ et al. (Cahn's famous equation for spinodal decomposition).

3) Simulation of crack tip temperature fields

Crack temperature rise has been dealt with by many authors. But the temperature fields obtained by conventional (uncoupled) solid mechanics were shown to be not realistic. The local temperature fields excert an important influence on the defect accumulation processes which are responsible for fatigue failure phenomena (reliability problem of structures containing cracks). The author deals with this problem applying a coupled field version of continuum thermodynamics for the steady-state case of crack propagation within viscoplastic strain-hardening materials. Convection, conduction, and heat dissipation have been taken into account. The influence of temperature-dependent material parameters is also considered. The temperature rise obtained seems to be more realistic than previous calculations. The reason is that a velocity-dependent constitutive law has been applied. The model can be further improved taking into account the finite structure of the process region /14-16/.

4) Evaluation of generalized J-integrals by holographic interferometry

In fracture mechanics there is an established and widely accepted fracture criterion based on a critical value of the J integral. Irrespective of stress-strain relations of the material considered the path- independent J-integral has physical significance with regard to the balance of momentum as a crack driving force. In order to analyse the J-integral with regard to a characteristic plane of a compact type specimen - such as 3-point bending - it is necessary to take into consideration the thickness effect of this specimen /17/. In ductile fracture there is moreover a finite plastic zone surrounding the crack tip. The generalized local conservation integral $\underline{J}$ with regard to a selected plane for 3-dimensional, ductile steady-state crack growth reads as /18/:

$$\underline{J} = \underline{J}_{Rice} + \underline{J}_{lateral} + \underline{J}_{cohesion} + \underline{J}_{Sliding\ friction}.$$

The expression $\underline{J}_{Rice} + \underline{J}_{lateral}$ was recently obtained by MIYAMOTO and KIKUCHI /19/ for the three-dimensional elastic case /2o/. The special case $\underline{J}_{Rice} + \underline{J}_{cohesion} + \underline{J}_{Sliding\ friction}$ was dealt with for plane strain conditions by AOKI, KISHIMOTO and SAKATA /21/.

Using piecewise numerical smoothing technique /22/ holographic interferometry was applied to determine the $\underline{J}$ integral immediately from displacement fields measured at the surface of cracked 3-point bending specimens in comparison with results from load-displacement curves /23/.

5) Electron states due to the presence of cracks ?

The method of so-called deformation potential in quantum mechanics leads to an approximation model for electron states in a distorted lattice of a solid. The modified SCHROEDINGER equation in the deformed region (such as around an edge dislocation) is written as follows:

$$(H_o + T' + V')\underline{\psi} = (E_o + E)\underline{\psi}$$

where H_o is the Hamiltonian of the unbounded medium wthout defects (Cracks). E_o- eigenvalue of H_o. E' - electron energy change due to the cracks. $\underline{\psi}$ - wave function (see /24,25/).

The most interesting fact is that the operators T' and V' (kinetic operator and potential operator of the perturbation term) contain the strain tensor ε and the deformation gradient β, respectively.

$$T' = \frac{\hbar^2}{m_{eff}} \sum_{i,j} \varepsilon_{ij} \frac{\partial}{\partial x_i} \frac{\partial}{\partial x_j} ,$$

$$V' = \sum_{i,j} \beta_{ij} \tilde{X}_{ij} + \varepsilon_I \tilde{X}_o .$$

Here $\hbar = h/2\pi$, h - PLANCK-s constant, m_{eff} is the effective mass of the electron gas. $\tilde{X}_{ij}$ and $\tilde{X}_o$ are known functions of position. x_i are the electron Cartesian coordinates. ε_I is the first invariant of the strain tensor ε . ε_I is immediately related to the elastic energy terms of the defects.

The conclusions drawn from this model with regard to crack-flank and crack-tip electron states are give in / 24,25/. The model leads to additional lines in the electromagnetic spectrum due to the cracks, the intensity of which could probably be obtained applying statistical methods (describing microcrack distribution, orientation etc.). This problem,however, has not yet been solved so far.

The problems discussed above are based upon the deterministic approach. One therefore could ask: Why to deal with them wthin this paper presented at the IUTAM Symposium on "Stochastic vibrations and reliability" at all?

Those quantities related with the defects provide good possibilities in the framework of an advanced theory of cumulative defect aggregations to get more detailed information on the distribution functions entering the various stochastic models of reliability connected with damage, failure,and strength of materials as well. The "gap" between the deterministic approach and the stochastic approach to reliability of mechanical systems can considerably be reduced by means of micromechanics, if the field theory approach (temperature field, strain field etc.) is coupled in a "suitable manner" with the particle theory approach (ensemble of defects), as this can lead to probability density functions necessary for the study of damage accumulation and related phenomena in quite a lot of models and concepts of stochastics as related to reliability.

REFERENCES:

/1/ MICHEL,B.: Arch.Mech.32(198o)1,73-79.

/2/ MICHEL,B.: Fortschr.d.Physik 3o(1982)5,233-31o.

/3/ MICHEL,B.: Acta Phys.Hung.44(1978)2,147-156.

/4/ MICHEL,B.: Phys.stat.sol.(b)89(1978)K179-K181.

/5/ MICHEL,B.: Cryst.Res.Technol.15(198o)12,1313-132o.

/6/ MICHEL,B.,SEIFERT,W.: FMC-Series,Academy of Sciences,Berlin, Institute of Mechanics,No3 (1982) (in print).

/7/ KALNIN,Y.,PIROGOV,V.P.:Phys.stat.sol.(b)84(1977)521.

/8/ MICHEL,B.,MICHAEL,D.:Proc.DIMETA 82,Budapest 1982 (in print).

/9/ MICHAEL,D.: Thesis, Univ.of Halle,1982.

/1o/ MICHEL,B.,MICHAEL,D.:Cryst.Res.Technol.18(1983)5.(in print).

/11/ SHEWMON,P.G.: Diffusion in Solids,MC Graw-Hill Book,New York 1967.

/12/ AIFANTIS,E.C.: Acta Mech. 37(198o)265.

/13/ CAHN,J.W.: Acta Met.9(1961)795.

/14/ MICHEL,B.,GRÜNDEMANN,H.: Exp.Techn.Phys.(in prep.).

/15/ MICHEL,B.,GrÜNDEMANN,H.:Cryst.Res.Technol.18(1983)5.(in print).

/16/ MICHEL,B.,GRÜNDEMANN,H.: FMC-Series No.2 (1982),Academy of Sciences, Institute of Mechanics, Berlin (in print).

/17/ KELLER,H.P.: Thesis, TH Aachen,1979.

/18/ WILL,P.: FMC-Series,Academy of Sciences,Inst.of.Mech,Berlin, No.2(1982) (in print).

/19/ MIYAMOTO,H.,KIKUCHI,M.: Proc.6.Int.Conf.Struct.Mech.Reactor Technol. vol.1 L6/4,1981.

/2o/ MICHEL,B.,WILL,P. et al. (in prep.)

/21/ AOKI,S.,KISHIMOTO,K.,SAKATA,M.: J.Appl.Mech.49(1982)75.

/22/ WILL,P.: (in print).

/23/ FRANKOWSKI,G.: Thesis,TH Magdeburg,Fac.Technol.Sci,1983.

/24/ MICHEL,B.: Wiss.Z.TH Karl-Marx-Stadt24(1983)1 (in print).

/25/ MICHEL,B.: FMC-Series No.5(1982) Academy of Sci.,Inst.of Mech. Berlin 1982.

/26/ GRÜNDEMANN,H.: Thesis B, Academy of Sciences, Inst of Mech.Berlin, 1983.

REASSESSMENT OF FIRST-PASSAGE TIME ESTIMATES FOR NORMAL STATIONARY RANDOM PROCESSES

Janusz W. Murzewski[1)]

Summary The Poissonian model of level upcrossings is analyzed for a not necessarily small time interval τ and a high level c. The exact formula for the bivariate normal distribution function is applied and then the mean number of level crossings λ_c is defined. Its derivative ν_c is taken as the decay rate of upcrossing probability. A modal value $\tilde{\nu}_c$ is defined for maxλ_c and it is calculated for four examplary correlation functions. The value $\tilde{\nu}_c$ may replace the limiting value ν_c^* $(\tau \to 0)$, known as the Rice estimate.

Introduction The system under consideration is subject to a random motion X(t). The quantity of interest is the probability that the random process X(t) will be kept below a proscribed bound c during the system operating time (0,t). The attention is restricted to ergodic stationary Gaussian processes X(t) with zero mean $\bar{x} = 0$, a finite variance μ^2 and a continuous autocorrelation function $q(\Delta t)$. The safety index $\beta = c/\mu$ is so large that the level upcrossings are rare events and the Poisson law is assumed [2]. Cramér [1] has shown that this assumption is in fact asymptotically exact when the threshold level increases to infinity. The probability of no upcrossing during the interval (0,t) has the exponential form in the Poissonian process X(t):

$$p(t) = \Phi(\beta) \exp(- \frac{\nu_c t}{2}) , \qquad (1)$$

in which $\Phi(\beta)$ denotes the probability of starting below the threshold c and ν_c = const is the decay rate of the level crossing probability.

Vanmarcke [6] considered the time the process actually spends in the unsafe domain. He assumed the exponential law for either state duration in the two-state description of the process and introduced a correlation factor as follows:

$$p(t) = \Phi(\beta) \exp\left[- \frac{\nu_c t}{2} / \Phi(\beta)\right] , \qquad (2)$$

1) Professor of Civil Engineering, Politechnika Krakowska, ul. Warszawska 24, 31-155 Kraków, Poland

where $\Phi(\beta) = \frac{1}{\sqrt{2\pi}\mu}\int_{-\infty}^{\beta} \exp(\frac{-x^2}{2\mu^2})\, dx$ is the Laplace integral.

The decay rate ν_c is usually taken from the Rice formula [4]

$$\nu_c^* = \nu_o^* \exp(-c^2 / 2\mu^2) , \tag{3}$$

where $\nu_o^* = \sqrt{-\varrho''(0)} / \pi = \sqrt{\lambda_2} / \pi$, $\varrho''(t) = d^2\varrho/dt^2$ represents the second derivative of the normed autocorrelation function and $\lambda_2 = \int_0^\infty \omega^2 s(\omega)\, d\omega$ the second moment of the one-sided unit-area spectral density function.

The validity of formula (3) is restricted to processes for which $\varrho'(0) = 0$ only [3]. But also in these cases the evaluation of the decay rate ν_c from formula (3) is not justified, in the author's opinion.

<u>The Poissonian model of level crossings</u> An odd number of level crossings will take place, if there is

$$\xi = \frac{x_2 - c}{x_1 - c} < 0 \tag{4}$$

after a time interval

$$\tau = t_2 - t_1 > 0, \tag{5}$$

where $x_1 = X(t_1)$, $x_2 = X(t_2)$, $c = \text{const}$. The probability of such an event in the Poissonian process $X(t)$ is as follows:

$$q_c = e^{-\lambda_c} (\frac{\lambda_c}{1!} + \frac{\lambda_c^3}{3!} + \frac{\lambda_c^5}{5!} + \ldots) = e^{-\lambda_c} \operatorname{sh} \lambda_c = \tfrac{1}{2}(1 - e^{-2\lambda_c}) , \tag{6}$$

where $\lambda_c = \nu_c t$ denotes the mean number of level crossings.

The same probability calculated for a two-dimensional Gaussian vector (x_1, x_2) may be expressed by means of the Owen integral $T(\alpha, \beta)$:

$$q_c = 4T(\alpha,\beta) = \frac{2}{\pi} \int_0^\alpha \exp \frac{-\beta^2(1+u^2)}{2} \frac{du}{1+u^2} , \tag{7}$$

where $\alpha = \sqrt{\frac{1 - \varrho(\tau)}{1 + \varrho(\tau)}}$, $\beta = c/\mu$.

The Laplace integrals may be taken into account in approximate calculations [5] instead of the Owen integral:

$$T(\alpha,\beta) \approx \Phi(-\beta) \left[\frac{1}{2} - \frac{\Phi(-\alpha\beta)}{1 + \alpha^2} \right] . \tag{8}$$

It gives a small error $\varepsilon < \exp\frac{-(\alpha\beta)^2}{2} / (\alpha\beta)^4$, if $\alpha > 0$, $\beta > 0$.

The number of mean level crossings follows from formulas (6) and (7)

$$\lambda_c = -\ln\sqrt{1 - 2q_c} = -\ln\sqrt{1 - 8T(\alpha,\beta)} \tag{9}$$

and the rate of level crossings per unit time is derived as follows:

$$\nu_c = \frac{d\lambda_c}{d\tau} = \frac{-4\frac{\partial T}{\partial\alpha}\frac{d\alpha}{d\varrho}\frac{d\varrho}{d\tau}}{1 - 8T(\alpha,\beta)} = \frac{-\varrho'(\tau)}{\pi\sqrt{1 - \varrho^2}}\frac{\exp\frac{-\beta^2}{1 + \varrho}}{1 - 8T(\alpha,\beta)} . \tag{10}$$

In case of $c = 0$, that means $\beta = 0$, there is $\nu_c = \nu_0$, where

$$\nu_0 = \frac{-\varrho'(\tau)}{\pi\sqrt{1 - \varrho^2(\tau)}}\frac{1}{1 - \frac{4}{\pi}\mathrm{arctg}\,\alpha} . \tag{11}$$

A reduction factor is taken when $c > 0$:

$$\frac{\nu_c}{\nu_0} = \frac{1 - \frac{4}{\pi}\mathrm{arctg}\,\alpha}{1 - 8T(\alpha,\beta)}\exp\frac{-\beta^2}{1 + \varrho} . \tag{12}$$

The basic value (11) must not be interpreted as the zero-crossing rate, because the Poissonian model does not apply for $\beta = 0$. It can not exist when $\tau \to 0$ since $\varrho(0) = 1$ for the normed correlation function $\varrho(\tau)$. If the derivative $\varrho'(\tau)$ is continuous, there will be also $\varrho'(0) = 0$ because of the symmetry condition $\varrho(\tau) = \varrho(-\tau)$. De l'Hôspital's rule may be applied to the quotient

$$\frac{\varrho'^2}{1 - \varrho^2} \to \frac{2\varrho'\varrho''}{-2\varrho\varrho'} = -\varrho''(0) , \tag{13}$$

and the limiting value ν_c^* in the case $\tau \to 0$ is as follows:

$$\nu_c = \frac{\sqrt{-\varrho''(0)}}{\pi}\exp\frac{-\beta^2}{2} . \tag{14}$$

So the Rice formula (3) is verified. The basic value ν_0^* and the reduction factor ν_c^*/ν_c can be easily separated in formula (14).

The other limiting case is when $\tau \to \infty$:

$$\alpha = \sqrt{\frac{1 - \varrho(\tau)}{1 + \varrho(\tau)}} \to 1, \qquad \delta = 1 - \frac{4}{\pi}\mathrm{arctg}\,\alpha(\tau) \to 0,$$

since $\varrho(\infty) = 0$ for the ergodic stationary function $X(t)$ holds

$$T(1,\beta) = \frac{1}{2\pi}\int_0^1 \exp\frac{-\beta^2(1+u^2)}{2}\frac{du}{1+u^2} = \frac{1}{2}\Phi(\beta)\Phi(-\beta) . \tag{15}$$

Application of de l'Hôspital's rule gives

$$\nu_o^{**} = -\frac{1}{\pi}\lim_{t\to\infty}\frac{\rho'(\ )}{\delta(\ )} = \frac{-\rho''(\infty)}{2\rho'(\infty)}\,, \tag{16}$$

and

$$\frac{\nu_c^{**}}{\nu_o^{**}} = \frac{\dot{\delta}(t)\,\exp(-\beta^2)}{1-4\,\Phi(\beta)\,\Phi(-\beta)} \longrightarrow 0\,. \tag{17}$$

The limiting values of ν_o for four exemplary correlation functions are shown in table 1, where $\Theta = \int_{-\infty}^{+}\rho(t)dt$ is the scale of fluctuation.

Table 1

No.	$\rho(t)$	The limiting rates	
		$\nu_o^{*} = \frac{1}{\pi}\sqrt{-\rho''(0)}$	$\nu_o^{**} = -\frac{1}{2}\frac{\rho''(\infty)}{\rho'(\infty)}$
1	$\exp\frac{-\pi t^2}{\Theta^2}$	$\sqrt{\frac{2}{\pi}}\,/\,\Theta$	$\pi\,/\,\Theta$
2	$\frac{\sin(\pi t/\Theta)}{t/\Theta}$	$\sqrt{\frac{1}{3}}\,/\,\Theta$	indefinite[2)]
3	$\exp\frac{-2\vert t\vert}{\Theta}$	1)	$1\,/\,\Theta$
4	$\begin{cases}\cos(2t/\Theta)\,, & \vert t\vert \le \frac{\pi\Theta}{4}\\ 0\,, & \vert t\vert > \frac{\pi\Theta}{4}\end{cases}$	$\frac{2}{\pi}\,/\,\Theta$	0[3)]

Remarks:

1) $-\rho'/\ 1-\rho^2 \quad 1/\sqrt{t\Theta}$ for $t \longrightarrow 0$

2) $(\pi/\Theta)\ \mathrm{tg}(\pi t/\Theta) \longrightarrow\ ?$ for $t \longrightarrow \infty$

3) $\nu_o = 0$ for $|t| \ge \pi\Theta/4$

<u>Modal decay rates</u> The probability decay rate ν_c (cf. formula (10)) depends on the time interval τ. The dependence $\nu_c(t)$ should not happen, if the level crossing process was exactly the Poissonian one. The lower bound estimates of probability q_c (cf. formula (6)) are obtained in the limiting cases, when $t \rightarrow 0$ and $t \rightarrow \infty$ and $\beta > 0$:

$$q_o = 0 \quad \text{for} \quad \lim_{t \to 0} \lambda_o(t) = 0 \quad \text{and} \quad \lim_{t \to \infty} \lambda_c(t) = 0 \, . \tag{18}$$

The upper bound of the probability of up-crossing can be estimated from the condition

$$t \, \nu_c(t) = \max \, . \tag{19}$$

Substituting (10) into (19) we get the equation

$$\frac{-t\varrho'(t)}{\pi\sqrt{1-\varrho^2(t)}} \; \frac{\exp \dfrac{-\beta^2}{1+\varrho(t)}}{1 - 8T\left[\sqrt{\dfrac{1 - \;(t)}{1 + \;(t)}}\, , \beta\right]} = \max(t) \, . \tag{20}$$

A modal value $t = \tilde{t}$ which satisfies condition (20) and the relative value $\tilde{\nu}_c = \nu_c(\tilde{t})$ can be estimated for every correlation function $\varrho(t)$. Not too small safety indices β will be assumed as it was indicated in the Introduction. Some numerically calculated results are presented in

Table 2

β	$\Phi(-\beta)$	No. of correlation function $\varrho(t)$			
		1	2	3	4
2	$2.28 \cdot 10^{-2}$	$6.83 \cdot 10^{-2}$	$5.12 \cdot 10^{-2}$	$6.59 \cdot 10^{-2}$	$6.08 \cdot 10^{-2}$
3	$1.35 \cdot 10^{-3}$	$5.34 \cdot 10^{-3}$	$3.97 \cdot 10^{-3}$	$7.20 \cdot 10^{-3}$	$4.25 \cdot 10^{-3}$
4	$0.32 \cdot 10^{-4}$	$1.62 \cdot 10^{-4}$	$1.19 \cdot 10^{-4}$	$2.75 \cdot 10^{-4}$	$1.32 \cdot 10^{-4}$

The values from Table 2 divided by specific scales of fluctuation Θ of stochastic functions give the modal values $\tilde{\nu}_c$. The probability

$$\Phi(\beta) = 1 - \Phi(-\beta) \tag{21}$$

is the correlation factor of formula (2). The modal values $\tilde{\nu}_c$ for correlation functions Nos. 1,2,4, which are differentiable at $\tau = 0$, appear to be 30 to 40 % smaller than the Rice limiting values ν_c^* calculated from formulas (3) or (14). The rates $\tilde{\nu}_c$ can be determined by means of formula (10) for $\tau = \tilde{t}$ also when the correlation functions do not satisfy the condition $\varrho'(0) = 0$, e.g. the No.3 from Table 1.

References

1 Cramér, H.: On the intersections between the trajectories of the normal stationary stochastic process and a high level, Arch. d. Math. 6 (1966),337

2 Murzewski, J.: Safety of Building Structures, Arkady, Warsaw 1970 (Polish ed.); VEB Verlag Bauwesen, Berlin 1974 (German ed.)

3 Papoulis, A.: Probability, Random Variables and Stochastic Processes, McGraw Hill Inc. 1965

4 Rice, S.O.: Mathematical analysis of random noise, Bell Syst. Techn. J. 23(1944),282 ; 24(1945),46

5 Smirnov, N.V.; Bol'shev, Z.N.: Tables for the Distribution of the Bivariate Normal Distribution Function (in Russian), Izdat. AN SSSR, Moscow 1962

6 Vanmarcke, E.H.: On the Distribution of the First-Passage Time for Normal Stationary Random Processes, J.Appl.Mech., ASME Paper No.75-APMW-12.

STOCHASTIC FRACTURE MODELS OF COMPOSITE MATERIALS

V. Tamuzs[1)]

The strength of unidirectional composite materials subjected to tension has been investigated taking into account the statistical fiber strength scattering. The obtained strength value of the extreme damageability and the distribution of lengths of broken fibers are compared with the experiment. The generalization of calculation considering durability in time has been proposed.

Reliability of construction materials depends first of all on the level of mean strength, strength scattering and the kind of fracture. In the fracture process of practically all materials two stages can be observed: 1) the so-called stage of volume or dispersed fracture when microcracks accumulate but their further enlargement is hampered by the heterogeneities of a material and 2) the germination of the macrocrack breaking a material and a construction into parts. Reliability of a material depends, of course, on the efficiency criterion itself, i.e. on the maximum admissible level of damage. If a definite level of volume damage is allowed, for example, due to leak-proofness, then a material which is subjected to volume fracture to a greater extent is more reliable as accidental outbursts of stresses over the strength surface will not cause a catastrophic germination of a macrocrack. Besides, the initial stage of fracture can be determined by acoustic emission. As was already said, practically all heterogeneous materials undergo the stage of volume fracture. This is relevant to polymers, rocks, metals, but perhaps more distinctly it is shown itself in composites [1].

It has long been observed that composites subjected to vibrational load irrevocably change their mechanical properties [2], which is the result of dispersed damage accumulation in them. Mechanical characteristics of composites are determined by their structure and the properties of constituents. In order to improve the strength and reliability of composites it is necessary to work out a theory of volume fracture which would give the answers to the following questions: 1) how do the mean strength and strength scattering of a composite depend on elastic properties and statistical strength parameters of initial fibers and 2) what extreme level of dispersed damage is admissible before unrestricted enlargement of defect takes place.

Apparently, the first statistical models of composite material fracture

1) Academy of Sciences of the Latvian SSR, Institute of Polymer Mechanics, Aizkraukles str. 23, 226006 Riga, USSR

are linked with works [3,4]. In [3] the composite material consists of a chain whose links were formed by a bundle of fibers of critical length. Application of the probable relation between the strength of the chain and strength of the individual elements led to an estimation on the probability of the accumulation of defects in each layer and the strength of the composite. In [4] the stress concentration on the fibers owing to the broken fibers was considered which included a study of the fracture sequence consisting of the breakage of adjacent fibers. When investigating a two dimensional model in [4] it has been obtained that even a small number (two, three) of the adjacent fibers can cause fracture of the entire reinforced composite.

In [5] the simplest kinetic model of fracture of a unidirectional composite has been investigated and in [6] the basis is given for an approximate analysis of the cumulative damage in a composite material by neglecting the effect of the fracture of one element on the strength of the other. A more general approach to the fracture problem of composite materials is based on the use of kinetic models. This approach permits within the limits of one model to take into account the non-stationary loading, temporary delay of fracture, accumulation of individual defects, their coalescence into a macrocrack and the macrocrack propagation - i.e. all the events of the actual fracture process of solids.

Calculation Model

To calculate the probability of fracture of a composite it is necessary to estimate the size of the structural element, i.e. of the element which can break once and only once in advance.

It is quite natural to link this size with the length of the fiber at which an imperturbable stress state is restored when a single fiber is broken. This length is determined according to a well-known formula

$$\delta = d_f\left(\frac{E_a}{G_m}\,\frac{1-V_a^{1/2}}{V_a^{1/2}}\right)^{1/2} \quad \text{arc ch}\left[\frac{1+(1-\varphi)^2}{2(1-\varphi)}\right] . \tag{1}$$

Here the parameters of the material: fiber modulus E_a, matrix modulus G_m, fiber diameter d_f and fiber volume V_a can be defined earlier; φ is the stress reset degree in distance δ from fiber break.

Secondly, the coefficient of the stress concentration next to a broken single fiber or a bundle of broken fibers will be determined. In order to obtain practically applicable dependences it is necessary to establish simplified assumptions. One of the hypotheses may be the following: it is assumed that the defect is always a penny-shaped crack and overstressed are only adjacent fibers; besides, the overstress along the fiber propagates to the length δ on both sides of the defect. Finally, it is necessary to know the statistical strength characteristics of fibers. It is usually assumed that the strength distribution of fibers is of the so-called Weibull distribution

$$F(\sigma) = 1 - \exp\left[- \frac{l}{l_o}\left(\frac{\sigma-\sigma_m}{o}\right)^{\beta}\right] . \quad (2)$$

Then the fracture calculation is reduced to the following approximate formulas [7]:

$$P_1 = F(\sigma) ; \quad P_2 = P_1 \cdot P_2 ; \; \ldots ; \quad P_n = P_{n-1} \cdot P_{n,n-1} \ldots$$

where P_1 is the probability of a single break appearance, $F(\sigma)$ is determined by formula (2) for $l = 2$; P_n denotes the probability of appearance of a group consisting of n broken elements. $P_{2,1}$ is the probability of breaking of at least one element next to an already broken one. It is determined in a similar way:

$$P_{2,1} = 1 - \left[\frac{1 - F(K_1\sigma)}{1 - F(\sigma)}\right]^{n_1} \quad (3)$$

where K_1 is the overstress coefficient, $\frac{F(K_1\sigma)-F(\sigma)}{1 - F(\sigma)}$ the probability of the adjacent element break after the appearance of the first defect, n the number of elements subjected to the overstress K_1;

$$\left[1 - \frac{F(K_1\sigma) - F(\sigma)}{1 - F(\sigma)}\right]^{n_1} = \left[\frac{1 - F(K_1\sigma)}{1 - F(\sigma)}\right]^{n_1} \quad (4)$$

is then the surviving probability of all the overstressed elements. The next probability is expressed by

$$P_{3,2} = 1 - \left[\frac{1 - F(K_2\sigma)}{1 - F(K_1\sigma)}\right]^{n_{21}}\left[\frac{1 - F(K_2\sigma)}{1 - F(\sigma)}\right]^{n_2-n_{21}} \quad (5)$$

where n_2 denotes the number of elements surrounding the double effect and n_2-n_{21} the number of adjacent elements which were not subjected to overstress before the break of a second fiber.

The probability of development in a specimen with N_o elements of at least one defect with size not smaller than n is determined by

$$W_n = 1 - \exp(-P_n N_o) . \quad (6)$$

At sufficiently large n all curves for W_m, $m > n$, merge. This limit curve determines the macrocrack appearance.

Calculation Results and Experimental Data

The parameters of Weibull's distribution were found out for real epoxy carbon fibers of three different types. Integral and differential strength distribution curves at different lengths are displayed in Fig. 1. The mean strength and dispersion are the following:

$$\langle \sigma_1 \rangle = \sigma_0 \left(\frac{l}{l_0}\right)^{-1/\beta} \Gamma\left(1 + \frac{1}{\beta}\right) + \sigma_m , \tag{7}$$

$$D = \sigma_0 \left(\frac{l}{l_0}\right)^{-1/\beta} \left[\Gamma\left(1 + \frac{2}{\beta}\right) - \Gamma\left(1 + \frac{1}{\beta}\right)^2\right]^{1/2} , \tag{8}$$

the variation coefficient is $K = \frac{D}{\langle \sigma_1 \rangle}$. The mean strength and dispersion are shown graphically in Fig. 2.

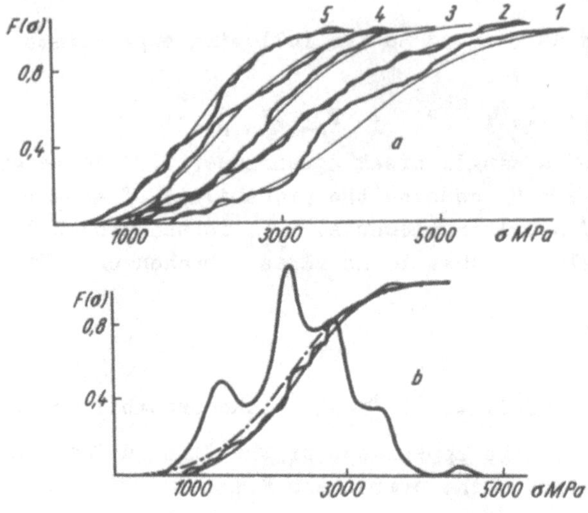

Fig. 1: Strength distribution functions of carbon fibers at lengths 1,2, 5,10,70 mm (curves 1 to 5 in Fig. a, respectively); distribution density for l = 25 mm (Fig. b)

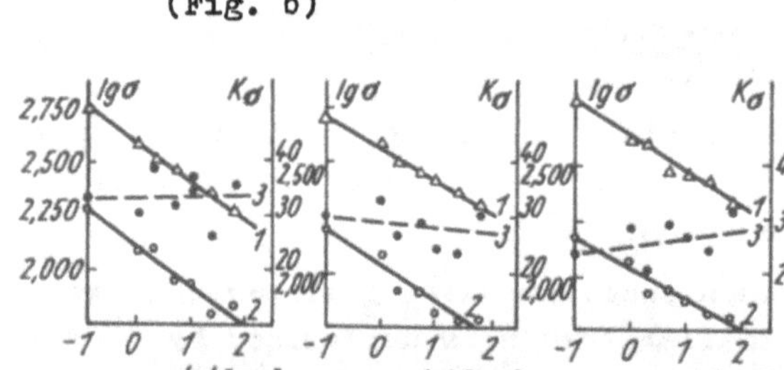

Fig. 2: Mean strength, dispersion and variation coefficient for three types of carbon fibers

The following states and numerical values are to be checked and calculated:

a) the multitudinous splitting of fibers;

b) the lengths distribution of fiber segments in a damaged composite and the level of extreme damage;

c) the calculated strength value of the composite and and its correspondence with the experiment.

Let us consider in detail the volume fracture of a unidirectional epoxy carbon composite. The count of fiber breaks and the damage analysis were carried out on specimens tested for statical tension and bending at load levels 0.2; 0.4; 0.6; 0.8; 0.93 from the maximum load. The broken specimens were tested, too. As is seen from the results presented in Fig. 3, the breakage of fibers starts at a load level approximately equal to a half of the maximum one. Then at load 0.9 - 0.95 from the maximum one the breakage turns into an avalanche-like process. Consequently, the fracture of an epoxy-carbon composite even at high filling levels has a cumulative nature.

The length distribution of broken fiber segments is shown in Fig. 4. It can be seen that the density distribution curve has two definite maxima at small lengths. As the fiber damage in segments shorter than the ineffective length δ is hardly probable, it can be assumed that the value of the first maximum corresponds to the size δ, according to formula (1).

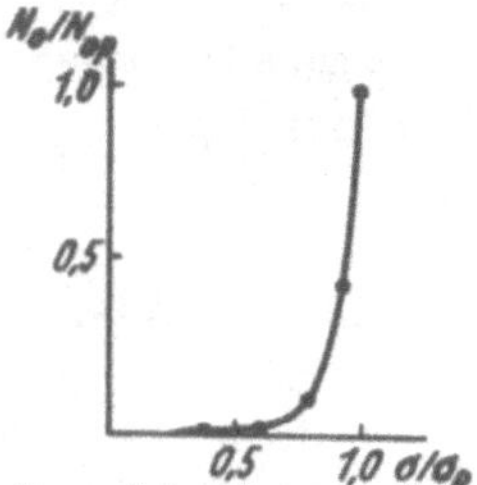

Fig. 3 (to the left): Break accumulation kinetics in an epoxy carbon composite

Fig. 4 (to the right): Length distribution density of broken fiber segments

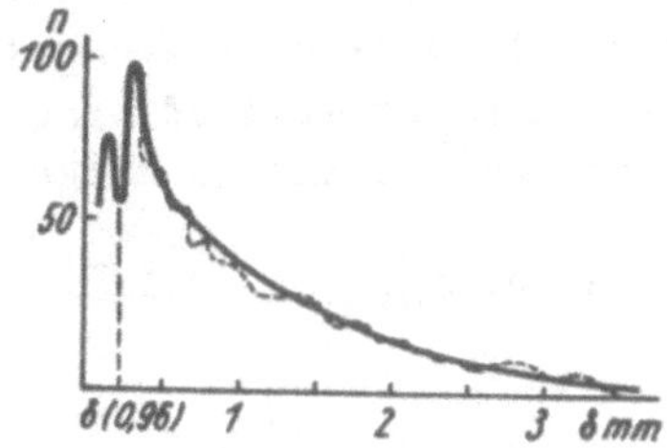

Considering σ in formula (2) as a parameter the expected length distribution of broken fiber segments can be obtained as

$$F(\lambda,\sigma) = 1 - \exp\left[-\frac{l-\delta}{l_0}\left(\frac{\sigma-\sigma_m}{\sigma_0}\right)^{\beta}\right]. \qquad (9)$$

As is seen from Fig. 4 the qualitative coincidence of the calculated and the real curve of the splitting is very good, quantitatively, however, there are differences.

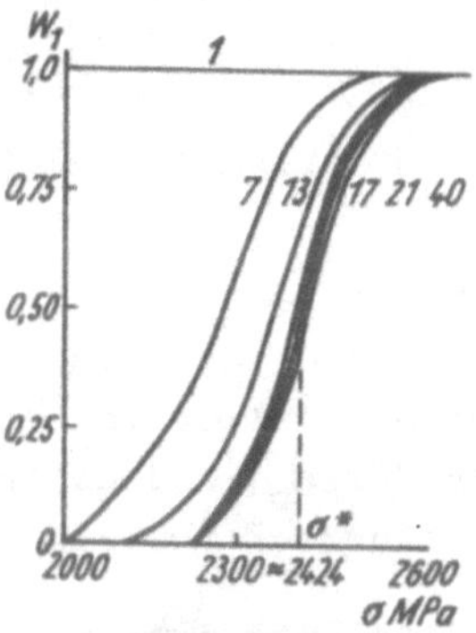

Fig. 5: Probability of macrocrack appearance in a composite

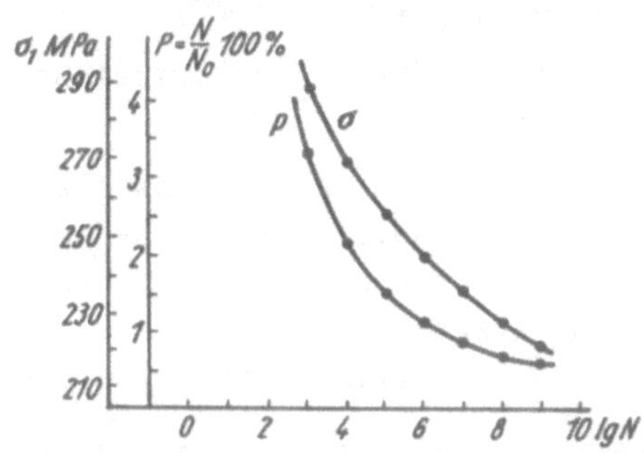

Fig. 6: Size effect on the strength value and extreme damage level

The calculation results of the fracture process are shown in Figs. 5 and 6. The merging of the curves W_n took place at $n \approx 30$ (Fig. 5). By the value W_{30} we obtain the level of maximum load σ on a fiber in a composite, then P_1^1, P_2^1 at the fracture moment are determined.

In Fig. 6 the dependence of the strength on the size of the specimen (scale effect) and, what is more essential, the dependence of the extreme level of damage on the size of a specimen is shown. This means that for the larger specimens the fracture takes place at a lower level of mean volume damage. The obtained theoretical values of ultimate damage coincide with the experimental ones.

It can be seen from Fig. 5 that the curves of probability of the large defect appearance are steeper. Consequently the more expressed volume fracture of material leads to a lower strength dispersion and to greater reliability of the material. The mean strength of the composite material is determined by a certain P % quantile of the fiber strength distribution extrapolated to the length 2δ:

$$\sigma(P) = \sigma_o\left[-\frac{l_o}{l}\ln(1-P)\right]^{1/\beta} + \sigma_m , \tag{10}$$

at $p \ll 1$

$$\sigma(P) \approx \sigma_o\left(\frac{l_o}{l}P\right)^{1/\beta} + \sigma_m . \tag{11}$$

The extreme amount of single defects as it is proved by calculation and experiment equals $P_1^1 \approx 1 - 4\%$.

Generalization for Long-term Strength

Let us assume that the distribution of durability of fibers follows the double Weibull distribution

$$F(\sigma,t,l) = 1 - \exp(-L\alpha t^{\gamma}\sigma^{\beta}) \tag{12}$$

where t, σ and L denote nondimensional time, stress and length of the fiber, respectively.

If the fracture of an element at its definite lifetime value $\tau(\sigma_o)$ is a deterministic quantity, then the probability of fracture of an element is $W[t,\tau(\sigma_o)=H\{t-\tau(\sigma_o)\}]$. But if due to scattering of properties the lifetime of an element is a random quantity with density $f(\tau,\sigma_o)$ and distribution function $F(\tau,\sigma_o)$, then the probability of fracture of the element for the time moment t is $W(t) = \int_0^\infty f(t)\, H(t-\tau)\, d\tau = F(t)$. The expected number of single defects in a specimen is $N_oF(t_o)$, where N_o is the number of elements in a specimen.

After the appearance of a single defect for the time moment $t = x$ an increased stress originates on its adjacent elements. If the loading history of the adjacent elements is not taken into account, the probability of fracture of such an element will be

$$W(t,\sigma_1) = \int_0^t f(\tau(\sigma_1))d\tau = F(t,\sigma_1) \tag{13}$$

where the time account has been restarted at the moment $t = x$. The probability of fracture of at least one adjacent element is

$$W_1^2 = 1 - [1 - F(t,\sigma_1)]^{n_1} \tag{14}$$

where n_1 is the number of the adjacent elements, while the density of this probability equals

$$P_1^2 = n_1\left[1 - F(t,\sigma_1)\right]^{n_1-1} f(t,\sigma_1) \ . \tag{15}$$

The probability of the development of a double defect W_2 can be determined by the formula

$$W_2 = \int_0^t \frac{dW_1(x)}{dx} W_1^2(t-x)\, dx = \int_0^t W_1(x)\, P_1^2(t-x)\, dx \ . \tag{16}$$

The probability of nucleation of a triple defect and defects of size j in general can be determined in a similar way:

$$W_j = \int_0^t W_{j-1}(x)\, P_{j-1}^j(t-x)\, dx \ . \tag{17}$$

In order to understand the loading history of the failing elements, let us examine a hexagonal packing of the fibers [9]. Till the moment of initial fracture all the elements are subjected to the stress σ_o. After the fracture of a single element at the moment x_1, six adjacent elements (the overstresses are considered for the immediate adjacent elements) have the loading history $\sigma = \sigma_o$ at $0 \leq t \leq x_1$, $\sigma = \sigma_1$ at $t > x_1$. For the initiation of a double defect the loading history of its neighbors is as follows: five elements are under $\sigma_o|_{o<t<x_1}$; $\sigma_1|_{x_1<t<x_2}$; $\sigma_2|_{t>x_2}$; three elements under $\sigma_o|_{o<t<x_2}$; $\sigma_2|_{t>x_2}$. Fracture of the third adjacent element entails a number of variations, which are shown conveniently in Table 1.

Table 1

Dimensions of defects	Number of neighboring elements	Stress in time periods			
		$0 < t < x_1$	$x_1 \leq t < x_2$	$x_2 \leq t < x_3$	$t \geq x_3$
1	6	σ_o	σ_1	-	-
2	5	σ_o	σ_1	σ_2	-
	3	σ_o	σ_o	σ_2	-
3 (one variant)	4	σ_o	σ_1	σ_2	σ_3
	3	σ_o	σ_o	σ_2	σ_3
	2	σ_o	σ_o	σ_o	σ_3
3 (another variant)	5	σ_o	σ_1	σ_2	σ_3
	2	σ_o	σ_o	σ_2	σ_3
	3	σ_o	σ_o	σ_o	σ_3
	etc.				

It is easy to see that the loading history of any element falls within the loading of all stress levels

$$t \rightarrow x_1 \rightarrow x_2 \dots \rightarrow x_{n-1} \rightarrow x_n \ ; \quad \sigma_0 \rightarrow \sigma_1 \rightarrow \sigma_2 \dots \rightarrow \sigma_{n-1} \rightarrow \sigma_n$$

and that of the lowest stress

$$t \rightarrow x_1 \rightarrow x_2 \dots \rightarrow x_{n-1} \rightarrow x_n \ ; \quad \sigma_0 \rightarrow \sigma_0 \rightarrow \sigma_0 \dots \rightarrow \sigma_0 \rightarrow \sigma_n \ . \tag{18}$$

Assigning the elements to one or the other extreme history of loading we will obtain two limits between lies and the true kinetics of cumulative defects. Knowing the dependence of the distribution function $F(t,\sigma)$ on the parameter σ the curves corresponding to the discrete stress levels σ_0, σ_1,... ..., σ_n (Fig. 7) can be found.

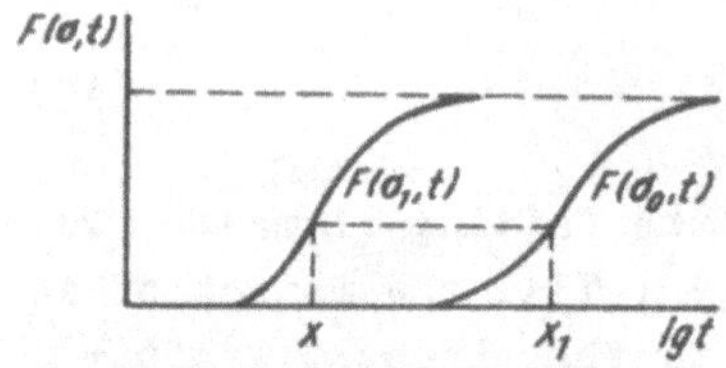

Fig. 7: Fracture probability of the structural element at two stress level loading

As an example the damage accumulation curves of a composite consisting of fibers of highly dispersed strength are shown in Fig. 8.

It can be seen that the majority of single defects appear at the initial loading but the amount of large defects increases in time till the final fracture taken place. The obtained data correspond to well-known experimental ones for damage accumulation in the materials with weak links [1].

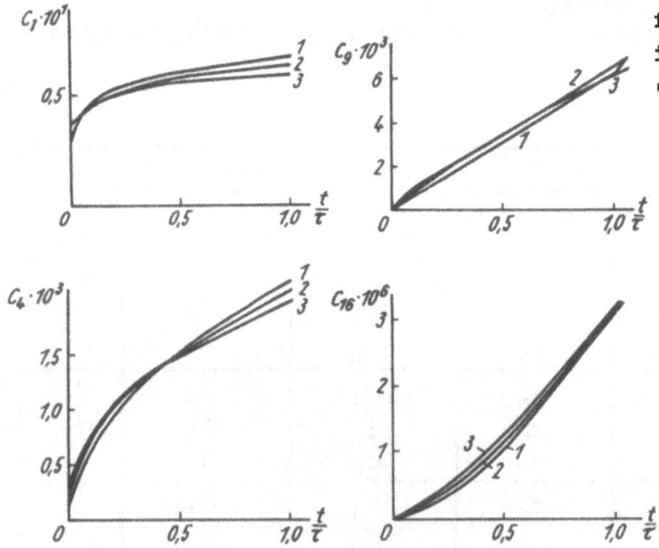

Fig. 8: Kinetics of cumulative defects of various sizes (sizes 1,4,9,16 in the four figures above); is the time of composite fracture. The curves 1,2,3 are obtained under different stress levels $\sigma_1 < \sigma_2 < \sigma_3$.

References

1 Kuksenko, V.S.; Tamuzs, V.P.; Fracture Micromechanics of Polymer Materials, Martinus Nijhoff Publishers, 1981, 311 pp.

2 Олдырев,П.П.; Тамуж,В.П.: Изменение свойтсв стеклотекстолита при циклическом растяжении-сжатии. -Механика полимеров, 1967,№5,с.864-872

3 Rosen, B.W.: Tensile failure of fibrous composites, AIAA J. (1964), 2,1985-1994

4 Zweben, C.: Tensile failure of composites, AIAA J. (1968),12,2325-2331

5 Болотин В.В. Некоторые математические и экспериментальные модели процессов разрушения. -Проблемы прочности, 1971, № 2, с.13-20

6 Болотин В.В. Статистическая теория накопления повреждений в композиционных материалах. -Механика полимеров, 1976, № 2, с.247-255.

7 Тамуж В.П. Особенности разрушения гетерогенных материалов. -Механика композитных материалов, 1982, № 3, с.406-409.

8 Тамуж В.П. и др. Разрушение однонаправленных углепластиков и реализация в них прочностных свойств волокон. -Механика композитных материалов, 1982, № 1, с.34-41.

9 Tamuzs, V.P.: Dispersed fracture of unidirectional composites. In: Fracture of Composite Materials (G.C. Sih; V.P. Tamuzs, eds.), Sijthoff & Noordhoff 1979, p.413.

RELIABILITY CONTROL OF MECHANICAL SYSTEMS

S.A. Timashev[1)]

The major reliability computation patterns for controled mechanical systems in cases of programmed two-level and differential control are considered. Solutions to the problem of controling reliability by programmed control of loads or impacts on a system and to the reliability control problem for a vibrating beam with a damper are presented.

1. Preliminary notes. The ultimate aim of every reliability computation is optimal control of quality variation processes in mechanical systems at the stages of their design, manufacture, and operation. Modern mechanical systems consist of three subsystems, viz.: 1) the structure proper, 2) simulation and control elements, 3) servomechanisms which alter the design scheme, boundary conditions, throw off or share load when it reaches a dangerous level. A built-in control system is employed when external impacts are dynamic or when human intervention is difficult to exercise or impossible. When the load is quasistatic and the operator can intervene readily the control function is in most cases performed by maintenance.

2. Reliability computation patterns for a controled mechanical system [4, 5]. A general reliability control circuit is shown in Fig. 1. As is seen, it is possible to influence external effects, the system operator, and the boundary conditions or to restore the structure according to a certain rule. Besides, each of the methods involved admits of programed (including feedback) and differential control. So let us consider the patterns of reliability computation with regard for control.

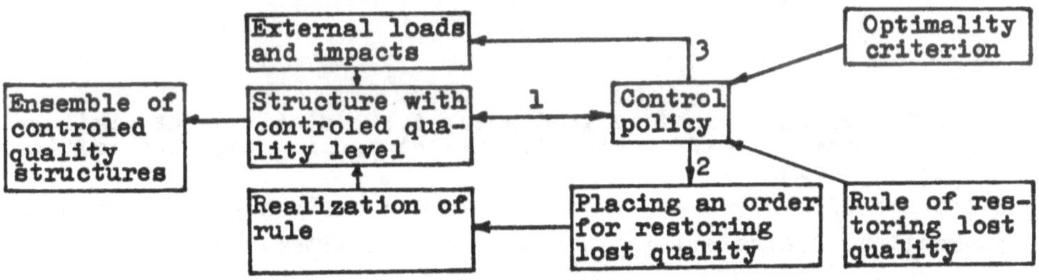

Fig. 1. General reliability control circuit of a mechanical system: 1 - quality level signal channel; 2 - system operator control channel; 3 - load vector control channel.

The programmed control case. Let the behavior of a system be described by the stochastic equation

$$Ls = q, \tag{1}$$

where the vector q is an element from the space Q of random input parameters (external effects), $q \in Q$; s is an element from the space S of

1) Dr. Sci., Prof. Department of Complex Problems of Machinery, Ural Science Research Center of the USSR Academy of Sciences, Sverdlovsk, USSR.

output parameters (specifying the behavior of the system), $s \in S$; $L(\varkappa, \varkappa_o, \varkappa_c, t)$ is some, generally speaking, nonlinear time - dependent operator; $\varkappa$, $\varkappa_o$ are, respectively, elements from the sets K and K_o of determinate properties of the system which are not and are subject to optimization; $\varkappa_c$ is an element of the set K_c of those properties of the system which are regarded as stochastic.

We define in the S set a quality subset V with the Γ_V boundary, which is characterized by the ultimate permissible value of quality (state) s_* of the system due to the warning failure. From it we derive the permissible region $\Omega s_*(\varkappa_c)$ in the Q set, which is found from the equation

$$\Omega s_*(\varkappa_c) = \left\{ q : q = Ls, \ s \in V \right\}. \qquad (2)$$

The operator L reflects the level of complexity in the formulation of a deterministic problem (since the vector $\varkappa_c$ is fixed).

Such an approach enables us to construct in Q space special permissible regions $\Omega_{i_j}(\varkappa_c)$ for each i_j-th quality criterion (Fig. 2).

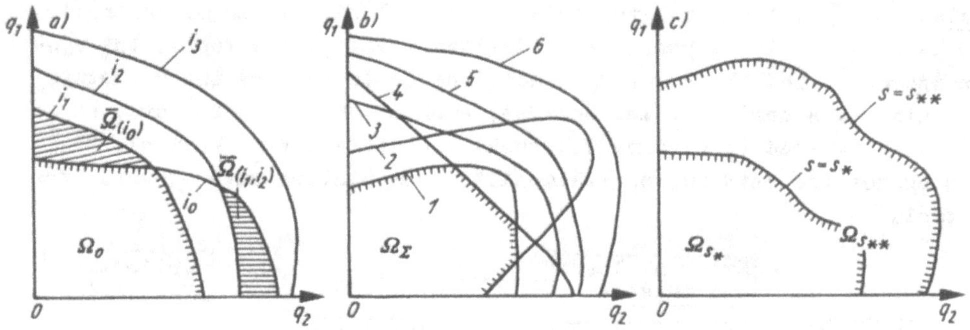

Fig. 2. Permissible regions in load space for: a) different failure criteria; b) different elements; c) warning (s_*) and critical (s_{**}) failures.

Their intersection gives a permissible region for a single element (cross-section) due to all quality criteria simultaneously (N of number)

$$\Omega_o = \bigcap_{j=1}^{N} \Omega_{(i_j)}(\varkappa_c). \qquad (3)$$

This way we can readily find the region $\bar{\Omega}_{i_j}(\varkappa_c)$ where one, two or more types of failure occur simultaneously, and hence take into account the interdependence of failures. For example, according to Fig. 2a,

$$\bar{\Omega}_{(i_o)}(\varkappa_c) = \Omega_{(i_1)} - \Omega_{(i_o)} \bigcap \Omega_{(i_1)}; \ \bar{\Omega}_{(i_1,i_2)}(\varkappa_c) = \Omega_{(i_o)} - \Omega_{(i_o)} \bigcap \Omega_{(i_2)} \quad \text{etc.}$$

If a system is multielement the permissible region for every element may be constructed according to some criterion. Their intersection yields the permissible region for the system as a whole with respect to that same criterion. Performing this procedure for all quality criteria the problem of obtaining the permissible region for the system with respect to all quality criteria is reduced to the pattern presented in Fig. 2a.

The technique outlined above permits, prior to computing the reliability function, to determine which of the quality criteria are the severest (i_0 and i_1 in Fig. 2a) and which elements do not participate in the formation of the permissible region (for instance, elements 3, 5, 6 in Fig. 2b). This enables us to single out excessive-reliability elements and to map out constructive steps to bring their reliability down to a level that does not affect the total reliability of the system. It should be noted that the load space dimension is in most cases essentially smaller than the state space dimension for a multicomponent system.

Since the admissible regions are plotted for a fixed value of the random vector $\varkappa_c$ it is no longer necessary to solve the corresponding statistical dynamics problem, but the reliability function obtained using these regions is conditional. Integration of this function with the weight (density) $f(\varkappa_c)$ yields an unconditional reliability function. The entire randomness is thus concentrated in the permissible region.

When $s_* < s < s_{**}$ (with s_{**} characterizing the onset of the critical failure) the system is described by one of the equations

$$L_1 s = q; \ Ls = q_1; \ L_2 s = q_2. \tag{4}$$

Here q_i, L_i, $i = 1, 2$ are respectively the vectors of loads on the system and the operators of the system after the warning failure.

Thus, what is essential here is to consider the reliability of the system in the case of two kinds of failure which necessarily follow one another and the fact that, after the primary (warning) failure, the system is controled due to a qualitative change in the design scheme of the system and/or in its loading and operating conditions (see Fig. 1). The permissible region with respect to critical failure, $\Omega_{s**}(\varkappa_c)$, is plotted similar to the region $\Omega_{s*}(\varkappa_c)$.

The conditional reliability function [1] with respect to critical failure can be found as

$$R_y(t) = 1 + \int_0^t \dot{P}[q(\tau) \in \Omega_{s*}(\varkappa_c)](1 - P[q_i(t-\tau) \in \Omega_{s**}(\varkappa_c)])\,d\tau, \quad 0 < \tau < t; \ i = 1,2 \tag{5}$$

(here and henceforth the dot above the symbol labels the differentiation of the reliability function with respect to time, and the complete function may be found according to the formula

$$R(t) = \int_{K_c} Ry(t) f(\varkappa_c)\,d\varkappa_c. \tag{6}$$

Thus the R (t) function is determined as a result of allowing for the most important environmental factors, properties of the system, for the process and operation requirements, and also for the restoration and control conditions. The control proves to be of two levels, thus corresponding best to real operating conditions.

The differential programmed control case. Several stages can be singled out here. Stage I deals with choosing a controled mechanical system design scheme

$$L(w) = q(t) + L_1\left[u(t)\right], \tag{7}$$

where L is the operator of the system controled; L_1 - operator of controls; q (t) - the vector of external forces (perturbations) from load space Q and u (t) - the vector of control actions from control space U; w (t) - the vector of states of the system from state space W. Stage II is the same as before singling out the K, K_o and Kc sets. Stage III consists in determining in state space W the quality subspace V and, from the latter, the permissible regions $\Omega_q(\varkappa_c)$, $\Omega_u \times (\varkappa_c)$ in spaces $Q(\varkappa_c)$ and $U(\varkappa_c)$.

Thus the permissible regions $\Omega_q(\varkappa_c)$ and $\Omega_u(\varkappa_c)$ turn out to be dependent since at any fixed value of q(t) the permissible value of the element u(t) is a function of the realized value of the element $\bar{q}(t)$,

$$u(t) = H_1\left[L(w_*) - \bar{q}(t)\right], \tag{8}$$

where w_* is the maximum permissible system quality value, $w_* \in \Gamma v$, Γv is the boundary of the permissible region in the quality space, and H_1 is an operator inverse to the operator L_1.

The permissible region $\Omega_u(\varkappa_c)$ is constructed according to the equation

$$\Omega_u(\varkappa_c) = \left\{u : u = H_1\left[L_1(w) - \bar{q}(t)\right],\ w \in V\right\}. \tag{9}$$

Stage IV is aimed at solving the optimal control problem subject to the formulated restrictions for the allowed function u(t). We seek such a function $u_o(t)$ (at fixed values of $\varkappa_o$ and $\varkappa_c$) which minimizes the maximal deviation of the loaded system from some 'normal' state. We understand that a 'normal' state is, for instance, a state corresponding to the mathematical expectation $\langle q(t) \rangle$ of the load vector. This problem can be solved on the basis of motion control theory [2]. Peculiarity of the problem is that it consists not in transferring a mechanical system from one phase space region to another but in retaining the system in the prescribed region (which can vary stochastically in size) and that the permissible control region sought from the standpoint of reliability theory differs substantially from the regions considered in mathematical control theory.

At stage V we seek the conditional optimal reliability (with $\varkappa_o$

and $\varkappa_c$ having fixed values)

$$R_y(t \mid \varkappa_o, \varkappa_c) = P\left[q(\tau) \in \Omega_q(\varkappa_o, \varkappa_c \mid u_o), \quad 0 \leq \tau \leq t\right], \tag{10}$$

and the conditional total reliability function for the structure

$$R_c(t|\varkappa_o) = \int_{K_c}\left\{1+\int_0^t \dot{R}_y(t|\varkappa_o,\varkappa_c)(1-R_m[q(\tau)\in\Omega_q(\varkappa_o,\varkappa_c|u_o)])d\tau\right\}f(\varkappa_c)d\varkappa_c, \quad 0\leq\tau\leq t. \tag{11}$$

The control system reliability function Rm(t) is sought separately, proceeding from particular features of the system.

At stage VI we write the efficiency function for the total cost of the system

$$C(u_o,\varkappa_o) = C_o^{(1)}(u_o,\varkappa_o)+C_o^{(2)}(u_o,\varkappa_o)-\int_0^T C_y^{(1)}(u_o,\varkappa_o,t)\dot{R}(u_o,\varkappa_o,t)\beta^{-t}dt - \\ -\int_0^T C_y^{(2)}(u_o,\varkappa_o,t)\dot{R}_m(u_o,\varkappa_o,t)\beta^{-t}dt + \int_0^T s(u_o,\varkappa_o,t)\beta^{-t}dt, \tag{12}$$

where $C_o^{(1)}(u_o, \varkappa_o)$ is the original cost of the structure; $C_o^{(2)}(u_o, \varkappa_o)$ - that of the control system; $Cy^{(1)}$ - the value of the damage incurred by the failure of the structure; $Cy^{(2)}$ - the counterpart for the control system; $s\ (u_o, \varkappa_o, t)$ - the rate of expenses for maintaining the quality of the entire system at the required level; β^{-t} - the discount factor.

By minimizing the cost function with respect to the parameters of the system (stage VII)

$$C\left[u_o(\varkappa_o), \varkappa_o\right] \longrightarrow \min_{\varkappa_o}, \tag{13}$$

we find the vector of optimal parameters of the structure, $\tilde{\varkappa}_o$. Substituting it into expression (11) we get the value of the total optimal reliability of the system

$$R_c(t) = R_c(t \mid \tilde{\varkappa}_o). \tag{14}$$

Let us consider some problems relating to the application of the reliability computation patterns outlined above.

3. Reliability control by programmed control of loads or impacts on the system. There exists a fairly broad class of structures operating under load that is a stepwise increasing stochastic process Q(t) (whose characteristics are known), it being possible to periodically remove the load from the structure when the latter reaches a dangerous level. In this case the engineer confronts the dilemma as to whether he should design the structure for a random load Q(t) or for a load of some level Q_B but with account to a possible periodic load throwoff. We have come to a problem of mechanical system reliability control theory which is typical in many respects [6] . It is solved in the following way.

Assume that the load Q(t) increases [passes from one state i to another (i+1)] at some random time intervals ξ_i having the same distribution function $K(\xi)$ and a finite mean $<\xi>$; at the moment the load, as it increases in magnitude, reaches some level Q_H an order is passed on to throw off the load by Q_H unities; this decrease occurs following a random time Ψ having the distribution function $L(\Psi)$ and a finite mean $<\Psi>$. A possible case is the situation when, after ordering the throwoff, the load rises to the level Q_B before being thrown off and the system fails. After failure, the system is restored within a random time τ and the load increases from zero again. The time τ has the distribution function $V(\tau)$ and a finite mean $<\tau>$.

Subject to the prerequisites formulated, the loading process Q(t) is a regenerative one, and the random quantities $\{\xi_i\}_{i=1}^{i=\infty}$, $\{\Psi_i\}_{i=1}^{i=\infty}$, $\{\tau_i\}_{i=1}^{i=\infty}$ are independent. The moments of initiating an order to share the load serve as regeneration points.

The main dependences required for further presentation are written below.

The mathematical expectation of the regeneration cycle Θ is

$$<\Theta> = p\left[(Q_H-Q_B)<\xi> - <\tau>\right] + Q_B<\xi> + <\tau> , \quad (15)$$

where $p = \int_0^\infty \left[1 - K_{B-H}(t)\right] dL(t)$; with $K_{(B-H)}(t)$ being a (Q_B-Q_H)-fold convolution of K(t).

The estimate of the mean time prior to the primary failure t_{0_1} is

$$<t_{0_1}> = Q_H<\xi>/q - (2Q_H - Q_B)<\xi> . \quad (16)$$

The controled system reliability function is obtained by applying the Laplace transformation. It has the form

$$R(t) \approx \exp\left[-qt/Q_H<\xi>\right] , \quad (17)$$

and explicitly connects the statistical characteristics of the load $\left[<\xi>, K(\xi)\right]$ and of the method of control with the parameters Q_H and Q_B.

The stationary value for the level of quality [in our case for the load Q(t)] , $\lim_{t\to\infty}<Q(t)> = <X(\Theta)>/<\Theta>$, where

$$<X(\Theta)> = <\xi> Q_H(4Q_B-3Q_H-1)/2 + <\xi> q\left[(Q_B-Q_H)^2-(Q_B-Q_H)\right]/2 -$$

$$- <\xi> (<\Psi> - r_{B-H});\ r_{B-H} = \int_0^\infty \left[1-L(v)\right] K_{B-H}(v)\, dv. \quad (18)$$

Introducing the cost of damage s_y per unit time due to failure of the system, taking the rate of expenses for maintaining the quality at a level i to be proportional to that level: $C_э = -k_э i+b$ ($k_э = b/Q^*$; Q^* is some basic value of the load) and the cost of restoring the lost quality to be equal to $C(Q_H)$, the stationary rate of losses in the reliability control of a single system will be

$$C = \left[s_y<\tau> - k_э <X(\Theta)> + b + C(Q_H)\right]/<\Theta> , \quad q=1-p. \quad (19)$$

Formulas (17) - (19) make it possible to compute and optimize the reliability function for a mechanical system controled with respect to load and to solve problems of selecting the optimal control policy and the infrastructure connected with the restoration of the system.

4. Reliability control of a vibrating beam with a damper. Let us consider the forced steady-state oscillations of a single beam (with the length l and rigidity EI) due to a distributed dynamic load of the form

$$Q\ (t,\ x) = Q_a\,\varphi(x)\ \sin\,\omega t, \tag{20}$$

with t being the time; ω - the load variation frequency; $x = \tilde{x}/l$ the dimensionless longitudinal axis of the beam; $\varphi(x)$ - the dimensionless law of variation of load intensity; Q_a - the load amplitude which is a random quantity with a distribution density p (Q_a).

It will be assumed that to suppress possible dangerous oscillations of the beam a damper exerting at every moment of time t a control action in the form of some force $\tilde{V}$ (t) is secured to the beam at a distance $\tilde{a}$ from the left-hand support.

We will assume that the beam fails when the maximal stresses exceed the yield point. In these conditions, let us find the reliability of the beam, R (t), on the assumption that the control force $\tilde{V}$ (t) is optimal in the sense that it ensures minimization of a possible deviation of the beam from the neutral nonloaded state

$$I = \max_{x,\tau} |z(x,\tau)| \longrightarrow \min_{\tilde{V}}, \tag{21}$$

where $\tau = \omega\tau$; $z = \tilde{y}\omega^2/\varphi_o$; $\varphi_o = Q_a/ml$; $\tilde{y}$ $(\tilde{x},\ t)$ is the deflection of the beam; m - the linear mass of the beam. Introduction of the notation: $b_k = \tilde{b}_k/l$; $a = \tilde{a}/l$; $u(\tau) = \tilde{V}(\tau)/m\varphi_o l$; $Q_k(\tau) = \tilde{Q}_k(\tau)/$ $/m\varphi_o l$; $k^2 = EI/m\omega^2 l^4$; b_k - the distance of the beam supports from the left-hand support (i.e. $b_1 = 0$; $b_2 = 1$); $\tilde{Q}_k$ - the response of the k-th support, yields the following equation for the forced oscillations of the beam

$$\frac{\partial^2 z}{\partial t^2} + k^2\,\frac{\partial^4 z}{\partial x^4} = \varphi(x)\ \sin\tau\ +\ \delta(x-a)u(\tau) + \sum_{k=1}^{2}(x-b_k)Q_k(\tau). \tag{22}$$

For a hinge-supported beam with the load obeying the law $\varphi(x) = \sin\pi x$, the optimal solution satisfying (22) and the optimal control satisfying the inequality $|u_n(\tau)| < 1$, respectively, are of the form [3]

$$u^o(\tau) = \begin{cases} u_a\ \mathrm{sign}\left([\eta(\pi/2)]\ \sin s\tau\ [\cos(s\pi/2)]^{-1}\right),\ \tau\in(0,\ \tau_1); \\ -\ \sin\tau\ +\ s^2\,\eta(\tau_1),\ \ \tau\in(\tau_1,\ \tau_2); \\ u_a\ \mathrm{sign}\left([\eta(\pi/2)]\ \sin s(\tau-\tau_2)\ [\cos s(\tau_2-\pi/2)]^{-1}\right),\ \tau\in(\tau_2,\pi), \end{cases} \tag{23}$$

where τ_1 and τ_2 are the moments at which the trajectory enters and escapes the constraint; u_n and u_a are the crest values of the controls involved in the subsystems of equations into which expression (22) breaks down;

$$z(x,\tau) = \sin(\pi \xi/l)\,\eta(\tau);\quad z_{max}(x,\tau)|_{\tau=\pi/2} = \sin(\pi\xi/l)\left[1-u_a/s^2\right];$$
$$u_a = \cos(s\tau)/(1-s^2). \tag{24}$$

The maximum deflection here (when $\tau = \pi/2 + 2\pi k$)

$$y_{max} = z_{max}\,\varphi_0/\omega^2 = \frac{Q_a}{m\omega^2 l}\sin(\pi\xi/l)\left[1-\cos(s\pi/2)/(1-s^2)s^2\right],$$
$$s = k\pi^2 = (\pi^2/\omega l^2)\sqrt{EI/m}; \tag{25}$$

2 h is the height of the beam.

Knowing the deflections, we find the bending moment $M = -EIy''$ and the maximum stresses σ_{max} in the most dangerous cross-section of the beam ($\xi = l/2$):

$$\sigma_{max} = \frac{Mh}{I}\Big|_{\xi=l/2} = Q_a(Eh\pi^2/\omega^2 ml^3)\left[1-\cos(s\pi/2)/(1-s^2)s^2\right] = AQ_o. \tag{26}$$

The distribution density $p(Q_a)$ being known, the stress distribution density is

$$p_1(\sigma_{max}) = |A|^{-1} p\left[Q_a(\sigma_{max})\right]$$

and reliability of the beam with respect to the type of failure assumed

$$R_c = P\{\sigma \leqslant \sigma_T\} = \int_0^{\sigma_T/A} p_1(Q_a)dQ_a. \tag{27}$$

Let us compare the reliability of the beam when with and without control. Setting $u = 0$ in Eq. (22) and following similar lines of argument we arrive at the result that

$$y_{max} = \frac{Q_a}{m\omega^2 l}\sin(\pi\xi/l)\left[1/s(1-s)\right];\quad \sigma_{max} = Q_a Eh\pi^2/\omega^2 ml^3 s(1-s) = BQ_a;$$

$$R = \int_0^{\sigma_T/B} p_1(Q_a)dQ_a.$$

Let us consider the difference

$$B-A = \frac{\cos(s\pi/2) + s(1+s)\left[1-s(1-s)\right]}{(1-s^2)s^2}\cdot\frac{Q_a Eh\pi^2}{ml^3\omega^2}.$$

When $s < 1$, $\cos(s\pi/2) > 0$ and $s(1-s) < 1$, whence $A-B > 0$ and $Rc > R$. If $s > 1$ then $B-A > 0$ and $Rc > R$ as before. So it has been proved that the reliability of a protected beam is higher than that of an unprotected one. This is so because the control action extends the permissible region in the load space. We will illustrate this conclusion on the example of the beam considered above. For convenience, we replace the load distributed according to (20) by the equivalent concentrated force $Q = 2Q_a/\pi$ applied to the middle of the beam (Fig.3,a).

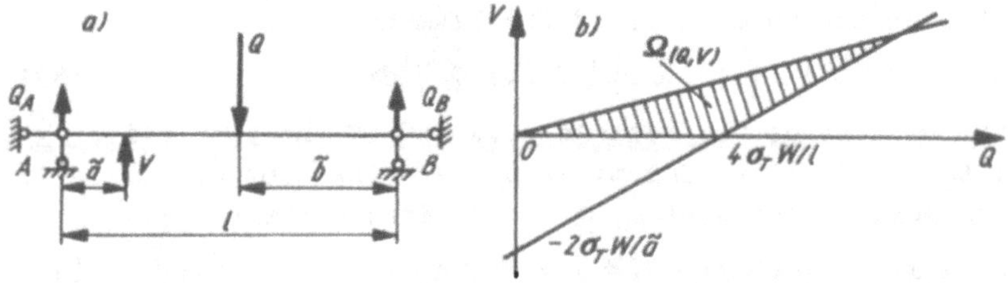

Fig. 3. The scheme of loading the beam (a) and the permissible region for it in the Q, V set (b)

The maximum bending moment is attained in the cross-section $\tilde{x} = l/2$ and the no-failure condition for the beam will become

$$M\ max/W \leqslant \sigma_T \ ; \ Ql-2V\tilde{a} \leqslant 4\sigma_T W \qquad (28)$$

(W is the section modulus).

To expression (28) we should add the condition $|u_n(\tau)| < 1$ which may be represented as

$$V < Q_a/2 \sin(\pi\tilde{a}/l) = Q\pi/4 \sin(\pi\tilde{a}/l). \qquad (29)$$

Considering the case $V > 0$, $Q \leqslant 0$ (for the time intervals t, when $V \leqslant 0$ and $Q > 0$, the lines of argument are similar) we obtain, as a result, the permissible region Ω in $\{Q, V\}$ space (Fig. 3, b) from which it is seen that the maximum permissible load Q value is equal to

$$Q_p = 8\sigma_T W \sin(\pi\tilde{a}/l)\left[2l \sin(\pi\tilde{a}/l) - \pi\tilde{a}\right]. \qquad (30)$$

If the control V is absent then $Q_p^* = 4\sigma_T W/l$ whence it follows that the control extends the range of permissible load values by the magnitude ΔQ,

$$\Delta Q = 4\sigma_T W\left\{2 \sin(\pi\tilde{a}/l)/\left[2 \sin(\pi\tilde{a}/l) - \pi\tilde{a}/l\right] -1\right\}/l. \qquad (31)$$

It is seen from the configuration of the permissible region shown in Fig. 3, b that for control V in mechanical system reliability problems there exists, in addition to a mathematical restriction from above, a physical restriction from below. This circumstance is crucial in the designing of control systems as such.

Determining the optimal control and the optimal solution of the system subject to such restrictions calls for the development of new algorithms. In solution (23) it turns out that $u_a = u_a(\Omega_{(Q,V)})$ and $u^\circ(\tau) = u^\circ[\Omega_{(Q,V)}, \tau]$ - depend on the form of the permissible region $\Omega_{(Q,V)}$. Since there may hold several failure criteria at a time the permissible region Ω (Q,V) in general is constructed as the intersection of admissible regions for each i_j-th kind of failure (see Fig. 2a).

Let us now determine the optimal reliability level for the steel beam considered. Assume that the load amplitude Q_a is distributed normally: $p(Q_a) = (\sqrt{2\pi}\,\sigma_{Q_a})^{-1} \exp\left[-(Q_a - m_{Q_a})^2 / 2\sigma_{Q_a}^2\right]$.

The function of the total cost of the beam is

$$C(T, \varkappa) = C_g(\varkappa) + \left[1 - R(\varkappa)\right] \int_0 C_y(t)\, \beta^{-t} dt, \tag{32}$$

where $C_g(\varkappa)$ is the cost of the beam in situ and of the protection system; $R(\varkappa)$ is the reliability function; $C_y(t)$ - the cost of damage incurred by the failure of the system; β^{-t} the discount factor;

$$C_g(\varkappa) = k_1 + k_2 G(\varkappa); \quad G(\varkappa) = Fl\gamma k, \tag{33}$$

where F is the cross-section area of the beam, k_1, k_2 are coefficients relating the cost of the structure to its mass $G(\varkappa)$, γ is the specific weight of steel; k is a coefficient allowing for the mass of additional structural elements of the beam. As the optimization parameter we can take the height of the beam section, h. Then we will obtain a one-parametric variational problem

$$C(T, \varkappa_o) = \min_{\varkappa} \left\{ C(T, \varkappa) \right\} \tag{34}$$

which may be reduced to an algebraic equation.

Determining the minimum $C(T, \varkappa)$ from the condition $dC(T, \varkappa)/d\varkappa = 0$ on the basis of the following initial data: $\alpha = 5 \times 10^{-3}$; $\gamma = 8 \times 10^{-3}$ kg/cm^3; $k = 1.1$; $k_1 = 0$; $k_2 = 5.13 \times 10^{-3}$ roubles per kilogram; $l = 1200$ cm; $\sigma_T = 230$ MPa; T = 20 Years; $\beta = 1.15$; $\sigma_{Q_a} = 7.48 = \times 10^4$ N; $\langle Q_a \rangle = 22.9 \times 10^4$ N; (Cy has been borrowed from Ref. 4), we have for the function a unique minimum at which $\varkappa_o = 54.4$ cm and $R_{opt} = 0.9989$.

References

1. Bolotin V.V. Primenenie teorii veroyatnostey i teorii nadezhnosti k raschetu sooruzheniy. M. Stroyizdat, 1971, 265 pp.
2. Krasovsky N.N. Teoriya upravleniya dvizheniem. M. Nauka, 1968, 675 pp.
3. Spiridonov V.V., Ulitskaya R.I. Upravlenie kolebaniyami i optimalnym gasheniem kolebaniy balok. Stroitelnaya mekhanika i raschet sooruzheniy. 1978, No 5, pp 56-60.
4. Timashev S.A. Sistemny podkhod k probleme nadezhnosti inzhenernykh sooruzheniy. Book: Issledovaniya v oblasti nadezhnosti inzhenernykh sooruzheniy. L., Leningradsky PSP, 1979, pp 5-24.
5. Timashev S.A. Postanovka nekotorykh klassov zadach upravleniya nadezhnostyu mekhanicheskikh sistem. Book: Prognozirovaniye i raschet individualnoy dolgovechnosti i nadezhnosti mekhanicheskikh sistem. Ural Science Research Center of the USSR Ac. Sci., Sverdlovsk, 1980, pp 16-21.
6. Timashev S.A. On a control problem in reliability. 3rd International Vilnius Conference on Theory of Probability and Mathematical Statistics, vol. 2, pp 186-187. Vilnius, 1981.

LEVEL EXCURSIONS AND EXTREMES OF HOMOGENEOUS RANDOM FIELDS

Erik H. Vanmarcke(*)

Summary

The paper discusses concepts underlying a new approach to representation and analysis of homogeneous random fields [5]. Attention is drawn to the problem of excessive sensitivity(to the choice of correlation model) of mean square derivatives upon which level excursion and extreme value statistics depend. This problem can be resolved for practical purposes by permitting some "local averaging" of a random field under study. This leads to the introduction of the variance function and the scale of fluctuation, key components of a random field theory which permits prediction of level excursion statistics and the(asymptotically exact)probability distribution of extreme values.

Introduction

Many physical "systems" have attributes which, viewed on an appropriate scale, exhibit complex patterns of variation in space and time. When the degree of disorder is sufficiently large, there is usually merit and economy in probabilistic rather than deterministic models. In particular,random field theory seeks to model complex patterns of variation (interdependence, correlation) in cases where deterministic treatment is inefficient and conventional statistics insufficient. An ideal random field model will capture the essential features of a complex random phenomenon in terms of a minimum number of physically meaningful and accessible parameters.

The question of scale is of paramount importance in practial applications of random field theory. A phenomenon which appears deterministic on the "microscale" may at a larger scale exhibit highly variable properties which call for probabilistic description. On an even larger scale (in this context, the "macroscale"), one may find these same structures embedded in objects amenable to deterministic description and characterization in terms of "average" or "composite" properties. In other words, there is usually a lower and an upper bound on the range of dimensions (distance or time) within which a random field model has practical value. Outside this range, a deterministic description or an entirely different stochastic description may be appropriate.

(*) Professor of Civil Engineering, Massachusetts Institute of Technology, Cambridge, Massachusetts 02139 USA

Much of the early literature on stochastic processes deals with variation in function of a single parameter, usually time. The classical approach to second-order analysis of stationary random processes is based on the Wiener-Khinchine relations between the auto-correlation function (a.c.f.) and the spectral density function (s.d.f.)[1,6,7]. Among the most important features of a random field are the patterns of correlation and persistence reflected in the second-order statistics, and much of the literature on applied stochastic processes is concerned with analytical modeling of the "correlation structure".

If a random process is Gaussian, first- and second-order information is sufficient for complete probabilistic characterization. An important related result due to S.O. Rice [4] is an expression for the mean rate of crossing of an arbitrary level by a one-dimensional random function. For stationary Gaussian processes, the mean threshold crossing rate is simply related to the second moment of the s.d.f. or equivalently, to the second derivative of the a.c.f. evaluated at zero distance separation ("lag").

Questions of "scale" motivate much of the work of Mandelbrot [2] who observed that temporal and spatial patterns of correlation and persistence occurring in nature often fail to obey common, simple correlation models. At whichever scale a random phenomenon is observed, it tends to contain "ripples" and "slow trends" which bear witness to the presence of components with lower and higher scales, respectively. Mandelbrot coined the term "self-similar" to refer to phenomena which display similar patterns of variation regardless of the scale at which they are observed. The spectral density function of such random processes appears to be inversely proportional to frequency, hence the term "1/f noise".

The limited purpose of this paper is to discuss the concepts and ideas underlying a simplified, yet rigorous approach to random fields developed in the writer's forthcoming book entitled "Random Fields: Analysis and Synthesis" [5]. Owing to the constraint on the paper's length, only a brief introductory survey is presented of specific new results.

2. Local Averages and Extremes of Local Averages

Central to the development of robust random field models is the concept of the "local average" of a random field. It is seldom useful or necessary to describe in detail the local "point-to-point" variation occurring on a "microscale" in time or space. Even if such information were desired, it may be impossible to obtain. Heisenberg's principle of uncertainty asserts that true patterns of point-to-point variation cannot be known: there is a basic tradeoff between the accuracy of a measurement and the (time or distance) interval within which the measurement is made. A pragmatic viewpoint on the question of scale in random field modeling is that there invariably exists a bound on the time of distance scale below which "micro-variations" of the attribute are:

(a) not observed or observable; and/or

(b) of no practical interest to the problem at hand; and or

(c) characterized by a deterministic microstructure

Much the same argument can be made with respect to very slow variations occurring on a "macroscale".

These observations constitute the basis for one of our central theses: random field models need only provide information about random variation on the "microscale" sufficient to represent the random field's behavior (under at least some amount of) local averaging. In fact, a more detailed description of the correlation structure on the microscale unnecessarily "overloads" the model. A similar principle in solid mechanics holds that erratic patterns of "local stresses" in a solid body may be replaced by locally averaged stresses for the purpose of overall force-deformation analysis. In other words, details of the microstructure of the material affect behavior on the macroscale only through their effect on local averages. Based on this principle, continuum mechanics becomes meaningful at the macroscale for "complex" materials such as granular soil, crystalline composites, or even living tissue.

The foregoing discussion about the relationship between "point" properties and "local averages" remains relevant when dealing with extremes and level excursions. Highly localized extreme values of a random field cause little concern if they do not affect the behavior of the system as a whole. However, such system-wide effects are often felt when the random field is in the aggregate high (or low) over a sufficiently large domain in space and/or time. The implication is that two quantities matter: (a) the size of the domain over which the random field is (locally) aggregated, and (b) the size of the deviation of the locally aggregated process from its expected value.

3. Stochastic Theory Based on Local Averages - An Introduction

3.1 Lack ot Stability of the Mean Square Derivative

Consider a one-dimensional continuous-parameter stationary random process X(t) with mean zero and unit variance. The correlation structure of X(t) is commonly represented either by the (auto) correlation function $\rho(\tau)$, where $\tau = t - t'$ is the "lag", or by the normalized (unit-area) spectral density function $g(\omega)$, defined for frequencies $\omega \geq 0$. These two functions constitute a Fourier transform pair [1,6,7].

Selection of an analytical model for $\rho(\tau)$ or $g(\omega)$ is usually based on the quality of fit in the range of observed (observable) values of τ or ω, which, for reasons mentioned earlier, does not include the microscale ($\tau \to 0$ or $\omega \to \infty$). The choice of a model for $\rho(\tau)$ or $g(\omega)$ does, however, imply an (essentially unverifiable) assumption about how the random process behaves on the microscale. Moreover, this assumption may profoundly affect predictions of quantities related to the pattern of fluctuation, e.g., the mean square derivative, mean threshold crossing rates and extreme values, as well as the behavior of the process under local averaging.

In particular, many common correlation models predict that the mean square derivative and the mean threshold crossing rates are non-existent (infinite), when in reality, in the sample paths these models represent, they are obviously finite. This "theoretical" problem, i.e., non-existence of the mean square derivative, disappears when a small amount of local averaging, enough to smooth the microscale fluctuations, is permitted.

3.2 One-Dimensional Variance Function

Consider the random process $X_T(t)$ obtained by averaging X(t) over the "window" T. The variance function $\gamma(T)$ is defined as the ratio of the variance of $X_T(t)$ over the "point" variance of the oroginal process X(t). $\gamma(T)$ decreases monotonically with T, its upper bound is $\gamma(0) = 1$, and for processes which are ergodic in the mean, $\gamma(\infty) = 0$. (This is a well-known "ergodic theorem" [3]). $\gamma(T)$ fully characterizes the correlation structure of X(t) and has the same information contents as $\rho(\tau)$ or $g(\omega)$.

3.3 Scale of Fluctuation

If a random process is ergodic in the mean, its scale of fluctuatuation θ can be expressed alternatively in terms of the variance function $\gamma(T)$, the correlation function $\rho(\tau)$, or the (unit-area) spectral density function $g(\omega)$, as stated below. The first statement serves as the definition of θ.

$$\theta = \begin{cases} T\ \gamma(T) & \text{for } T \to \infty \\ \int_{-\infty}^{+\infty} \rho(\tau)\ d\tau & \\ \pi\ g(\omega_o) & \text{for } \omega_o \to 0 \end{cases} \tag{1}$$

The condition for validity of the three interpretations of θ is $\dot{g}(0) = 0$, where $\dot{g}(\omega)$ is the first derivative of $g(\omega)$, or $|\theta_{2,X}| < \infty$, where $\theta_{2,X}$ is the second moment of the correlation function $\rho(\tau)$.

For wide-band processes, as encountered in the study of isotropic random media, θ captures information not only about low frequency content (since $\theta = \pi\ g(0)$ and $\dot{g}(0) = 0$), but also about high frequency content, since contributions to the variance associated with frequencies above $(2\pi/\theta)$ become negligible (owing to the fact that $g(\omega)$ has unit area). Random processes for which $\theta = 0$ have no "long-term memory" and their variance decays in inverse proportion to T^2 (rather than T) under extended local averaging.

When a process with variance σ^2 and scale θ is integrated over a window of finite length T, the variance of the integral converges to $\sigma^2\ \theta\, T$ when $T >> \theta$. The product $\sigma^2\theta$ is itself an important parameter which in some applications (e.g., diffusion problems where only the variance of the displacement matters) provides sufficient characterization of a random phenomenon under study (i.e., σ^2 and θ need not be known separately).

3.4 Statistics of Level Excursions and Extremes

In the one-dimensional case, the variance of $X_T(t)$ is Var $[X_T] = \sigma^2\ \gamma(T)$, and the ratio of the variances of $\dot{X}_T(t)$ and $X_T(t)$ is

$$\frac{\mathrm{Var}[\dot{X}_T]}{\mathrm{Var}[X_T]} = \frac{2}{T^2} \left(\frac{1 - \rho(T)}{\gamma(T)} \right) \xrightarrow[T >> \theta]{\text{For}} \frac{2}{\theta T} \tag{2}$$

For $T \to 0$, $\mathrm{Var}[\dot{X}_T]$ reduces to $\sigma_{\dot{X}}^2$ provided $X(t)$ is mean square differentiable. In general, the parameter θ controls the behavior of a random process under long-term averaging (when $T >> \theta$). For wideband processes, however, even if the window T is only a fraction of the "scale" θ , the stochastic behavior of the family of "local average" processes can be described satisfactorily in terms of the mean m, the variance σ^2 and scale of fluctuation θ of the original process. Details of the "correlation structure" are of secondary importance (unless $T \to 0$), and their inport diminishes fast as the averaging interval grows.

The three-parameter (m, σ^2, θ) description suffices to obtain exact asymptotic results (for $T \gg \theta$) of appealing analytical simplicity for the mean threshold crossing rate and the probability distribution of extreme values for the family of derived "local average" processes. The original process need not be Gaussian, and the theory applies to discrete-parameter and discrete-state processes as well.

3.5 Generalization

The concept of the variance function and the idea of defining a correlation parameter in terms of the asymptotic form of the variance function remain valid in multidimensional situations. In the two-dimensional case, the variance function $\gamma(T_1,T_2)$ is defined in terms of the local average of $X(t_1,t_2)$ over a rectangular window $A = T_1T_2$ (with sides parallel to the two axes). The "correlation area" α has basic properties similar to those of θ in the 1-D case; in particular, the 2-D variance function converges to the expression α/A when the dimensions T_1 and T_2 are much larger than their respective (direction-dependent) scales of fluctuation, θ_1 and θ_2. Of course, for an isotropic field, $\theta_1 = \theta_2 = \theta$.

The 2-D variance function $\gamma(T_1,T_2)$ can be expressed as the product of two 1-D variance functions, one "marginal" and the other "conditional" (much in the same way as one expresses the joint probability density function of two dependent random variables). The conditional variance function $\gamma(T_2 \mid T_1)$ depends on the conditional scale of fluctuation $\theta_2(T_1)$. These quantities characterize the 1-D random process obtained by averaging a 2-D random field within a band of prescribed width T_1 parallel to the t_2-axis. When $T_1 = 0$, there is no averaging, and the conditional scale of fluctuation reduces to the scale θ_1. As the width T_1 of the averaging band increases, $\theta_2(T_1)$ tends to increase slowly and it approaches the asymptotic value $\alpha/\theta_2 = c\theta_1$, where the dimensionless constant $c = \alpha/\theta_1\theta_2$ characterizes the interaction (as regards correlation) between the principal coordinates. If the 2-D correlation structure is "separable", i.e., $\rho(\tau_1, \tau_2) = \rho(\tau_1)\,\rho(\tau_2)$, then $\alpha = \theta_1\theta_2$ and $c = 1$.

The properties of a homogeneous n-dimensional random field can be described by an n-dimensional variance function which can be expressed as a product of n one-dimensional variance functions. Additional correlation parameters (and conditional scales of fluctuation) are introduced in the same way as in the two-dimensional case. An explicit analytical solution is derived for the mean rate of maxima above a high level, and the mean size of regions of excursion above a high level, for the family of local average fields derived from homogeneous (not necessarily Gaussian) n-dimensional random fields. By assuming that high level excursions are characterized by a multidimensional Poisson process (with known mean occurrence rate), a closed-form expression can be found for the probability distribution of extreme values.

3.6 Some related results

The three-parameter representation (mean m, variance σ^2, scale θ) provides a flexible tool for practical stochastic modeling: a random process with "complex" correlation structure can be modeled as a sum of two or more uncorrelated "simple" processes, each characterized by parameters m_i, σ_i^2 and θ_i which can easily be related to the corresponding composite parameters. This permits modeling and analysis of patterns of random variation which display the characteristics of 1/f noise . The composite model (which extends easily to the multidimensional case) can accommodate uncorrelated "systematic" errors (for which "$\theta = 0$"), perfectly correlated "systematic" errors (for which $\theta = \infty$), as well as "$\theta = 0$" narrow band components.

Finite element (or finite difference) analysis of systems with distributed random properties requires as input the local averages of the random property within each finite element. The proposed methodology permits efficient evaluation of the matrix of covariances between local averages associated with pairs of finite elements. These results may also be used to update predictions of random fields based on limited, discrete observations, and to assess alternative strategies for data acquisition (i.e., the frequency of sampling in time and/or in space). A variety of approaches are also considered to the important problem of how to estimate the scale of fluctuation (and related correlation parameters) from actual measurements of random field.

Conclusion

The proposed approach based on local averages resolves a long-standing issue of how to deal with the non-existence of mean square derivatives and level excursion statistics of homogeneous random functions. As regards specific new results, this paper has given no more than a very brief introductory account of the random field methodology proposed in the writer's forthcoming book [5].

Acknowledgement

This research was sponsored in part by the National Science Foundation under Grant No. PFR-7815898 "Risk-Based Assessment of the Safety of Dams". Any opinions, findings, conclusions or recommendations expressed herein are those of the writer and do not necessarily reflect the views of N.S.F.

References

1.
KHINCHINE, A.I., "Korrelationstheorie der stationaren Stochastischen Prozesse", Math. Ann, vol. 109, pp. 605-616, 1934.

2.
MANDELBROT, B.B., "Une classe de processus homothétiques à soi; application à la loi climatologique de H.E. Hurst", C.R. Acad. Sci., 260, Paris, 1965, p. 3274.

3.
PUGACHEV, V.S., Theory of Random Functions and its Application to Control Problems, Pergamon Press, 1965, p. 302.

4.
RICE, S.O., "Mathematical Analysis of Random Noise", Bell Systems Tech Journal, 22, 1944, p. 282; 25, p.46, 1945.

5.
VANMARCKE, E.H., Random Fields: Analysis and Synthesis, forthcoming at the M.I.T. Press, Cambridge, Massachusetts and London , England, January 1983.

6.
WIENER, N., "Generalized Harmonic Analysis", Acta Mathematica, Vol. 55, pp. 117-258, 1930.

7.
WIENER, N., Extrapolation, Interpolation and Smoothing of Stationary Time Series, The M.I.T. Press, 1950.

SECTION I D E N T I F I C A T I O N

ON EXPERIENCES IN THE APPLICATION OF SYSTEM IDENTIFICATION PROCEDURES TO MECHANICAL ENGINEERING

Beer, H.-J.; Hardtke, H.-J.; Holzweißig, F.[1)]

A complete system of procedures for the experimental system identification is presented, which are implemented in computer programs. Adaptive methods are emphasized and an example for the application of the overall conception is given.

In system identification from experimental data for problems of machine dynamics three questions must be answered in advance:

1. Which class of model should be employed?
   (model with constant or with time-variant parameters, linear or nonlinear equations of motion)
2. Which structure for the model must be taken as a basis?
   (discrete structure, continuum, selection of coordinates, type of coupling)
3. In which way the model shall be described?
   (by characteristic functions in the time or frequency domain or by differential equations)

After having decided upon these basic questions the problem of system identification comprises the confirmation of the present mathematical structure as well as the determination of their parameter values.
In engineering practice such models are preferred that permit a direct correspondence of system parameters (e. g. masses) to model parameters.
In many cases a clear structure is selected and their parameters are evaluated. For this approach adaptive methods, which reduce the identification problem to an optimization, have proved their worth. By means of these methods the parameters of a structure are optimized in a manner, that an error function (e. g. weighted difference between model and system behaviour) is minimized. Prerequisite is a thorough a-priori knowledge of the structure and of start values for the parameters. These demands lead to an overall conception into which adaptive methods are imbedded (fig. 1). It has proved its worth in the treatment of many practical problems, which are tackled in three steps. The method is bases on measured input and output time functions of the system under test (often: excitation of a single coordinate and response of several ones). In the first step the digitized measured data are processed and characteristic functions in the time domain

1) Dipl.-Ing. H.-J. Beer, Dr.-Ing. H.-J. Hardtke, Prof. Dr.-Ing. habil. F. Holzweißig, TU Dresden, DDR 8027 Dresden, Mommsenstr. 13

(impulse functions $g_{kl}(t)$) as well as in the frequency domain (transfer functions $G_{kl}(j\omega)$ are computed. This is implemented in display-controlled computer programs. At now there are two program systems, "KENNWERTE" and "SYSKEN". They include 20 subprograms. 2048(4096) real data can be processed in an minimum sampling interval of about 1 ms (analogue input).

From one data record characteristics of the time function and characteristic functions of the system under test can be determined. The following subprograms are available: analogue data input (single ~and multichannel), Fourier Transformation (FFT), power spectral density, level and peak counting (one parametric~, two parametric~), band-limeted Fourier Transformation, digital filter, impulse response and transfer function, frequency response locus plot, approximation, convolution, exciting force time function.

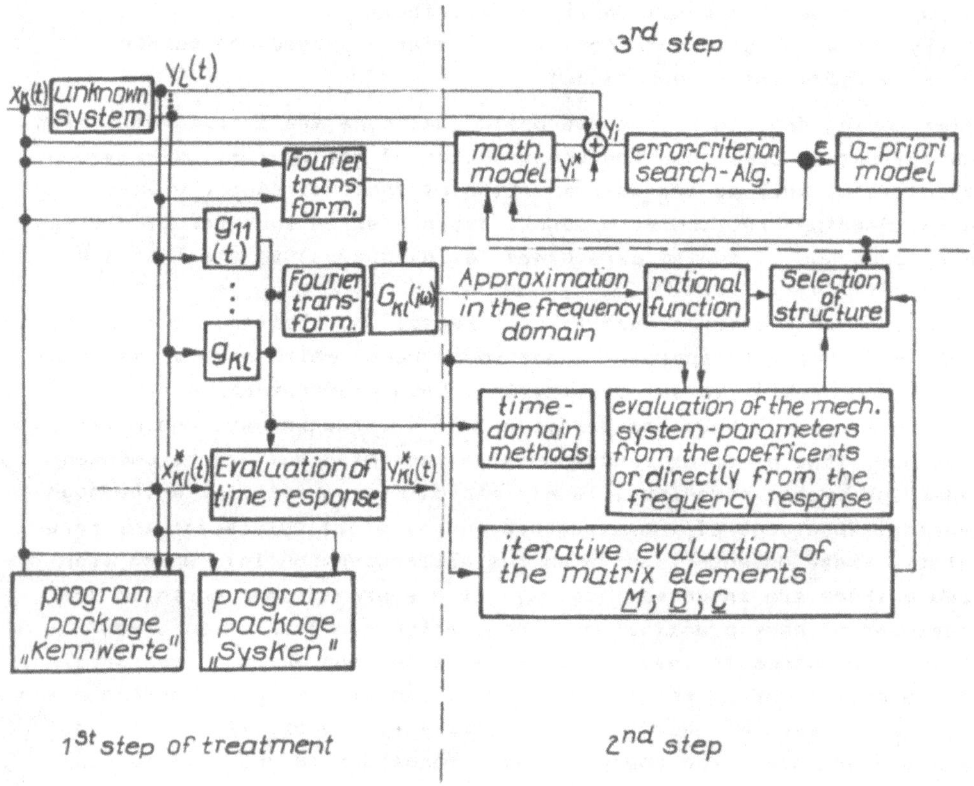

fig. 1

The programs provide an approximated transfer locus, which is the main part of the second step. Transfer loci permit to find conclusions for the structure (number of degrees of freedom) and they give hints on a-priori parameter values.

The third step of the treatmant comprises the adaptive system identification. Since 1975 our department has been engaged in this field.

A summary of the arising problems and the state of the art is shown in fig. 2. First of all shall be reported on the optimization strategy and on the model description (fig. 3). In general a complete identification is only feasible, if a row or a column of the transfer matrix $\underline{G}$ is known ($\underline{y} = \underline{G} \cdot \underline{x}$, $\underline{x}$ = input vector, $\underline{y}$ = output vector). In the field of machine dynamics the evaluation of a column is in general not too difficult (measurement of all responses to a single excitation).

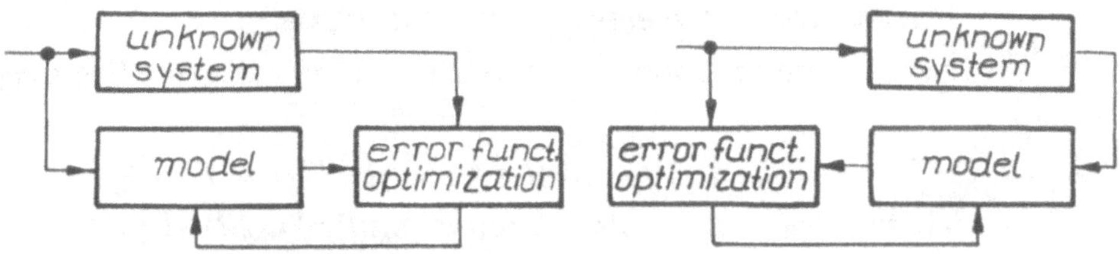

optimization of the output — optimization of the input

problems : 1. structure of optimization (input or output)
2. strategy of optimization (complete or incomplete identification)
3. selection of model (e.g. differential equation, impulse response, transfer function)
4. evaluation method
5. quality criterion, error function
6. optimization algorithm (method of Zettel; ~ Rosenbrock)

model description	output optimization		input optimization	
	solution of model equations by		solution of model equations by	
	numerically	hybrid comp.	numerically	hybrid comp.
polynom of the order 2n			F,T	T
system of different. equat.	T	T	F,T	
impulse response				
transfer-function	F,T			

F: optimization in the frequency domain ; T: optimization in the time domain

problems and state of treatment of the adaptive identification approach

fig. 2

Often one must be satisfied with a single or only a few responses G_{ik}, because not all outputs may be accessible. Strictly speaking the model then represents only the relationship between measured inputs and outputs.

As a model description the differential equations are favourable for their direct correspondence between model and system parameters.

The computation of the model has particular significance. Because of the large number of numerical operations required in the optimization process the computing time for evaluating the model should be as short as possible. This suggested the application of an analogue computer. Comprehensive studies on this subject, however, indicated that the advantage of a short computation gets lost owing to a huge effort of

model < *structure, parameters* — *minimal model*
model description — *set of equations in the frequency or time domain*

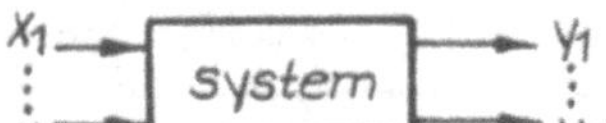

1. *differential equations:* $\underline{M}\ddot{y}+\underline{B}\dot{y}+\underline{C}y=\underline{X}$
2. *impulse function:* $g_{kl}(t)=\sum_{i=1}^{n} c_{ikl}e^{s_i t};\ s_i=\sigma_i \pm j\omega_i$
3. *transfer function:* $G_{kl}(s)=\dfrac{a_0+a_1s+\ldots a_{m-1}s^m}{b_0+b_1s+\ldots b_{n-1}s^n}=\sum_{i=1}^{n}\dfrac{c_{ikl}}{s-s_i}$

Complete model description: $\underline{y}=\underline{G}\cdot\underline{X}$

incomplete model identification: – *evaluation of the parameters from* $y_k=G_{kl}\cdot x_l$; G_{kl} *contains all system parameters*
– *model, described by* G_{kl}, *represents only the functional relation between* y_l *and* x_k.

Complete model identification: $y_k=\sum_{i=1}^{n}G_{ki}\,x_i$ — *evaluation of the* G_{ki} *from a single measured response* y_k, *excited by* $x_i\ (i=1,\ldots n)$; *i.e. one output measured, all inputs excited (one row of matrix* $\underline{G}$*)*

or: $y_i=G_{il}\cdot x_l$ — *evaluation of* G_{il} *from measured data of all* y_i, *that are excited by only one input* $x_l\ (i=1,\ldots,n)$; *i.e. all outputs measured, only one input excited (one column of matrix* $\underline{G}$*)*

in praxi: complete identification is not feasible
strategy: selection of proper input and output coordinates

selection of model and optimization strategy

fig. 3

preparation and to accuracy demands. The optimization strategy can be considered as decisive for the decrease of computing time. According to fig. 2 you can distinguish between input and output optimization. If the error function is evaluated from the output time functions of the system and the model, a system of differential equations must be solved. On the other hand the input optimization requires only multiplications. This approach is therefore qualified for nonlinear problems and for the incomplete (partial) identification (fig. 4). But it is supposed, that all outputs of the model are available from measured system outputs.

Studies indicated, that either the direct measurement (absolute phase-correct measurements are demanded) or measurement and subsequent inte-

advantages: model computation by multiplication without solution of differential equations
sinted for: nonlinear systems; complete model identification (if all outputs are measured only one input contributes to the error function)

problems : correct phase evaluation of y_K; $\dot{y}_K$; $\ddot{y}_K$; $(\dot{y} - \frac{dy}{dt})$

1. selection of measured time functions: investigations of the errors in the presence of noise indicate, that only the cases a; b; f are possible (no numerical derivation)

	a	b	c	d	e	f	g
y_K	X	O	O	X	O	X	X
$\dot{y}_K$	X	O	X	O	X	O	X
$\ddot{y}_K$	X	X	O	O	X	X	O

X - measured value ; O - computed value

2. selection of the sampling interval:

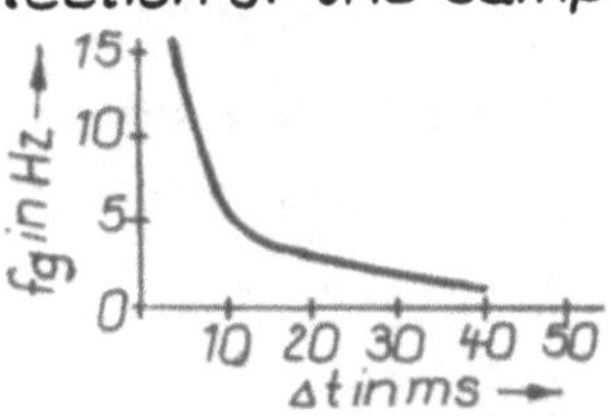

sufficient small sampling interval is necessary

for optimization a greater sampling interval is admissible

3. Start synchronisation of input and output measurements

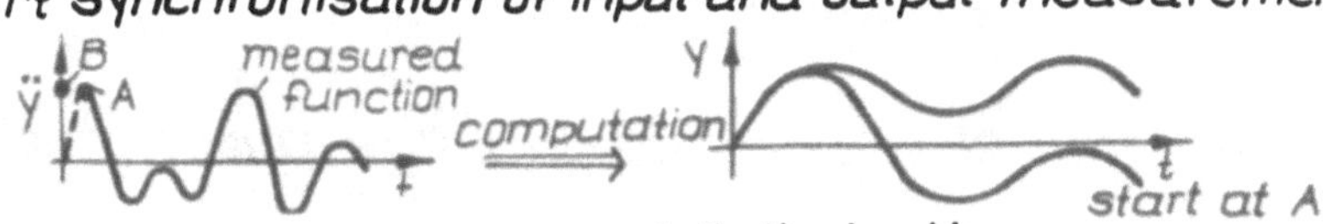

Problems of Input Optimization

fig. 4

gration are reasonable. Sufficiently small sampling intervals and synchronization of input and output measurements play here a significant part.
As an example for the application of the different steps of treatment we consider the determination of the hydraulic axial force of a pumped-storage power station. Figure 5 illustrates the construction. Response measurements at the turbine in operation have been only feasible at the upper bearing. The rotating parts are axially forced at the turbine wheel. For a treatment of the problem the following steps have been necessary.

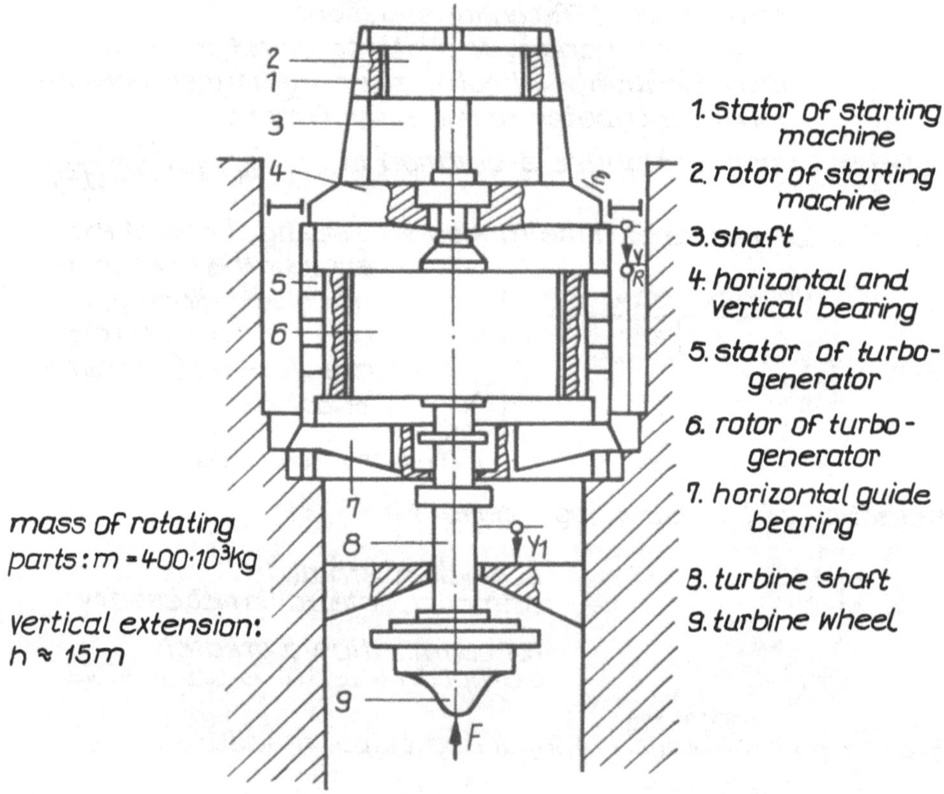

turbogenerator of a pumped-storage power station

fig. 5

Preliminary experiments: In order to determine the transfer function a force formed by a step function (about 500 kN) was applied to the turbine wheel and the resulting strain was measured.

Evaluation of the transfer locus: From response and excitation the transfer function was determined and the transfer locus was approximated (fig. 6). The measurement indicated, that only three degrees of freedom were in evidence.

Determination of the structure and their parameter values: By means of the adaptive methods with output optimization the parameter values of a model with three degrees of freedom have been computed. The number of degrees of freedom has been deduced from the transfer locus and as structure a multiple-degree-of-freedom system has been selected.

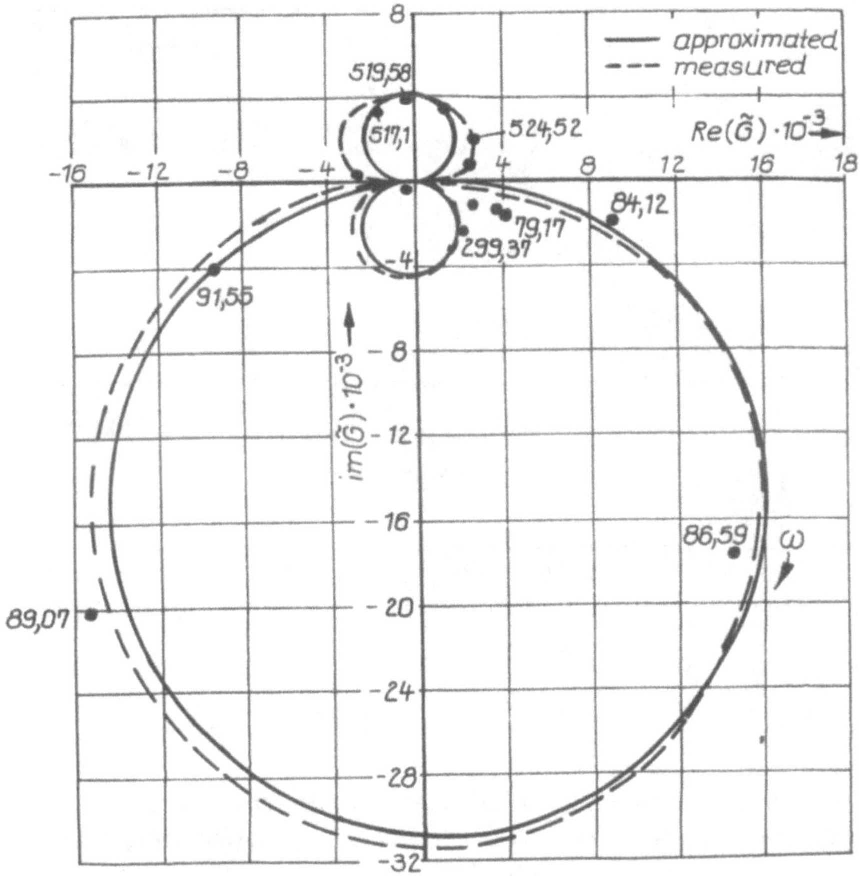

frequency response locus of the strain of vertical bearing $\tilde{G}_{\varepsilon,F}(j\omega)$

fig. 6

Evaluation of the exciting force: The strain time function had to be processed by a digital filter, which consisted of a low pass and two band passes, the pass bands of which have been fitted to the resonance ranges of the transfer function. The time function of the exciting force has been computed by this filtered strain time series and by the approximated transfer function (summary see fig. 7).

As a result we have achieved a describing model and the dynamic axial force in different operational states of the turbine.

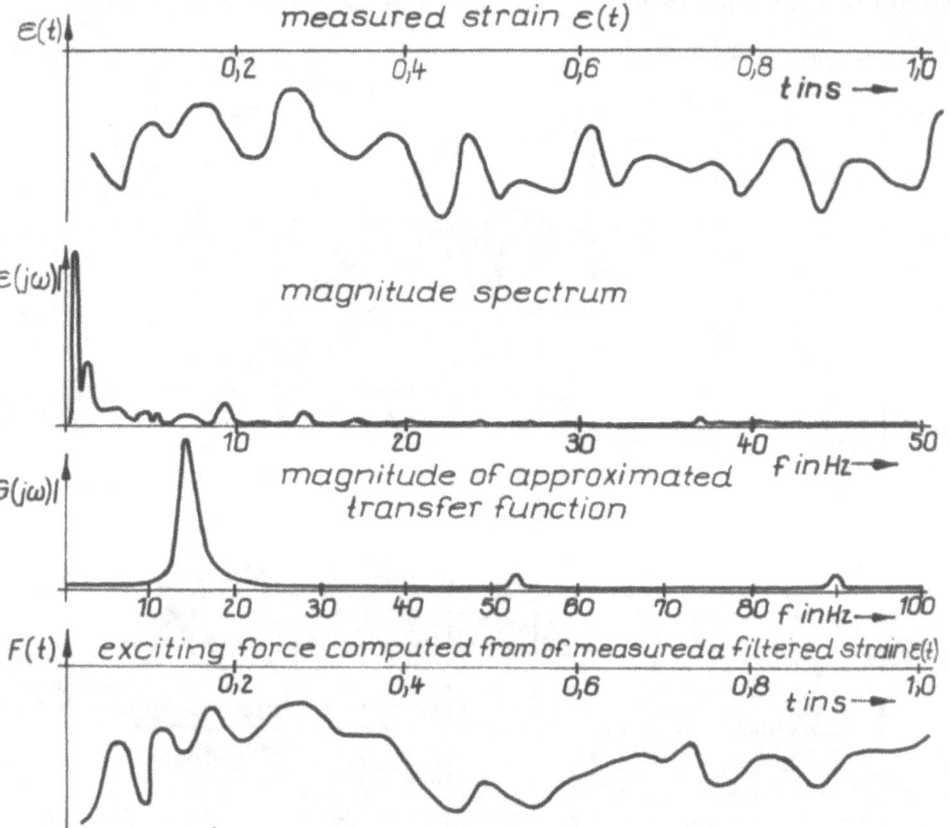

steps in the evaluation of the exiting force from measured time function of strain ε(t) and approximated transfer function

fig. 7

Literature: Following papers on experimental system identification have been published at the Bereich Dynamik und Betriebsfestigkeit der Sektion Grundlagen des Maschinenwesens der TU Dresden:

AWAD, T.: Spezielle Probleme der adaptiven Modellfindung in der Maschinendynamik. Diss. 1978, TU Dresden

BACKHAUS, E.: Bestimmung von Feder-Dämpfer-Kennlinien aus Schwingungsmessungen. Diss. 1972, TU Dresden

BEER, H.-J.: Approximation von Systemkennfunktionen mit Hilfe eines Optimierungsverfahrens. Beitrag BXX in Forschungsberichte 1979. Leipzig, VEB Fachbuchverlag 1979

BEER, H.-J.; HARDTKE, H.-J.; RICHTER, F.: Programmsystem zur experimentellen Bestimmung dynamischer Systemkennwerte. Maschinenbautechnik 27 (1978) 3, S. 103 - 105

DÖGE, S.: Parameterbestimmung für elastisch gelagerte Starrkörper mit Hilfe eines adaptiven Modellfindungsverfahren. Diss. 1983, TU Dresden

DUNG, PHUNG DUY: Beiträge zur adaptiven Modellfindung linearer und nichtlinearer mechanischer Systeme. Diss. 1981, TU Dresden

HARDTKE, H.-J.: Beiträge zur Identifikation mechanischer Systeme. Diss. 1977, TU Dresden

HOLZWEISSIG, F.; HARDTKE, H.-J.: Verfahren zur experimentellen Modellfindung der Maschinendynamik. Maschinenbautechnik 26 (1977) 11, S. 500 - 504

HOLZWEISSIG, F.: Zur Entwicklung der Maschinendynamik in der DDR von 1969 bis 1979. Maschinenbautechnik 28 (1979) 10, S. 442 - 447

RENATUS, H.-J.: Beitrag zur Bestimmung von Rückstellfunktionen für spezielle Hysteresewirkung. Diss. 1983, TU Dresden

STRÜBING, W.: Beiträge zur experimentellen Identifikation von mechanischen Schwingungssystemen bei vorgegebenen Strukturen. Diss. 1978, TU Dresden

PARAMETER ESTIMATORS IN LINEAR STOCHASTIC DIFFERENTIAL EQUATIONS AND THEIR PROPERTIES

Bärbel Bellach[1)]

In this paper a summary of parameter estimators in linear stochastic differential equations and their properties is given. In the case of white noise excitations the consistency of the Maximum Likelihood Estimator (MLE) holds without assuming the stability of the system. Although the MLE, which can also be interpreted as Least Squares Estimator, has some optimality properties under white noise disturbances, an example is given that for coloured noise the same estimator has bad properties. Therefore a special construction principle of Instrumental Variables Estimators is given, which yields consistent estimators in the case of coloured noise excitations in the stable case.

1. Introduction

We are concerned with dynamic systems which can be described by a linear stochastic differential equation of the form

$$\underline{\dot{Y}(t)} = A\underline{Y(t)} + BX(t) + \underline{\varepsilon(t)}, \qquad t \in [o,T], ^{2)} \tag{1.1.}$$

where $\{\underline{\varepsilon(t)}\}$ is a d-dimensional noise process with $E\underline{\varepsilon(t)} = 0$, A and B are unknown (d x d) and (d x p) respectivelly parameter matrices and $X(\cdot) : [0,T] \to R^p$ is a known p-dimensional function. Usually the noise process $\underline{\varepsilon(t)}$ has the property to be "nearly" uncorrelated at different time points t_1 and t_2, with a large distance. The Gaussian white noise is often used as a good mathematical idealization of such processes, although it has the property to have realizations which are nowhere continuous.

In the first part of our paper we assume that $\underline{\varepsilon(t)}$ is a Gaussian white noise process, so that (1.1.) becomes a stochastic differential equation from Itô-type:

$$d\,\underline{Y(t)} = [A\underline{Y(t)} + BX(t)]\,dt + d\underline{W(t)}, \quad t \in [0,T], \tag{1.2.}$$

where $\underline{W(t)}$ is a d-dimensional Wiener process.

After observing our process $\underline{Y(t)}$ over a time interval $[0,T]$ we are interested in estimating the unknown parameter matrices A and B.

---

2) We will always underline random variables for distinguishing them from deterministic variables as X(t) or from realizations of a random variable.

---

1) Dr. B. Bellach
Humboldt-Universität Berlin, Sektion Mathematik
PSF 1297, DDR-1086 Berlin

One possibility for deriving an estimator is the principle of "maximum likelihood ": It is known that the processes $\{\underline{Y(t)}, t \in [0,T]\}$ and $\{\underline{W(t)}, t \in [0,T]\}$ induce Gaussian measures P^Y resp. P^W on the space $C^d[0,T]$ of the d-dimensional continuous functions, which are equivalent iff $\int_0^T \|X(t)\|^2 dt < \infty$. Using the explicit form of the density $\frac{dP^Y}{dP^W}(Y(t), t \in [0,T])$ (see Arato (1)) we obtain the Maximum Likelihood Estimator (MLE) $(A(T) \vdots B(T))$ by taking that value of the matrix $(A \vdots B)$ that maximizes the density relative to our observed realization $\{Y(t), t \in [0,T]\}$ (See also Kozin (4)) .

In our model (1.2.) the MLE $(\underline{A(T)} \vdots \underline{B(T)})$ has the form

$$(\underline{A(T)} \vdots \underline{B(T)}) = \left[\int_0^T \begin{pmatrix} \underline{Y(t)} \\ X(t) \end{pmatrix} d\underline{Y'(t)} \right]' \left[\int_0^T \begin{pmatrix} \underline{Y(t)} \\ X(t) \end{pmatrix} (\underline{Y'(t)} \vdots X'(t)) dt \right]^{-1} . \quad (1.3.)$$

2. Properties of the MLE in the Itô-differential equation model

Before applying such an estimator it is necessary to know, what properties it has. This question will be answered by the following theorems:

Theorem 1

Let the process $\{\underline{Y(t)}, t \in [0,T]\}$ be defined by (1.2.). Under the assumptions

A 1 The real parts of the characteristic roots of the matrix A are all different from zero. If λ_i and λ_j are any characteristic roots of the matrix A, the real parts of which are positive, then λ_i and λ_j are different.

A 2 X(t) fulfills some weak regularity conditions.

the MLE $(\underline{A(T)} \vdots \underline{B(T)})$ defined by (1.3.) is a consistent estimator for $(A \vdots B)$, that means for T tending to infinity the estimator tends to the true parameter matrices (in probability).

It follows from this theorem that stationarity of the process $\{\underline{Y(t)}\}$ is not a necessary condition for the consistency of our MLE and even in the explosive case the MLE tends to the true parameters.

The asymptotic distribution of our MLE is known only in the stable case, that means $\mathrm{Re}\,\lambda(A) < 0$, but also with a deterministic input X(t). The following theorem holds:

Theorem 2

Let the process $\{\underline{Y}(t),\ t \in [0,T]\}$ be defined by (1.2.) . Under the assumptions A 2 from theorem 1 and

A 1 The real parts of the characteristic roots of the matrix A are all negative

the MLE $(\underline{A(T)} \vdots \underline{B(T)})$ is asymptotically normally distributed. It holds

$$\mathcal{L}(T^{1/2}(\overrightarrow{(\underline{A(T)} \vdots \underline{B(T)})} - \overrightarrow{(A \vdots B)})) \xrightarrow[T \to \infty]{} N(0, I_d \otimes C_{AB}) \text{ with}$$

$$C_{AB} := \left(\begin{array}{c|c} P_1+P_2 & L \\ \hline L & P_3 \end{array} \right) \quad \text{where} \quad P_1 := \int_0^\infty e^{At} e^{A't} dt,$$

$$P_2 := \lim_{T \to \infty} \frac{1}{T} \int_0^T \int_0^t e^{A(t-\tau)} BX(\tau) d\tau \left(\int_0^t e^{A(t-\tau)} BX(\tau) d\tau \right)' dt,$$

$$P_3 := \lim_{T \to \infty} \frac{1}{T} \int_0^T X(t) X(t)' dt.$$

It is also possible to prove that our MLE is a best estimator in the class of all asymptotically normally distributed estimators in the following sense:

Theorem 3

Under the assumptions of theorem 2 it holds for all other estimators $\underline{S(T)}$ of the unknown parameter matrices $(A \vdots B)$ with

$$\mathcal{L}(T^{1/2}(\overrightarrow{\underline{S(T)}} - \overrightarrow{(A \vdots B)})) \xrightarrow[T \to \infty]{} N(0, C^S_{AB})$$

that $C^S_{AB} \geqslant C_{AB}$,

that means $(A(T) \vdots B(T))$ is the best asymptotically normally distributed estimator (BAN).

For the proofs of these three theorems the reader is referred to Bellach (2).

So we have seen that the MLE in model (1.2.) has theoretically very good properties. But although the Gaussian white noise seems to be a good approximation for "nearly" uncorrelated disturbance processes, there arise difficulties, since the process $\underline{Y}(t)$ defined by (1.2.) has realizations of unbounded variation, while real physical processes usually are of bounded variation. Our model (1.2.) with the stationary Gaussian white noise can only be an idealized model of real processes, but unfortunately the good properties of our MLE, which can also be interpreted as Least Squares Estimator, depend strongly on this idealization. This can be illustrated by the following example:

Let $\underline{Y(t)}$ be defined by

$\underline{\dot{Y}(t)} = a\underline{Y(t)} + \underline{\varepsilon(t)}$, $Y(0) = 0$, with $a < 0$, where

$\underline{\varepsilon(t)}$ is a stationary Gaussian process which has a continuous spectral density and the covariance structure $E\{\varepsilon(t)\,\varepsilon(\tau)\} = Ke^{\gamma|t-\tau|}$.[1] It is known that such processes are also a good approximation for "nearly" uncorrelated processes, if we choose K and $|\gamma|$ large enough.

In this model the estimator (which is MLE in model 1.2.)

$$\underline{a(T)} = \frac{\int_0^T \underline{Y(t)}d\underline{Y(t)}}{\int_0^T \underline{Y(t)}^2 dt}$$ converges to zero for all parameters a.

(See Bellach (3))

So this estimator may be not suitable for practical purposes and we try to construct estimators in another, more practicable model.

3. Parameter estimators in linear stochastic differential equations with dependent disturbances

In the following we will confine ourselves to the univariate case, since the underlying ideas become clearer. In principle all considerations and theorems hold analogously in the multivariate case.

We consider processes described by

$$\underline{\dot{Y}(t)} = a\underline{Y(t)} + bX(t) + \underline{\varepsilon(t)}, \quad Y(0) = 0, \qquad (3.1.)$$

where $a < 0$ and b are unknown parameters and $\underline{\varepsilon(t)}$ has continuous realizations with probability 1, $E\,\underline{\varepsilon(t)} = 0$.

Now we try to find estimators which are consistent in this model. We cannot use Maximum Likelihood Estimators in this case, because we don't know densities of the induced measure P^Y with respect to some other measures. Therefore we will construct a special form of 'optimal' instrumental variables estimators, where lagged values of the exogeneous input variable $X(t)$ are the instruments.

For illustrating our concept of instrumental variables estimators let us consider the following general model:

$\underline{V_T} = b\underline{U_T} + \underline{\varepsilon_T}$, where $\underline{V_T}$, $\underline{U_T}$ and $\underline{\varepsilon_T}$ are random variables

---

1) $K \in R^+$, $\gamma \in R^-$

(for example $Y_T = (Y_0, Y_{t_1} \ldots Y_T)'$ or $Y_T = \{Y_t, t \in [0,T]\}$).

b is an unknown parameter $\in R$ and it holds $E\underline{\varepsilon_T} = 0$.

The Least Squares Estimator b_T for b is then defined by

b_T: $\min_b \| V_T - bU_T \|_T^2$ which is $b_T = \langle U_T, U_T \rangle_T^{-1} \langle U_T, V_T \rangle_T$,

where $\langle U_T, V_T \rangle_T$ may be $\begin{cases} \frac{1}{T} \sum_{i=1}^{T} U_{t_i} V_{t_i} & \text{for discrete observations} \\ \frac{1}{T} \int_0^T V(t)U(t)dt & \text{for continuous observat.} \end{cases}$

Then it is $\lim_{T \to \infty} b_T = b + \lim_{T \to \infty} \langle U_T, U_T \rangle_T^{-1} \langle U_T, \varepsilon_T \rangle_T$.

If $\underline{U_T}$ and $\underline{\varepsilon_T}$ are correlated, then $\langle U_T, \varepsilon_T \rangle_T$ must not tend to zero for T going to infinity. Therefore, we replace $\underline{U_T}$ by some $\underline{Z_T}$, so that we obtain a new estimator

$$\underline{b_T} = \langle \underline{Z_T}, \underline{Z_T} \rangle_T^{-1} \langle \underline{Z_T}, \underline{U_T} \rangle_T$$

$$= b + \langle \underline{Z_T}, \underline{Z_T} \rangle_T^{-1} \langle \underline{Z_T}, b(E\underline{U_T} + \underline{\eta_T} - \underline{Z_T}) + \underline{\varepsilon_T} \rangle_T ,$$

where $\underline{\eta_T} := \underline{U_T} - E\underline{U_T}$.

For deterministic Z_T a sufficient condition for the consistency of $\underline{b_T}$ is $\lim_{T \to \infty} \langle Z_T, EU_T - Z_T \rangle_T = 0$.

Furthermore, $\langle Z_T, E\underline{U_T} - Z_T \rangle_T = 0$ is sufficient for the unbiasedness of b_T. From the projection theorem it follows that this condition is fulfilled iff $Z_T = \mathcal{P}_L \{ EU_T \}$, where $\mathcal{P}_L$ is the projection operator on any subspace L. (3.2.)

Now the question arises , how we can choose the subspace L in the best way. For answering this questions we try to get the variance of $\underline{b_T}$ as small as possible.

It is
$$E(\sqrt{T}(\underline{b_T} - b))^2 = \langle \underline{Z_T}, \underline{Z_T} \rangle^{-2} E(\langle Z_T, b\underline{\eta_T} + \underline{\varepsilon_T} \rangle_T)^2$$
$$= \langle Z_T, Z_T \rangle_T^{-2} \langle Z_T, R_T Z_T \rangle ,$$

where R_T is the covariance operator of $(b\underline{\eta_T} + \underline{\varepsilon_T})$.

Now it holds

$$E(\sqrt{T}(\underline{b_T} - b))^2 \leq \lambda_{\max}(R_T) \| Z_T \|_T^{-2}.$$

The upper bound becomes minimal, if we choose Z_T with maximal norm. Since $\|EU_T\|_T \geq \|\mathcal{P}_L\{EU_T\}\|_T$ for all L, it would be the best to take

$$Z_T = E\underline{U}_T . \tag{3.3.}$$

In our model(3.1.) it is

$$\underline{Y(t)} = \int_0^t e^{a(t-\tau)} bX(\tau)d\tau + \int_0^t e^{a(t-\tau)} \underline{\varepsilon}(\tau)\, d\tau \quad \text{and hence}$$

$$E\underline{Y(t)} = \int_0^t e^{a(t-\tau)} bX(\tau)d\tau . \tag{3.4.}$$

Since the expectation depends on the unknown parameters a and b, we cannot use it as an instrument and it is also difficult to estimate $\{E\underline{Y(t)},\ t \in [0,T]\}$ after observing a realization $\{Y(t),\ t \in [0,T]\}$.

Therefore we go the following way: Taking into consideration only a finite past $[t-h,t]$ we approximate $\{E\underline{Y(t)},\ t \in [0,T]\}$ because of (3.4.) by a real function $\{\overline{Y(t)},\ t \in [0,T]\}$ which is the projection on one of the linear Hilbert spaces $L \in \mathfrak{M}$ with

$$\mathfrak{M} := \{L_{h,C},\ L_{K,(\tau_1,\dots,\tau_K)} \quad K = 1,2\dots \text{ and } 0 \leq \tau_1 < \tau_2 \dots < \tau_K \leq h\}$$

where $L_{h,C} := \left\{ \int_0^h g(s)X(t-s)ds,\ t \in [0,T],\ g(\cdot) \in L^2\, [0,h] \right\}$ and

$$L_{K,(\tau_1 \dots \tau_K)} = \left\{ \sum_{i=1}^{K} g_i\, X(t-\tau_i),\ t \in [0,T],\ g_i \in R \right\}.$$

This means, we approximate $E\underline{Y(t)}$ by $\overline{Y}_T(t)$ which has either the form

$$\overline{Y_T}(t) = \int_0^h \overline{g_T}(s)X(t-s)ds \qquad \text{or}$$

$$\overline{Y_T}(t) = \sum_{i=1}^{K} \overline{g_{i,T}}X(t-\tau_i).$$

The instrumental variables estimators we consider are obtained now in two steps:

1. We compute an estimate $\{\hat{Y}_T(t),\ t \in [0,T]\}$ for $\{\overline{Y_T}(t),\ t \in [0,T]\}$ by projecting the realization $\{Y(t),\ t \in [0,T]\}$ on L
2. As instrumental variables estimator of a and b we denote the estimator

$$\begin{pmatrix} \hat{a}_T \\ \hat{b}_T \end{pmatrix} = \left[\int_0^T \begin{pmatrix} \hat{Y}_T(t) \\ X(t) \end{pmatrix} \left(\hat{Y}_T(t) \vdots X(t) \right) dt \right]^{-1} \int_0^T \begin{pmatrix} \hat{Y}_T(t) \\ X(t) \end{pmatrix} \dot{Y}(t)dt. \tag{3.5.}$$

Before investigating the properties of these estimators we have to decide, which linear subspace $L \in \mathcal{M}$ we will choose for our projection. This depends on several theoretical considerations and on the "feeling" of the practicioner, who has to deal with certain inputs $X(t)$ and who possesses the concrete realizations $\{Y(t),\ t \in [0,T]\}$. But the following theorem tells us, that the asymptotic properties of the so constructed instrumental variables estimator do not depend on the choice of the projection space L:

Theorem 4

Let $\underline{Y(t)}$ be a stochastic process defined by

$$\dot{\underline{Y(t)}} = a\underline{Y(t)} + bX(t) + \underline{\varepsilon(t)}, \quad Y(0) = 0,$$

with $a \in R^-$, $b \in R$, and $\{\underline{\varepsilon(t)},\ t \in R^+\}$ have continuous realizations with probability 1. Furthermore, let be $E\underline{\varepsilon(t)} = 0$ and $X(\cdot): R^+ \longrightarrow R$ measurable and uniformly bounded. Under some regularity conditions and the assumption that there exist constants $K > 0$ and $\gamma < 0$ so that

$$|E\underline{\varepsilon(t)}\,\underline{\varepsilon(\tau)}| \leq Ke^{\gamma|t-\tau|} \text{ for all } t,\tau \in R^+ \qquad (3.6.)$$

the instrumental variables estimator $(\underline{a_T} \vdots \underline{b_T})'$ defined by (3.5.) is a consistent estimator for $(a \vdots b)'$ and it holds

$$\sqrt{T}\left[\begin{pmatrix} \underline{a_T} \\ \underline{b_T} \end{pmatrix} - \begin{pmatrix} a \\ b \end{pmatrix}\right] = O_p(1).$$

(Proof see Bellach (3))

So we possess now a consistent estimator for our unknown parameters in a more general model than that of type (1.2.). And the class of disturbance processes we have used here is very practicable as the following considerations will show:

The Gaussian white noise was used to describe a disturbance process $\{\underline{\varepsilon(t)},\ t \in [0,T]\}$ which should have the property to be "nearly" uncorrelated at different time points t_1 and t_2 with large distance, and to be "strongly" correlated at time points t_1 and t_2 in a small neighbourhood. But these properties has for example the diffusion process $\{\underline{\varepsilon(t)},\ t \in [0,T]\}$ which is the solution of

$$d\underline{\varepsilon(t)} = \varrho\,\underline{\varepsilon(t)}\,dt + c\,d\underline{W(t)}, \quad \varepsilon(0) = 0,\ \varrho < 0,\ c > 0. \qquad (3.7.)$$

It is $E\underline{\varepsilon(t)}^2 = \frac{-c^2}{2\varrho} + \frac{c^2}{2\varrho}e^{2\varrho t}$,

$$E\underline{\varepsilon(t_1)}\,\underline{\varepsilon(t_2)} = \frac{-c^2}{2\varrho} e^{\varrho|t_1-t_2|} + \frac{c^2}{2\varrho} e^{\varrho(t_1+t_2)} . \qquad (3.8.)$$

If c and ϱ tend to infinity in such a way that $-\frac{c}{\varrho}$ converges to a constant k, then the process $\{\underline{\varepsilon(t)}\}$ tends to a Gaussian white noise process.

Because of (3.8.) $\underline{\varepsilon(t)}$ fulfills the condition (3.6.) of theorem 4 and hence it is possible to construct a consistent instrumental variables estimator. This estimator can again be used as an initial estimator for a "One Step Gauss-Newton-Estimator" which can be shown to be a best asymptotically normally distributed estimator. (See Bellach (3))

References

1. Arato, M.
   Точные формулы для плотностей мер элементарных Гауссовских процессов
   (Density formulas for the measures of elementary Gaussian processes)
   Studia Scientiarum Mathematicarum Hungaria 5/1970, 17-21

2. Bellach, B.
   Consistency, Asymptotic Normality and Asymptotic Efficiency of the Maximum-Likelihood-Estimator in Linear Stochastic Differential Equations
   Math. Operationsforsch. Statist., Ser. Statistics
   Vol 11 (1980) 227-277

3. Bellach, B.
   Parameter Estimators in Linear Stochastic Differential Equations and their Asymptotic Properties
   Math. Operationsforsch. Statist., Ser. Statistics
   Vol 14 (1983), Heft 1

4. Kozin, F.
   Estimation of parameters for systems driven by white noise excitations
   Proceedings of the IUTAM-Symposium on "Random Vibrations and Reliability"
   Akademie-Verlag, Berlin 1983, pp.163-172

5. Lipzer/Schirjajew
   Статистика случайных процессов
   (Statistics of random processes)
   Moskau 1973

GUARANTEED ELLIPSOIDAL ESTIMATES OF STATE FOR DYNAMIC SYSTEMS

F.L.Chernousko[1)]

The guaranteed approach is developed to state estimation of dynamic systems subjected to disturbances or control forces. The approach is based on approximation of sets to which state vectors of the systems belong by means of ellipsoids. Two-sided (external and internal) optimal ellipsoidal bounds for attainable sets are obtained. The evolution of approximating ellipsoids is described by specific non-linear systems of ordinary differential equations. Some properties of these equations are established. The applications of the method of ellipsoidal estimates are discussed.

ESTIMATION OF ATTAINABLE SETS

We consider a dynamic system

$$\dot{x} = f(x,u,t), \qquad u(t) \in U(t,x),$$
$$x(s) \in M, \qquad t \geqslant s, \qquad (\dot{\ }) = d(\ )/dt. \tag{1}$$

Here t is time, x is an n-dimensional vector of state, u is an m-dimensional vector of disturbances or control functions, a given closed set $U(t,x)$ describes imposed constraints, f is a given vector function, s is an initial time moment, and M is a given initial set. We denote by $X(x,t)$ the set of values of $f(x,u,t)$ for arbitrary $u(t) \in U(t,x)$. Then (1) can be rewritten as a differential inclusion

$$\dot{x} \in X(x,t), \quad x(s) \in M, \qquad t \geqslant s. \tag{2}$$

The <u>attainable set</u> $D(t,s,M)$ of the system (2) for $t \geqslant s$ is a set of all vectors $x(t)$ which are the values of all functions $x(\tau)$ satisfying (2) for $\tau \in [s,t]$. The attainable sets are important for a number of problems in the theory of control and state estimation, see, for example, books [6 - 10]. However, the actual construction of attainable sets is connected with many difficulties, especially for large n. Therefore it is useful to have simple and efficient estimates of these sets.

The method described below was proposed in papers [1,2], see also [3 - 5].

We denote by $E(a,Q)$ the following n-dimensional ellipsoid

1) Professor, Doctor of Phys.-Math. Sciences, Institute for Problems of Mechanics, USSR Academy of Sciences, prospekt Vernadskogo, 101, 117526, USSR.

$$E(a,Q) = \left\{x: (x-a)^T Q^{-1}(x-a) < 1\right\}. \tag{3}$$

Here a is an n-dimensional vector of the centre of the ellipsoid, and Q is a symmetric positive-definite $n \times n$ matrix. We assume that the sets X and M in (2) satisfy the following ellipsoidal estimates

$$E(A^-(t)x + f^-(t), G^-(t)) \subset X(x,t) \subset E(A^+(t)x + f^+(t), G^+(t)), \quad t \geqslant s,$$
$$E(a_0^-, Q_0^-) \subset M \subset E(a_0^+, Q_0^+). \tag{4}$$

Here $A^{\pm}$ and $G^{\pm}$ are given $n \times n$ matrices depending on t; $f^{\pm}$ are given n-dimensional vector functions; $a_0^{\pm}$ are given n-dimensional vectors; and $Q_0^{\pm}$ are given $n \times n$ matrices. The matrices $G^{\pm}$, $Q_0^{\pm}$ are symmetric and positive-definite. The estimates (4) correspond to two linear systems with ellipsoidal constraints imposed on control and initial data

$$\dot{x} = A^{\pm}(t)x + f^{\pm}(t) + u, \qquad t \geqslant s,$$
$$u(t) \in E(0, G^{\pm}(t)), \qquad x(s) \in E(a_0^{\pm}, Q_0^{\pm}). \tag{5}$$

We denote by $D^{\pm}(t,s,M)$ the attainable sets of the systems (5). It follows from (4)

$$D^-(t,s,E(a_0^-,Q_0^-)) \subset D(t,s,M) \subset D^+(t,s,E(a_0^+,Q_0^+)). \tag{6}$$

The sets D^- and D^+ are not ellipsoids in the general case. We consider the ellipsoidal approximations $E(a^{\pm}(t), Q^{\pm}(t))$ of these sets satisfying the following conditions.

1) Initial conditions:

$$a^-(s) = a_0^-, \qquad Q^-(s) = Q_0^-,$$
$$a^+(s) = a_0^+, \qquad Q^+(s) = Q_0^+. \tag{7}$$

2) Inclusion conditions for all $\tau \in [s,t]$:

$$E(a^-(t),Q^-(t)) \subset D^-(t,\tau,E(a^-(\tau),Q^-(\tau)),$$
$$E(a^+(t),Q^+(t)) \supset D^+(t,\tau,E(a^+(\tau),Q^+(\tau)). \tag{8}$$

3) Optimality conditions for volumes of ellipsoids:

$$dv^-/dt \longrightarrow \max, \qquad dv^+/dt \longrightarrow \min. \tag{9}$$

The conditions (8) mean that the approximation is evolutional: if at any time moment $\tau \geqslant s$ we replace the attainable set by an approximating ellipsoid then the inclusion will take place for all $t \geqslant \tau$. The conditions (9) mean that the volumes $v^{\pm}$ of ellipsoids $E(a^{\pm}(t),Q^{\pm}(t))$ change with maximal (for v^-) or minimal (for v^+) velocities possible under conditions (7), (8).

EQUATIONS OF ELLIPSOIDS

We present here some theorems stating the properties of approximating ellipsoids. The proofs are not given, some of them see in [1, 2].

Theorem 1. Ellipsoids satisfying conditions (7) - (9) are unique, and their parameters $a^{\pm}$ and $Q^{\pm}$ satisfy the following differential equations and initial conditions

$$\dot{a} = A(t)a + f(t), \qquad a(s) = a_0, \tag{10}$$

$$\dot{Q}^- = AQ^- + Q^-A^T + 2R^{-1}(RQ^-R^T)^{1/2}(RGR^T)^{1/2}(R^{-1})^T,$$
$$Q^-(s) = Q_0, \tag{11}$$

$$\dot{Q}^+ = AQ^+ + Q^+A^T + hQ^+ + h^{-1}G,$$
$$h = \left\{n^{-1}\,\mathrm{Tr}\left[(Q^+)^{-1}G\right]\right\}^{1/2}, \quad Q^+(s) = Q_0. \tag{12}$$

Here the dependence of A,f,G,R, and h on t, and indices $^-$ in (11) and $^+$ in (12) after A,G, and Q are omitted. The linear equation (10) is true for a^- and a^+ if we put appropriate indices $^-$ and $^+$ after A, f, and a_0. The matrix $R(t)$ in (11) is such that both matrices RQ^-R^T and RGR^T are diagonal for all $t \geqslant s$.

If we solve the initial problems (10) - (12) for vectors $a^{\pm}(t)$ and symmetric positive-definite matrices $Q^{\pm}(t)$ then we obtain, according to (6), (8), the estimates

$$E(a^-(t),Q^-(t)) \subset D(t,s,M) \subset E(a^+(t),Q^+(t)). \tag{13}$$

If we have only internal (external) estimates (4) then we can obtain only internal (external) estimates (13). If the system (1) is linear and has the form

$$\begin{aligned} &\dot{x} = A(t)x + f(t) + u, \qquad t \geq s, \\ &u(t) \in E(0,G(t)), \quad x(s) \in E(a_0,Q_0), \end{aligned} \tag{14}$$

then both systems (5) coincide with (14). In this case $a^-(t) = a^+(t) = a(t)$, and the indices $\pm$ after A, f, G, a_0, and Q_0 in (10) - (12) should not be put.

The following theorem shows when both ellipsoids (13) coincide with the attainable set.

Theorem 2. Ellipsoids (13) coincide ($a^- = a^+$, $Q^- = Q^+$) for all $t \geq s$, if and only if the system (1) has the form (14), and matrices A, G and Q_0 satisfy the equalities

$$\begin{aligned} &AG + GA^T - \dot{G} = \mu(t)G, \qquad t \geq s, \\ &Q_0 = \lambda_0^2 G(s), \qquad \lambda_0 = \text{const} \geq 0. \end{aligned} \tag{15}$$

Here $\mu(t)$ is a scalar function. Under these conditions the common centre $a(t)$ of ellipsoids (13) satisfies (10), and their matrix $Q(t)$ is given by

$$\begin{aligned} Q(t) = \lambda^2(t)G(t), \quad \lambda(t) = &\int_s^t \exp\left[\frac{1}{2}\int_{t_2}^t \mu(t_1)dt_1\right]dt_2 + \\ &+ \lambda_0 \exp\left[\frac{1}{2}\int_s^t \mu(t_1)dt_1\right]. \end{aligned} \tag{16}$$

The systems (11), (12) can be simplified by the change of variables

$$Q^{\pm} = V^{\pm} Z^{\pm}(V^{\pm})^T \tag{17}$$

where $Z^{\pm}$ are new variables and $V^{\pm}(t)$ are non-degenerate $n \times n$ matrices.

Theorem 3. If $V^{\pm}$ are fundamental matrices of the systems (5)

$$\dot{V}^{\pm} = A^{\pm}(t)V^{\pm}, \quad t \geqslant s, \quad V^{\pm}(s) = I, \tag{18}$$

where I is an identity $n \times n$ matrix, then the change of variables (17) reduces the systems (11), (12) to the form (11), (12) with $A^{\pm} \equiv 0$. If $V^{\pm}=G^{1/2}$ then the change of variables (17) reduces (11), (12) to the form

$$\begin{aligned}
&\dot{Z}^{-} = A_1Z^{-} + Z^{-}A_1^{T} + 2(Z^{-})^{1/2}, \\
&\dot{Z}^{+} = A_1Z^{+} + Z^{+}A_1^{T} + hZ^{+} + h^{-1}I, \\
&h = \left\{ n^{-1} \operatorname{Tr}\left[(Z^{+})^{-1}\right]\right\}^{1/2}, \quad A_1 = G^{-1/2}\left[AG^{1/2} - \right. \\
&\left. - \frac{d}{dt}(G^{1/2})\right], \quad Z^{-}(s) = Z^{+}(s) = Z_o = G^{-1/2}(s)Q_oG^{-1/2}(s),
\end{aligned} \tag{19}$$

Instead of equations (11), (12) we can deal further with equivalent but simpler equations (19).

In the important special case of fixed initial point we have $x(s) = a_o$ in (1). Here the initial ellipsoids $E(a_o^{\pm}, Q_o^{\pm})$ in (4) degenerate into the point a_o, and the initial conditions (11), (12) and (19) for matrices $Q^{\pm}$ and $Z^{\pm}$ become zero: $Q_o = Z_o = 0$. The equations (11), (12), and (19) have singularities at $Q^{\pm} = Z^{\pm} = 0$. The asymptotic solutions of equations (19) near the singular initial point are given by the following theorem.

Theorem 4. If the matrix A_1 in (19) has the expansion

$$A_1(t) = A_{10} + A_{11}\theta + O(\theta^2), \quad \theta = t - s \geqslant 0, \tag{20}$$

then the solutions of (19) for $Z_o = 0$ have the expansions

$$\begin{aligned}
&Z^{-}(t) = \theta^2 I + \theta^3 D_o + \theta^4\left(\frac{7}{12} D_o^2 + \frac{2}{3} D_1\right) + O(\theta^5), \\
&Z^{+}(t) = \theta^2 I + \theta^3 D_o + \theta^4\left[\frac{2}{3} D_o^2 + \frac{2}{3} D_1 + \frac{(\operatorname{Tr}D_o)^2}{12n^2} I - \frac{D_o \operatorname{Tr}D_o}{6n}\right] + O(\theta^5),
\end{aligned} \tag{21}$$

$$D_o = (A_{10} + A_{10}^{T})/2, \quad D_1 = (A_{11} + A_{11}^{T})/2.$$

The expansions (21) for Z^-, Z^+ coincide up to the terms $O(\theta^3)$. The asymptotic solutions (21) can be used for starting numerical integration.

Some other properties of approximating ellipsoids are also established, for example, their asymptotic behaviour for $t \to \infty$, bounds for their volumes etc. We shall present here one of these results.

Theorem 5. The volumes $v^{\pm}$ of ellipsoids (13) for $A^{\pm} = 0$ satisfy the inequalities

$$v^+(t) \geqslant v^-(t), \quad \frac{v^+(t)}{v_0^+} \leqslant \left[\frac{v^-(t)}{v_0^-}\right]^{\sqrt{n}}, \quad t \geqslant s. \tag{22}$$

Here $v_0^{\pm}$ are the volumes of the initial ellipsoids $E(a_0^{\pm}, Q_0^{\pm})$ in (5). Note that according to Theorem 2, the general case can be reduced to the case $A^{\pm} = 0$.

By means of numerical integration of the systems (10) - (12) or (19) it is possible to obtain estimates (13) practically. A number of numerical examples as well as some exact analytical solutions of these systems are given in [1 - 5].

APPLICATIONS

The estimates (13) can be used in various problems of control and estimation. Let u be some unknown disturbance bound by the constraint (1); its stochastic properties may be arbitrary. Then the external ellipsoid (13) gives the guaranteed outer bound for all possible motions of the system (1) under the disturbance. On the other hand, if u is a control, then the internal ellipsoid (13) gives the guaranteed inner bound for all controlled motions.

The obtained bounds can be used for two-sided estimation of the optimal value of the cost functional in optimal control problems. Thus, for the problem of minimizing the terminal cost functional $J = F(x(T)) \longrightarrow \min$ for the system (1), where $F(x)$ is a given scalar function and $T \geqslant s$ is a given terminal time moment, we have the evident two-sided bounds

$$\min_{x \in E^+} F(x) \leqslant J^* \leqslant \min_{x \in E^-} F(x),$$
$$J^* = \min_{x \in D(T,s,M)} F(x), \quad E^{\pm} = E(a^{\pm}(T), Q^{\pm}(T)). \tag{23}$$

Similar bounds are true for other classes of optimal control problems.

Let u in (1) be a vector of unknown disturbances, and the following observations be available

$$y_i = N_i x(t_i) + w_i, \quad w_i \in E(0, L_i)$$
$$t_o < t_1 < \ldots, \qquad i = 1, 2, \ldots \tag{24}$$

Here t_i are given times of observation, k-dimensional vectors y_i are the observation results, k-dimensional vectors v_i are the observation errors which are arbitrary but bounded by ellipsoids, and N_i and L_i are given matrices of dimensions $k \times n$ and $k \times k$, respectively. The matrices L_i are symmetric and positive-definite. Let $P(t)$ be the uncertainty set to which the state vector $x(t)$ belongs if the observations (24) with $t_i < t$ are taken into account. The problem of guaranteed state estimation (guaranteed filtering) consists in estimation of the set $P(t)$. The following ellipsoidal bound for $P(t)$ was derived

$$x \in P(t) \subset E(a(t), Q(t)). \tag{25}$$

Here the vector $a(t)$ and the matrix $Q(t)$ are piecewise continuous functions of time satisfying systems (10), (12) for a^+, Q^+ within the intervals (t_i, t_{i+1}) between observations (24), $i = 0, 1, \ldots$. At the times t_i the ellipsoid $E(a(t_i-0), Q(t_i-0))$ representing all the past information intersects with the ellipsoid of the observation (24); the last ellipsoid can be degenerate and have infinite axes. The intersection of these two ellipsoids is a convex set, the ellipsoid containing this set was obtained in [1] . Taking its parameters $a(t_i+0)$, $Q(t_i+0)$ as initial data, we continue integration of the systems (10), (12) till the next observation at t_{i+1} and so on. The procedure described briefly above realizes the guaranteed method of filtering for dynamic systems in the presence of external disturbances and observation errors. At each time moment the state of the system is estimated by the ellipsoid (25).

CONCLUSIONS

The method described in the paper gives external and internal ellipsoidal bounds for the set of possible states of dynamic systems in the presence of disturbances and control. These bounds are obtained for linear as well as for nonlinear systems.

Note that the guaranteed approach to estimation of disturbances used in this paper has some advantages compared with the stochastic approach. The guaranteed approach does not require any knowledge of stochastic properties of disturbances (these properties are seldom well-known in practical problems); only the bounds for disturbances are assumed to be known. The estimates obtained by the guaranteed approach are true for each individual realization of disturbances. On the other hand, the guaranteed estimates may be too "wary", especially when we have a big number of realizations.

The qualitative analysis and numerical examples showed that the two-sided bounds presented above give quite satisfactory approximation of attainable sets.

The method of optimal ellipsoidal estimates can be used for guaranteed estimating of the influence of disturbances acting upon the dynamic system, for obtaining two-sided bounds in optimal control and differential games, for guaranteed filtering in the presence of bounded observation errors.

More details, proofs and examples see in [1 - 5].

REFERENCES

1. Chernousko, F.L. "Optimal guaranteed estimates of uncertainties by means of ellipsoids". Izvestiya of the USSR Academy of Sciences, Engineering Cybernetics, I - 3, 3 - 11; II - 4, 3 - 11; III - 5, 5 - 11, 1980 (in Russian).
2. Chernousko, F.L. "Ellipsoidal estimates for the attainable set of a controlled systems". Applied Mathematics and Mechanics, 45, 1, 11 - 19, 1981.
3. Chernousko, F.L. "Guaranteed ellipsoidal estimates of uncertainties in control problems". Technische Hochschule Leipzig, Wissenschaftliche Zeitschrift, 4, 6, 325 - 329, 1980.
4. Chernousko, F.L. "Guaranteed ellipsoidal estimates of uncertainties in control problems". Proceedings of the 8th IFAC Congress, Kyoto, vol. YI, 203 - 208, 1981.
5. Chernousko, F.L. "Ellipsoidal bounds for sets of attainability and uncertainty in control problems". Optimal Control Applications and Methods, vol. 3, N 2, 181 - 202, 1982.
6. Chernousko, F.L. and Melikyan, A.A. Game Problems of Control and Search, Nauka, Moscow, 1978 (in Russian).
7. Krasovsky, N.N. Theory of Control of Motion. Nauka, Moscow, 1968 (in Russian).
8. Kurzhansky, A.B. Control and Observation in Conditions of Uncertainty. Nauka, Moscow, 1977 (in Russian).
9. Lee, E.B.and Markus L. Foundations of Optimal Control Theory. New York, John Wiley, 1967.
10. Schweppe, F.C. Uncertain Dynamic Systems. Prentice Hall, 1973.

LEAST SQUARE ESTIMATION OF VIBRATION SYSTEMS

H. Diesing [1]), B. Heimann [1]), W. Tischer [1])

Summary. This paper is a contribution to identification problems of vibration systems. The criterion used for identification is a quadratic cost functional of the system output with respect to the unknown quantities like the impulse-response-function of linear systems or the system parameters. The proposed identification methods are derived from the generalized WIENER-HOPF equation. The results presented here for multiinput-multioutput continuous-time and discrete-time systems are applied to special vibration systems with few degrees of freedom. Some numerical results are given.

Introduction

The importance of system identification has been recognized for a long time, although a unified mathematical treatment has emerged in the last decade. Some of earlier attempts were based on frequency domain methods and engineering judgement. An interesting discussion of the relative advantages and disadvantages of frequency domain methods and time domain methods in application to mechanical vibration systems can be observed in the literature. In the field of mechanics a large amount of literature exists on frequency methods with harmonic and periodic inputs. Using nonstationary or random inputs in order to develop recursive methods for parameter identification, then time domain methods have some advantages. Therefore, the presented work is devoted to these methods and contains some applications to special mechanical vibration systems.

1. Problem Statement

Consider a time-invariant vibration system described by a nonlinear second order vector differential equation

$$\underline{M}\,\ddot{y} + \underline{G}(y,\dot{y}) + \underline{G}_0(y) = f(t) \tag{1}$$

($\underline{M}$ mass matrix; $\underline{G}(y,\dot{y})$ matrix of nonconservative forces; $\underline{G}_0(y)$ matrix of conservative forces; $f(t)$ excitation; $f, y \in R^n$).

The special case of linear vibration systems is important both from the theoretical and practical point of view

$$\underline{M}\,\ddot{y} + \underline{D}\,\dot{y} + \underline{C}\,y = f \tag{2}$$

[1]) Academy of Sciences of GDR, Institute of Mechanics, GDR-1199 Berlin, Rudower Chaussee 5.

($\underline{D}$ damping matrix; $\underline{C}$ stiffness matrix). The equivalent first order system is given by

$$\dot{x} = \underline{A}\, x + b \tag{3}$$

where $x := (y,\dot{y})^T \in R^{2n}$; $b := (0,f(t))^T \in R^{2n}$;

$$\underline{A} = \left[\begin{array}{c|c} \underline{0} & \underline{I} \\ \hline -\underline{M}^{-1}\underline{C} & -\underline{M}^{-1}\underline{D} \end{array}\right].$$

Assuming that f(t) consists of a deterministic excitation u(t) and a noise sequence w(t) we may write

$$f(t) = \underline{B}\, u(t) + \underline{Q}\, w(t) \tag{4}$$

($u \in R^m$, $w \in R^l$; $m,l \leq 2n$).
The properties of w(t) are as follows:

$$E[w(t)] = 0 ;$$
$$E[w(t)w^T(\tau)] = \underline{R}_w \delta(t-\tau) \; ; \; \underline{R}_w < 0$$
$$E[u(t)w^T(\tau)] = 0 \; , \forall t,\tau .$$

For model building by data analysis the input and output signals will be measured in a time-discrete manner. Therefore, it is necessary to transform the equation of motion into the following time-discrete form. In the case of linear systems

$$x(t_{k+1}) = e^{\underline{A}(t_{k+1}-t_k)} x(t_k) + \int_{t_k}^{t_{k+1}} \exp\{\underline{A}(t_{k+1}-\tau)\}\, b(\tau)d\tau \tag{5}$$

(k=1,...,N; see also /1/,/2/).
Under the assumptions

i) $h = t_{k+1} - t_k = \text{const} \; , \forall k$

ii) $b(\tau) = b(t_k) = \text{const} \; , \forall \tau \in [t_k, t_{k+1}]$

it follows from equation (5)

$$x_{k+1} = \underline{G}\, x_k + \underline{H}\, b_k \tag{6}$$

where $x_k := x(kh)$; $b_k := b(kh)$;

$\underline{G} := \exp\{\underline{A}h\}$; $\underline{H} := \underline{A}^{-1}(\exp\{\underline{A}h\} - \underline{I})$.

The condition ii) can be approximately fulfilled by choosing the frequency $f=1/h \gg f_n$ (f_n highest system eigenfrequency).
Another possibility for transforming into a time-discrete system - also suitable for nonlinear systems - consists in replacing the derivations by difference quotients

$$\dot{y}_k = \frac{1}{2h}(y_{k+1}-y_{k-1}),$$

$$\ddot{y}_k = \frac{1}{h^2}(y_{k+1}-2y_k+y_{k-1}).$$

In the case of linear systems we get

$$\underline{\overline{M}}\, y_{k+2} + \underline{\overline{D}}\, y_{k+1} + \underline{\overline{C}}\, y_k = \underline{\overline{F}}^{(1)}_{k+1} + \underline{\overline{F}}^{(2)}_k . \tag{6a}$$

<u>Remarks</u>

a) In order to reduce the numerical effort in identification we take into account the special structure of vibration systems, for instance by using structure matrices

$$\sum_i m_i \underline{M}_i\, y + \sum_j d_j \underline{D}_j\, y + \sum_k c_k \underline{C}_k\, y + \sum_l a_l h(y,\dot{y}) = f(t) \tag{7}$$

where $\Theta := (m,d,c,a)$ are the parameters to be estimated ($m:=\{m_i\}$, $d:=\{d_j\}$, $c:=\{c_k\}$, $a:=\{a_l\}$) and, $\underline{M}_i$, $\underline{D}_j$, $\underline{C}_k$ the matrices describing the system structure, $h(y,\dot{y})$ a given nonlinear vector function.

b) For backward transformation from (6) or (6a) into a time-continuous system in order to get physically interpretable parameters the following possibilities are known:

- a simple linear transformation in the case of using difference quotients, for instance $A \approx \frac{1}{h}(G-I)$;
- numerical computation of $A = \frac{1}{h} \ln G$;
- computation of A using the transformation (6):

$$A = [H(G-I)^{-1}]^{-1}.$$

2. WIENER-HOPF-Equation

Denote by P the set of n-dimensional random vector processes (RVP) and by R the set of linear and bounded operators.

<u>Definition.</u> The estimation is said to be optimal, if the estimator

$$\hat{y}(t) = \underline{G}(t,t_0)\, y_0 + \underline{A}_\lambda(t,\lambda)\, f(\lambda) \tag{8}$$

fulfills the condition

$$E[\|\hat{y}(t)-y(t)\|^2_{L_2(T)}] = \min! \tag{9}$$

($f,y,\hat{y} \in P$; $\underline{A}_\lambda(t,\lambda) \in R$; $\underline{G}(t,t_0)$ n,n-matrix; $y(t_0)=y_0$).

It is a well-known fact that the principle of orthogonality

$$E[(\hat{y}(t)-y(t),v(t))_{L_2(T)}] = 0 \;, \forall\, v \in P \tag{10}$$

is a necessary and sufficient condition for the existence of a minimum of (9) (see for instance /5/).

Applying this principle to equ. (8) and assuming that v(t) can be represented by $v(t)=\underline{B}_s(t,s)f(s)$, ($\underline{B}_s(t,s) \in R$, arbitrary) we get

$$\underline{G}(t,t_0)\underline{K}_{\hat{y}f}(t_0,s) + \underline{A}_\lambda(t,\lambda)K_{ff}(\lambda,s) - \underline{K}_{yf}(t,s) = 0 \quad (11)$$

($\underline{K}_{\hat{y}f}(t_0,s):=E[\hat{y}(t_0)f^*(s)]$; $\underline{K}_{ff}(\lambda,s):=E[f(\lambda)f^*(s)]$; $\underline{K}_{yf}(t,s):=E[y(t)f^*(s)]$).

Equ. (11) is the so-called generalized WIENER-HOPF-equation, because the known WIENER-HOPF-integral equation follows from it as a special case.

We use now the generalized WIENER-HOPF-equation in order to derive some identification procedures for linear estimation problems and their application to vibration systems.

Remarks

a) According to equ. (8) we set $\underline{A}_{\tau'}(t,\tau')f(\tau') := \int_0^{t-t_0} \underline{K}(\tau')f(t-\tau')d\tau'$ and in the case of stationary RVP we find from equ. (11) the WIENER-HOPF-integral equation

$$\underline{K}(T)\underline{K}_{\hat{y}f}(\beta-T) + \int_0^T \underline{K}(\tau')\underline{K}_{ff}(\beta-\tau')d\tau' - \underline{K}_{yf}(\beta) = 0 \quad (12)$$

($\tau':=t-\tau, \beta=t-s,\ T:=t-t_0,\ 0 \leq \beta \leq T$).

b) Consider the time-discrete equation. For $t_0=0, \tau'=\frac{p}{N}T; \beta=\frac{l}{N}T;$ $p,l=0,1,\ldots,N$, and with the abbreviations $\underline{K}(p) := \underline{K}(\frac{p}{N}T)$, $\underline{K}_{ff}(l-p) := \underline{K}_{ff}(\frac{l-p}{N}T)$, $\underline{K}_{yf}(l) := \underline{K}_{yf}(\frac{l}{N}T)$ we find for L_2-stable systems and for sufficient large T the approximative relations

$$\sum_{p=0}^{N} \alpha_p \underline{K}(p)\, \underline{K}_{ff}(l-p) = \underline{K}_{yf}(l) \quad (13)$$

($l=0,1,\ldots,N$; α_k coefficients of numerical integration procedure).

c) By rearranging of equ. (13) we get the well-known equation of the linear estimation theory

$$\underline{Y} = \underline{X}\cdot\underline{\Theta} \quad (14)$$

where

$$\underline{Y} := [\underline{K}_{yf}(0),\ldots,\underline{K}_{yf}(N)]^T,$$

$$\underline{\Theta} := [\underline{K}(0),\ldots,\underline{K}(N)]^T,$$

$$\underline{X} := \begin{bmatrix} \alpha_1\underline{K}_{ff}(0) & \ldots & \alpha_N\underline{K}_{ff}(-N) \\ \vdots & & \vdots \\ \alpha_1\underline{K}_{ff}(N) & \ldots & \alpha_N\underline{K}_{ff}(0) \end{bmatrix}.$$

3. Identification procedure

Using the pseudoinverse of MOORE-PENROSE the solution of equ. (14) can be written as follows

$$\underline{\Theta} = (\underline{X}^T\underline{X})^{-1}\, \underline{X}^T\underline{Y} \,. \tag{15}$$

This equation is the basic relation for the least-square-estimation.

3.1. Estimation of impulse-response-functions

In order to reduce the dimension of equ. (14) we use time-discrete orthonormal polynomials defined on the set of equidistant points $\{0,1,\dots,N\}$

$$\sum_{p=0}^{N} P_{q,N}(p)P_{j,N}(p) = \delta_{q,j} \tag{16}$$

($\delta_{q,j}$ KRONECKER-symbol).

Approximating $\underline{K}(p)$ by these polynomials

$$\underline{K}(p) \doteq \underline{K}_r(p) = \sum_{q=0}^{r} \underline{B}_q\, P_{q,N}(p)$$

we find for the unknown coefficients $\underline{B}_q$ the following relations

$$\sum_{q=0}^{r} \underline{B}_q\, \underline{A}_{jq} = \underline{C}_j \;;\; j=0,1,\dots,r, \tag{17}$$

as a special form of equ. (15) where

$$\underline{A}_{jq} = \underline{A}_{qj} = \sum_{k=0}^{N}\sum_{l=0}^{N} \alpha_k\, \underline{K}_1(k,l)\, P_{q,N}(k)\, P_{j,N}(l),$$

$$\underline{C}_q = \sum_{p=0}^{N} \underline{F}_1(p)\, P_{q,N}(p),$$

$$\underline{K}_1(k,l)= \sum_{p=0}^{N} \alpha_p\, \underline{K}_{ff}(k-p)\, \underline{K}_{ff}^T(l-p),$$

$$F_1(p)= \sum_{p=0}^{N} \alpha_p\, \underline{K}_{yf}(p)\, \underline{K}_{ff}(l-p).$$

The advantage of equ. (17) in comparison to (15) consists in its lower dimension, because in the most practical cases it is possible to choose $r \ll N$.

In order to approximate the impuls-response-function of L_2-stable and time-invariant systems we use

$$\varphi_i(p) = p^i \exp\left\{-\frac{\lambda T}{2N}\, p\right\}$$

($i=0,1,\dots,r, \lambda>0, p=0,1,\dots,N$) and get by using E.SCHMIDT's orthonormalization procedure the system $\{P_{q,N}(p)\}$.

3.2. Parameter estimation

Provided that the sequence w(t) is not significant the simplest form of least-square estimation is applicable to model building. Remember that

$$\text{i)}\quad E[Y] = \underline{X}\theta;$$

$$\text{ii)}\quad s = (Y-\underline{X}\theta)^T (Y-\underline{X}\theta) = \min!$$

are the assumptions for getting equ. (15).

Denoting by x_k^T the k-th row of $\underline{X}_N$,

y_k the k-th component of Y_N,

$$\underline{P}_k := (\underline{X}_k^T \underline{X}_k)^{-1} \; ; \; k=0,1,\ldots,N-1,$$

$$\underline{P}_{k+1} = (\underline{X}_k^T \underline{X}_k + x_{k+1}x_{k+1}^T)^{-1} = (\underline{I} - \underline{P}_k \frac{x_{k+1}x_{k+1}^T}{1 + x_{k+1}^T \underline{P}_k x_{k+1}})\underline{P}_k, \quad (18)$$

we get the following recursive estimation equations

$$\hat{\theta}_{k+1} = \hat{\theta}_k + K_{k+1} (y_{k+1} - x_{k+1}^T \hat{\theta}_k),$$
$$K_{k+1} = \underline{P}_k x_{k+1} (1 + x_{k+1}^T \underline{P}_k x_{k+1})^{-1}, \quad (19)$$

where $k=0,1,\ldots,N-1$; K_{k+1} correction vector. **As starting parameters we** use equ. (18) or $\underline{P}_0 = \alpha^{-1}\, \underline{I}$, α small.

a) Estimation by using complete information,

i.e., all components of state and excitation vectors are measurable. Then equs. (15) or (19) are directly applicable to parameter estimation. Assuming that the mass matrix is known and that the excitation has the form $\overline{\underline{F}}_{k+1}^{(1)} = \overline{\underline{B}}^{(1)} u_{k+1}$, $\overline{\underline{F}}_k^{(2)} = \overline{\underline{B}}^{(2)} u_k$ we get in the case of a linear time-discrete system the following least square notation

$$x_k^T := [-y_{k+1}^T, -y_k^T, u_{k+1}^T, u_k^T] \; ; \; \underline{\theta}^T := [\, \overline{\underline{D}}, \overline{\underline{C}}, \overline{\underline{B}}^{(1)}, \overline{\underline{B}}^{(2)}]$$
$$\underline{Y}_N := \begin{bmatrix} y_2^T \\ \vdots \\ y_{N+1}^T \end{bmatrix} \; ; \; \underline{X}_N := \begin{bmatrix} x_0^T \\ \vdots \\ x_{N-1}^T \end{bmatrix} \quad (20)$$

Note that $\underline{\theta}$ and $\underline{Y}_N$ are matrices and therefore the cost functional to be minimized becomes $s = \operatorname{tr}\{(\underline{Y}-\underline{X}\underline{\theta})^T (\underline{Y}-\underline{X}\underline{\theta})\}$. The result of minimization is the same as in equ. (15).

b) Estimation by using incomplete information

In this case we have to introduce a measurement equation

$$z_k = \underline{U}\, x_k + v_k \quad (21)$$

($\underline{U}$ rectangular observation matrix, v_k noise sequence of measurements). In order to estimate the complete state vector and the unknown parameters we combine the recursive least square method with KALMAN-filter approach (assuming that v_k and b_k are white noise sequences with covariance matrices R_v and R_b, respectively):

Extrapolation: $\quad x_{k+1}^* = \hat{\underline{G}}_k \hat{x}_k \quad ; \quad \underline{S}_{k+1}^* = \hat{\underline{G}}_k \underline{S}_k \hat{\underline{G}}_k^T + \underline{R}_b.$

Correction: $\hat{x}_{k+1} = x^*_k + \underline{L}_{k+1}\,(z_{k+1} - \underline{U}\,x^*_{k+1})$,

$$\underline{L}_{k+1} = \underline{S}^*_{k+1}\,\underline{U}^T\,(\underline{U}\,\underline{S}^*_{k+1}\,\underline{U}^T + \underline{R}_v)^{-1}, \tag{22b}$$

$$\underline{S}_{k+1} = (\underline{I} - \underline{L}_{k+1}\,\underline{U})\,\underline{S}^*_{k+1}.$$

The following **algorithm** is proposed:

Step 1: Set $\hat{x}(0) = \hat{x}_0$, $S(0) = S_0$ and choose $\hat{\Theta}_0^T = [\,\hat{G}_0, \hat{H}_0]$ and P_0.

Step 2: Compute $\hat{x}_k$ using (22) with $\hat{\Theta}_{k-1}$.

Step 3: Compute $\hat{\Theta}_k$ using (19) with $\hat{x}_{k-1}$.

Step 4: If converged - stop , otherwise set k:= k+1 and go to step 2.

It is clear, that the convergence of the algorithm essentially depends on the starting values $\hat{x}_0$ and $\hat{\Theta}_0$.

4. Examples

a) $\ddot{y} + 2D\omega_0\dot{y} + \omega_0^2 y = z$, where $K_{zz}(\alpha-\beta) = e^{-1.5|\alpha-\beta|}$ and $K_{yz}(\alpha) = 3/70\,\exp\{-0.5\alpha\}(8/Q\,\sin Q\alpha - \cos Q\alpha) + 1/7\,\exp\{-1.5\alpha\}$, $Q=2.5\sqrt{0.96}$. Fig.1 shows the influence of the approximation order r.

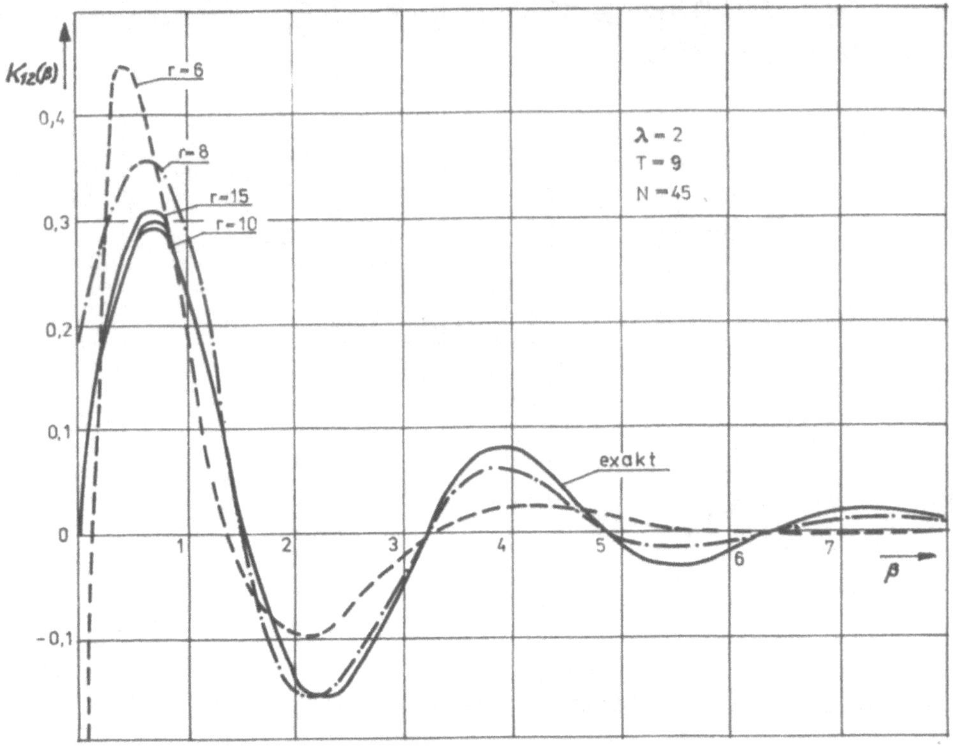

Fig.1. Approximation of impulse-response-function

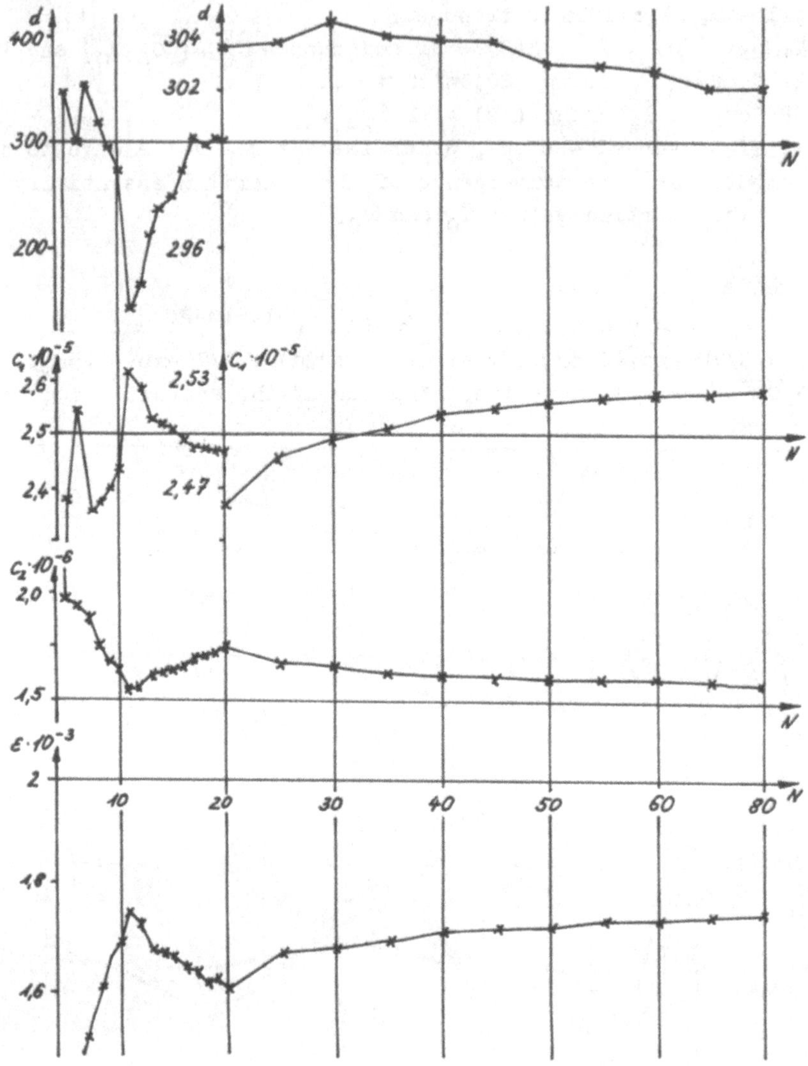

Fig.2. Results of parameter estimation in dependence on time points

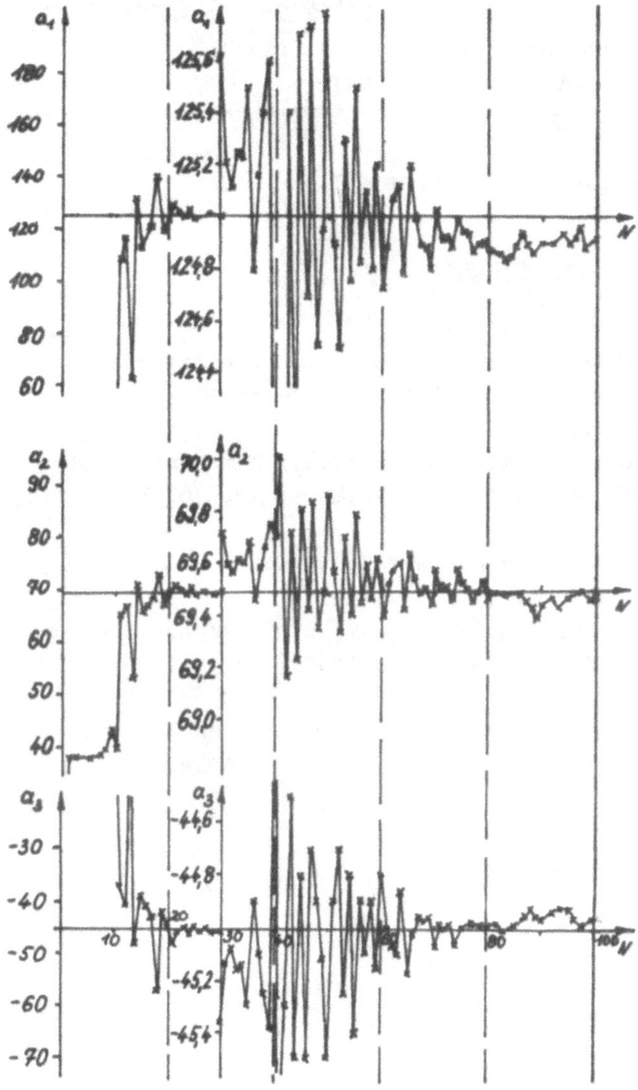

Fig.3. Estimation of a_1, a_2, a_3 in the case of plane robot motion (see example c))

b) Two degree-of-freedom system

$$m_1\ddot{x}_1 + d(\dot{x}_1-\dot{x}_2) + c_1(x_1-x_2) = 0,$$

$$m_2\ddot{x}_2 - d(\dot{x}_1-\dot{x}_2) - c_1(x_1-x_2) + c_2(x_2-s) + \varepsilon(x_2-s)^3 = 0.$$

Results of parameter estimation are given in fig.2.

c) Plane motion of a robot (structure DSS)

$$a_1\ddot{s} - (sa_1 + a_3)\dot{\varphi}^2 = F_1,$$

$$(a_2 + 2a_3 s + a_1 s^2)\ddot{\varphi} + 2(sa_1 + a_3)\,\dot{s}\dot{\varphi} = M_2$$

(see fig.3).

REFERENCES.

/1/ GOODWIN, G.C., PAYNE, R.L.: Dynamic system identification. Experiment design and data analysis, Academic Press, New York, San Francisco, London, 1977.

/2/ SCHWARZ, R.G.: Identifikation mechanischer Mehrkörpersysteme, Fortschr.-Ber. VDI-Zeitschriften, Reihe 8, Nr. 30, VDI-Verlag GmbH, Düsseldorf, 1980.

/3/ ASTRÖM, K.J., EYKHOFF, R.: System identification- A survey, Automatica, Vol. 7, Pergamon Press, 1971.

/4/ SÖDERSTRÖM, T., LJUNG, L., GUSTAVSSON, I.: A theoretical analysis of recursive identification methods, Automatica, Oxford, 14(1978)3.

/5/ SCHNEEWEIß, W.G.: Zufallsprozesse in dynamischen Systemen, Springer-Verlag, Berlin, Heidelberg, New York, 1974.

ESTIMATION OF PARAMETERS FOR SYSTEMS DRIVEN BY WHITE NOISE EXCITATIONS

Frank Kozin[1)2)]

Parameter estimates for real structures are becoming increasingly important for determining true dynamical characteristics, for assessing damage due to severe loadings such as strong motion earthquakes, and for purposes of structural control. In this paper we discuss maximum likelihood estimates of parameters for linear or non-linear systems subjected to white noise excitations. We establish the form of the estimates, as well as their asymptotic properties and apply the results to two examples. In particular, we establish an often observed result that for the stable damped oscillators the spring constant is estimated with smaller relative error than is the damping constant.

I. Introduction

In this paper we shall discuss and apply the theoretical results that were presented in [1]. Parameter estimation has become significant in structural engineering, and the results contained in [1] are of sufficient generality to allow the structural engineer to study the parameters for linear as well as non-linear systems. For a more recent study that applies to non-stationary as well as stationary statistics see [2].

We are concerned with systems of the form

$$\dot{x}_t = \sum_{j=1}^{k} a_j^o \phi_j(x_t) + W_t \tag{1.1}$$

where x_t is an N-vector, $\phi_j(x_t)$ are given N-vector functions and W_t is an N-vector of statistically independent gaussian white noise components. The $\{a_j^o\}$ denote the "true" scalar parameters of the structure.

The question we ask is the following: given that we observe the state vector x_t, under what conditions (if any) can we be assured that an estimator of the coefficients, as a functional of the observations, which we denote by $\hat{a}(T)$, will converge to the true coefficients a^o as the observation time T increases to infinity. That is, under what conditions can we be assured that in some limiting sense we can write

$$\lim_{T\uparrow\infty} \hat{a}(T) = a^o \tag{1.2}$$

Notice that we are talking about the limit properties of random variables. As such, therefore, there are a number of modes of convergence that we might consider such as, convergence in probability, convergence in the mean, and, finally, almost sure convergence.

The most practical form of convergence from the engineering point of view is almost sure convergence, since it tells us that for almost any time history of observed values of x_t the estimated coefficients $\hat{a}(T)$ will converge to the true coefficients a^o as the observation time approaches infinity.

The estimator that we shall be concerned with is the maximum likelihood estimator which we denote as $\hat{a}_{ML}$ and we wish to determine the conditions for which

$$\lim_{T\uparrow\infty} \hat{a}_{ML}(T) = a^o, \text{ almost surely.} \tag{1.3}$$

Then, it will allow us to determine the parameters of the structure being subjected to white noise inputs. We will have a firm statistical foundation upon which to base our estimates. Furthermore, as we shall see, we can even obtain an idea of the range within which the estimate lies after a given observation time T, which will allow us to determine the error bounds for our estimated parameters. We recall that the likelihood function for a given set of discrete observations $x_1,\ldots,x_N$, with parameter vector $\vec{a}$, is given by the joint density of the random variables $x_1,\ldots,x_N$,

---

1) Polytechnic Institute of New York, Brooklyn, New York, USA

2) The author gratefully acknowledges the support of the National Science Foundation under research grant CEE7825430.

$$f(x_1,\dots,x_N : \vec{a}) = L(\vec{x} : \vec{a}) \tag{1.4}$$

The maximum likelihood estimate of $\vec{a}$ is that value of $\vec{a}$ that maximizes (1.4) relative to the given observations. Thus a_{ML} satisfies

$$\max_{\vec{a}} L(\vec{x} : \vec{a}) = L(\vec{x} : \hat{a}_{ML}) \tag{1.5}$$

This usually may be obtained simply by taking the partial derivatives of L with respect to the unknown parameters and solving

$$\frac{\partial L(\vec{x} : \vec{a})}{\partial \vec{a}} = 0 \tag{1.6}$$

We may recall that in the gaussian case, the maximum likelihood estimators are identical to least squares estimators.

In order to proceed, we must determine the likelihood function for the system (1.1). We interpret the equation (1.1) in its proper Ito form as

$$dx_t = \sum_{j=1}^{k} a_j^o \phi_j(x_t)dt + dB_t \tag{1.7}$$

where the B_t is a vector of independent standard Brownian motion processes. That is, each component of B_t is a gaussian process with stationary independent increments for which $E\{B_t\} = 0$ and $E\{B_t B_t'\} = \text{diag}(t)$.

How much of the ideas above carry over into our case? Note that we are concerned with a continuous time parameter, possibly non-linear, formulation of the system dynamics. We shall obtain the likelihood function for the system (1.7) in a highly heuristic fashion which contains the essential ideas but was first made rigorous by Girsanov, [3], in 1960 and is fully discussed in [4].

We shall consider a simple scalar version of the Ito equation (1.7). On the interval $[0,T]$ we form the partition $t_i = i(\frac{T}{N}) \equiv i\Delta_N$. Therefore, $\Delta_i t = t_{i+1} - t_i = \Delta_n$, $x_i = x_{t_i}$, $\Delta_i x = x_{i+1} - x_i$, and finally $\Delta_i B = B_{i+1} - B_i$. Therefore, for large N we can interpret (1.7) for arbitrary a_j simply as,

$$\Delta_i x = \sum_{j=1}^{k} a_j \phi_j(x_i)\Delta_N + \Delta_i B \tag{1.8}$$

But since $\Delta_i B$, $\Delta_j B$ are on disjoint intervals for $i \neq j$, it follows that they are independent random variables, and their joint density can be written as

$$\prod_{i=0}^{N-1} \frac{1}{\sqrt{2\pi\Delta_n}} e^{-\frac{1}{2\Delta_N}(\Delta_i B)^2} = \frac{1}{(2\pi\Delta_N)^{N/2}} e^{-\frac{1}{2\Delta_N}\sum_{i=0}^{N-1}(\Delta_i B)^2} = p_B \tag{1.9}$$

Upon substitution of the variable x from (1.7) into (1.9) will give us the joint density

$$\frac{1}{(2\pi\Delta_N)^{N/2}} e^{-\frac{1}{2\Delta_N}\sum_{i=0}^{N-1}\left(\Delta_i x - \sum_{j=1}^{k} a_j\phi_j(x_i)\Delta_N\right)^2} \tag{1.10}$$

$$= \frac{1}{(2\pi\Delta_N)^{N/2}} e^{-\left[\frac{1}{2\Delta_N}\sum_{i=0}^{N-1}(\Delta_i x)^2 - \sum_{i=1}^{N-1}\Delta x_i\left(\sum_{j=1}^{k} a_j\phi_j(x_i)\right) + \frac{1}{2}\sum_{i=0}^{N-1}\left(\sum_{j=1}^{k} a_j\phi_j(x_i)\right)^2\Delta_N\right]} = p_x.$$

The so-called likelihood ratio of the (discrete) x-process with respect to the Brownian motion process is the ratio of (1.10) to (1.9),

$$\frac{P_x}{P_B} = e^{-\left\{\frac{1}{2\Delta_N}\sum_{i=0}^{N-1}\left[(\Delta_i x)^2 - (\Delta_i B)^2\right] - \sum_{i=1}^{N-1}\Delta x_i\left(\sum_{j=1}^{k} a_j\phi_j(x_i)\right) + \frac{1}{2}\sum_{i=0}^{N-1}\left(\sum_{j=1}^{k} a_j\phi_j(x_i)\right)^2 \Delta_N\right\}} .$$

Now, as $N\to\infty$, $(\Delta_i x)^2 = \Delta_N$, $(\Delta_i B)^2 = \Delta_N$ (see [4]) and $\sum_{i=0}^{N-1}[(\Delta_i x)^2 - (\Delta_i B)^2] = o(\Delta_N)$.

Furthermore, the second two sums can be interpreted as stochastic integrals to yield in the limiting case (for continuous time parameter), the proper likelihood function,

$$L(a_1,\dots,a_K : T) = e^{-\frac{1}{2}\int_0^T\left(\sum_{j=1}^{k} a_j\phi_j(x_t)\right)^2 dt + \int_0^T\left(\sum_{j=1}^{k} a_j\phi_j(x_t)\right) dx_t} . \tag{1.11}$$

Again we repeat, though correct in the sequence of steps, this derivation is only formal. All limiting operations must be made rigorous. Modern Martingale representation theorems [4] make this derivation quite clean and straightforward.

If we set $\mu(x_t) \equiv \sum a_j\phi_j(x_t)$ in (1.7) for the general vector case, we have

$$L(\vec{a},T) = e^{-\frac{1}{2}\int_0^T \mu'(x_t)\mu(x_t)d_t + \int_0^T \mu'(x_t)dx_t} \tag{1.12}$$

$\equiv e^{\ell(\vec{a},T)}$, where ℓ is the log likelihood and " ' " denotes transpose.

In order to obtain our asymptotic convergence theorem, we will require the existence of a stationary distribution (or invariant measure) for the system (1.7). The problem of existence of invariant measure for Ito systems was studied rigorously by Khazminskii [5]. Using Lyapunov ideas, Wonham [6] modified the conditions in [5] somewhat.

It is known that for certain regularity conditions on $\phi(x)$ the solution to (3.1) is a Markov diffusion process. In terms of notation for Markov processes we write the transition probability function as

$$P(t,x,S) = \text{Prob}\{x(t)\varepsilon S \mid x(0) = x\} .$$

If $\mu_o(\cdot)$ is the distribution of the initial value $x(0) = x_o$, then we write

$$\mu_t(S) = \int_{R^n} P(t,x,S)\mu_o(dx) \equiv (U_t\mu_o)(S) ,$$

which is the probability distribution for $x(t)$.

If μ_o satisfies the property,

$$U_t\mu_o = \mu_o \qquad \text{for all} \quad t \geq 0 ,$$

then μ_o is called an invariant measure.

In the usual terms of transition densities, if there is a density $f_o(x)$, for the Markov diffusion process with transition probability density, $p(x,t|\xi,s)$, $s < t$, that satisfies,

$$f_o(x) = \int_{R^n} p(x,t|\xi,0)f_o(\xi)d\xi \quad , \tag{1.13}$$

then $f_o(x)$ is a stationary density for the process.

For example, the scalar linear system (Ornstein-Uhlenbeck process)

$$dx = -\beta x dt + dB_t \ , \ B > 0$$

possesses the stationary density

$$f_o(x) = \frac{1}{\sqrt{2\pi\ \sigma^2/2\beta}}\ e^{-\frac{x^2}{2(\sigma^2/2\beta)}} \quad .$$

The stable non-linear system

$$\begin{cases} dx_1 = x_2 dt \\ dx_2 = (-x_2 - h(x_1))dt + dB_t \end{cases}$$

possesses the stationary density

$$f_o(x_1,x_2) = C\ e^{-\left[\frac{1}{2}x_2^2 + H(x_1)\right]}$$

where C is the normalizing constant and $H(x_1) = \int_0^{x_1} h(x)dx.$.

For the arbitrary Ito system, $dx(t) = m(x(t))dt + G(x(t))dB_t$, where x,m are n-vectors, G is nxq, and B_t is a q-vector of independent Brownian motions, the associated backward operator $\mathcal{L}$ is given by

$$\mathcal{L} = \sum_{i=1}^{n} m_i(x)\frac{\partial}{\partial x_q} + \frac{1}{2}\sum_{i,j=1}^{n} g_{ij}(x)\frac{\partial^2}{\partial x_i \partial x_j} \tag{1.14}$$

where

$$g_{ij} = (GG')_{ij} \quad .$$

The Fokker-Planck equation $\partial u/\partial t = \mathcal{L}^* u$, where $\mathcal{L}^*$ is the adjoint operator of $\mathcal{L}$ (usually referred to as the forward operator) can often be used to obtain the stationary density as the normalized positive solution to $\mathcal{L}^* u = 0$.

The result of Wonham is as follows: Let V(x) be a function (analogous to Lyapunov functions) satisfying

(1) V is defined for all x in $\overline{D}_R$, $D_R = \{x: \|x\| > R\}$ for any $R > 0$.

(2) V is continuous in $\overline{D}_R$, and twice continuously differentiable in D_R.

(3) $V(x) \geq 0$, $x\epsilon\overline{D}_R$ and $V(x)\uparrow\infty$ as $\|x\| \uparrow\infty$.

Theorem [6]

If there exists a function V satisfying (1) - (3) for which

$$\mathcal{L}V(x) \leq -1, \ x\epsilon D_R \quad ,$$

then the x-process possesses an invariant measure.

It is also known that the almost sure, ergodic property

$$\text{Prob}\left\{\lim_{T\to\infty} \frac{1}{T}\int_0^T g(x_t)dt = \int_{R^n} g(x)f_o(x)dx\right\} = 1 \quad , \tag{1.15}$$

holds.

There is one more result that is required, one can show that

$$\lim_{T\to\infty} \frac{1}{T}\int_0^T g'(x_t)dB_t = 0 \tag{1.16}$$

almost surely for any function g, satisfying $E\{g'(x_t)g(x_t)\} < \infty$, for the x_t-process defined by (1.7).

The result follows from simple Borel-Cantelli arguments for the integral martingale (1.16).

With these preliminaries at hand, we can now proceed to our asymptotic convergence result.

II. Main Results

In this section we derive our main almost sure convergence result.

Theorem 1.

Let $\hat{a}(T) = (\hat{a}_1(T),\ldots,\hat{a}_k(T))'$ be the estimate of $(a_1^o,\ldots,a_k^o)'$ that is obtained by maximizing (1.11) for the observed sample solution to (1.7). If the system (1.7) possesses a solution process with an invariant measure which must satisfy (1.15), and if the vector $\mu(x_t)$ satisfies $E\{\mu'(x_t)\mu(x_t)\} < \infty$, so that (1.16) holds, then it follows that

$$\lim_{T\to\infty} \hat{a}_{ML}(T) = (a_1^o,\ldots,a_k^o)' \text{ , with probability one.}$$

Proof.

$$\ell(a,T) = -\frac{1}{2}\int_0^T \mu'(x_s)\mu(x_s)ds + \int_0^T \mu'(x_s)dx_s$$

$$= -\frac{1}{2}\int_0^T \left[\sum_{j=1}^k a_j\phi_j(x_s)\right]'\left[\sum_{j=1}^k a_j\phi_j(x_s)\right]ds + \sum_{j=1}^k a_j\int_0^T \phi'_j(x_s)dx_s$$

$$= -\frac{1}{2}\sum_{i,j=1}^k a_ia_j\int_0^T \phi_i'(x_s)\phi_j(x_s)ds + \sum_{j=1}^k a_j\int_0^T \phi_j'(x_s)dx_s$$

$$= -\frac{1}{2}a'Ca + a'\rho \quad , \tag{2.1}$$

where

$$\left.\begin{aligned} c_{ij} &= \int_0^T \phi_i'(x_s)\phi_j(x_s)ds \text{ , } C = (c_{ij}) \\ \rho_j &= \int_0^T \phi_j'(x_s)dx_s \quad . \end{aligned}\right\} \tag{2.2}$$

The matrix C, above, is non-singular for sufficiently large T with probability one, since the $\{\phi_j(x)\}$ are linearly independent. Hence, from (2.1) we have

$$\hat{a}_{ML}(T) = C^{-1}\rho \tag{2.3}$$

$$\rho_i = \int_0^T \phi_i'(x_s)dx_s = \int_0^T \phi_i'(x_s) \sum_{j=1}^{k} a_j^o \phi_j(x_s)\, ds + \int_0^T \phi_i'(x_s)dB_s$$

$$= \sum_{j=1}^{k} a_j^o \int_0^T \phi_i'(x_s)\phi_j(x_s)ds + \int_0^T \phi_i'(x_s)dB_s \quad ;$$

or

$$\rho = Ca^o + Z \quad , \tag{2.4}$$

where Z_i is defined as

$$Z_i = \int_0^T \phi_i'(x_s)dB_s \quad ,$$

Hence, from (2.4) we have

$$a^o = C^{-1}p - C^{-1}Z \quad ,$$

and combining this result with (2.3) yields

$$\hat{a}_{ML}(T) - a^o = C^{-1}Z \quad . \tag{2.5}$$

Now, since we assume that there is a stationary solution to the stochastic equation (1.7) with stationary density $f_o(x)$, it is known that

$$\lim_{T\to\infty} \frac{1}{T} \int_0^T \mu(x_t)dt = \int \mu(x)f_o(x)dx$$

almost surely, therefore,

$$\lim_{T\to\infty}(C/T) = \Gamma \text{ , with probability one,} \tag{2.6}$$

where

$$\Gamma_{ij} = \int_{-\infty}^{\infty} \phi_i'(x)\phi_j(x)f_o(x)dx \quad . \tag{2.7}$$

It is shown below that Γ is non-singular; thus Γ^{-1} exists and

$$\lim_{T\to\infty} TC^{-1} = \Gamma^{-1} \text{ , with probability one} \quad . \tag{2.8}$$

Finally, from (2.5) we have, by application of (1.16),

$$\lim_{T\to\infty}(\hat{a}_{ML}(T) - a^o) = \lim_{T\to\infty}(TC^{-1}(Z/T))$$

$$= \Gamma^{-1} \lim_{T\to\infty}(Z/T) = 0. \tag{2.9}$$

This completes the proof.

To show that Γ is non-singular, we simply note that the kxk matrix $G(x)$, with elements

$$G_{ij}(x) = \phi_i'(x)\phi_j(x) \quad ,$$

is positive definite.

This follows from the fact that $\phi_1(x),\ldots,\phi_k(x)$ are linearly independent vectors. Therefore, there is no non-trivial relationship of the form

$$\sum_{i=1}^{k} c_i\phi_i(x) = 0 \quad ,$$

for any non-identically zero vector of constants $(c_1,\ldots,c_k)$.

Hence, the magnitude

$$\| \sum c_i\phi_i(x) \| = \left(\sum_{i=1}^{k} c_i\phi_i(x) \ , \ \sum_{i=1}^{k} c_j\phi_j(x) \right)^{1/2}$$

$$= c^T(\phi_i'(x)\phi_j(x))c \geq 0 \quad ,$$

is only equal to zero for $c \equiv 0$.

Hence the matrix $G(x) = (\phi_i'(x)\phi_j(x))$, usually called the Gram matrix, is non-negative definite.

Thus, for any non-zero k-vector, b, we have for the probability density $f_o(x)$,

$$0 < \int dxb'G(x)bf_o(x) = b'\left[\int dxG(x)f_o(x)\right]b$$

$$= b'\Gamma b \quad ,$$

and the non-singularity follows.

We have established that the maximum likelihood estimates of the unknown coefficients must approach their true values as the observation time approaches infinity. Of course, in real applications, we may observe over a long period of time, but it still is finite. Thus, we must address the question of the error in our estimated values for finite intervals of observation. A measure of the error is obtained by study of the asymptotic statistical properties of our estimates. We shall only state the major result here, for a discussion and references see [1]. Upon denoting a normal distribution with mean m and variance σ^2 as $N(m,\sigma^2)$, we have the asymptotic result.

Theorem 2.

Assuming the hypotheses of Theorem 1, the estimator $\hat{a}_{ML}(T)$ defined by (2.3) satisfies,

$$T^{1/2}(\hat{a}_{ML}(T) - a^o) \xrightarrow{D} N(0,\Gamma^{-1}) \quad ,$$

where the covariance matrix Γ is defined by (2.7), and D denotes convergence in distribution.

We can now consider examples.

III. Applications

In this section, we shall look at parameter estimation for a simulated non-linear system, and use Theorem 2 to discuss the error in estimates of spring and damping coefficients for the damped linear oscillator.

Example 1.

The non-linear system, written in Ito differential form,

$$\begin{pmatrix} dx_1 \\ dx_2 \end{pmatrix} = \left\{ a^o \begin{pmatrix} x_2 \\ x_1 + \sin x_1 \end{pmatrix} + b^o \begin{pmatrix} 0 \\ 2x_1 + x_2 \end{pmatrix} \right\} dt + \begin{pmatrix} dB_1 \\ dB_2 \end{pmatrix} \qquad (3.1)$$

with $a^o = 1$, $b^o = -2$ can be shown to satisfy Wonham's theorem [6] for existence of a stationary solution.

Furthermore, (3.1) possesses the backward operator

$$\mathcal{L} = x_2 \frac{\partial}{\partial x_1} + (\sin x_1 - 3x_1 - 2x_2) \frac{\partial}{\partial x_2} + \frac{1}{2}\left(\frac{\partial^2}{\partial x_1^2} + \frac{\partial^2}{\partial x_2^2}\right) \tag{3.2}$$

and Γ is given by the matrix

$$\Gamma = \begin{pmatrix} E\{x_2^2 + (x_1 + \sin x_1)^2\} & E\{(2x_1 + x_2)(x_1 + \sin x_1)\} \\ E\{(2x_1 + x_2)(x_1 + \sin x_1)\} & E\{(2x_1 + x_2)^2\} \end{pmatrix}, \tag{3.3}$$

The system (3.1) was simulated by

$$\begin{aligned} \begin{pmatrix} x_1(t+\Delta t) \\ x_2(t+\Delta t) \end{pmatrix} &= \begin{pmatrix} x_1(t) \\ x_2(t) \end{pmatrix} + \left\{ a^o \begin{pmatrix} x_2(t) \\ x_1(t) + \sin x_1(t) \end{pmatrix} + b^o \begin{pmatrix} 0 \\ 2x_1(t) + x_2(t) \end{pmatrix} \right\} \Delta t \\ &+ \begin{pmatrix} B_1(t+\Delta t) - B_1(t) \\ B_2(t+\Delta t) - B_2(t) \end{pmatrix} \end{aligned} \tag{3.4}$$

where $\Delta t = 0.01$, $E\{(B_i(t+\Delta t) - B(t))^2\} = \Delta t$, $i = 1,2$.

The following table contains the estimates and errors for 500 samples (T=5), 1000 samples (T=10), 1500 samples (T=15).

T	$\hat{a}_{ML}(T)$	$\hat{b}_{ML}(T)$	$T^{1/2}(\hat{a}_{ML}(T)-a^o)$	$T^{1/2}(\hat{b}_{ML}(T)-b^o)$
5	1.060	-1.689	0.134	-0.695
10	1.633	-2.293	2.00	-0.927
15	1.141	-2.006	0.546	-0.023

Since, we cannot explicitly calculate the matrix Γ in (3.3) we applied simulation to estimate the various expectations for T = 100, 200, 300, 400. At T=400, for example, we find the estimated value of $E\{x_1x_2\}$, with true value -0.5, to be -.495. We used T=400 for all estimated moments to yield

$$\Gamma = \begin{pmatrix} 3.124 & 1.851 \\ 1.851 & 2.057 \end{pmatrix}, \quad \Gamma^{-1} = \begin{pmatrix} 0.686 & -0.617 \\ -0.617 & 1.041 \end{pmatrix}.$$

Thus the standard deviations σ_a, σ_b of $T^{1/2}(\hat{a}_{ML}(T) - a_o)$, $T^{1/2}(\hat{b}_{ML}(T) - b_o)$ respectively are approximately $\sqrt{0.686} = 0.828$, $\sqrt{1.041} = 1.02$. Comparing this with the table, we see that for T = 5, 15, the estimates $\hat{a}_{ML}(T)$, $\hat{b}_{ML}(T)$ are well within one standard deviation. For T = 10, $\hat{a}_{ML}(T)$ is within three standard deviations. Except for T = 10, the estimators are quite close to the true values, as we should expect.

It is generally observed that the estimators for damping coefficients exhibit larger errors than the estimates for spring coefficients. This is certainly reasonable for the real structure, since the nature of the damping is subject to interpretation. On the other hand, we find greater errors even in simulated systems, where the damping term is of a given form. Based upon the asymptotic variance properties of the estimators given by Theorem 2, we see in the following example that the larger damping estimator errors are predicted.

Example 2.

In this example we consider the simple damped oscillator driven by gaussian white noise which, in scalar form, we write as

$$\ddot{x}_t + 2\zeta\omega\dot{x}_t + \omega^2 x_t = W_t \tag{3.5}$$

where as usual, ζ is the damping ratio, ω is the natural frequency and W is the gaussian white noise. It is easily shown that for positive ζ, a stationary solution to (3.5) exists.

We can write (3.5) in the vector Ito form (1.7) as

$$\begin{pmatrix} dx_1 \\ dx_2 \end{pmatrix} = \left[-\omega^2 \begin{pmatrix} 0 \\ x_1 \end{pmatrix} - 2\zeta\omega \begin{pmatrix} 0 \\ x_2 \end{pmatrix} + \begin{pmatrix} x_2 \\ 0 \end{pmatrix} \right] dt + \begin{pmatrix} 0 \\ dB \end{pmatrix}. \tag{3.6}$$

We estimate the parameters, $(\omega^2, \zeta\omega)$. Notice that the coefficient of $\binom{x_2}{0}$ is known (unity) and is not to be estimated.

We find from (2.1), (2.3), for finite observation time T, the estimators to be

$$\begin{pmatrix} \hat{\omega}^2 \\ 2\widehat{\zeta\omega} \end{pmatrix}_{ML} = \begin{pmatrix} -\int_0^T x_1^2(s)ds & 0 \\ 0 & -\int_0^T x_2^2(s)ds \end{pmatrix}^{-1} \begin{pmatrix} \int_0^T x_1(s)dx_2(s) \\ \int_0^T x_2(s)dx_2(s) \end{pmatrix}. \tag{3.7}$$

The asymptotic error variances as given by Theorem 2 are

$$\Gamma^{-1} = \begin{pmatrix} E\{x_1^2\}^{-1} & 0 \\ 0 & E\{x_2^2\}^{-1} \end{pmatrix}. \tag{3.8}$$

The constant moments in (3.8) may be evaluated by Dynkin's Theorem through the backward operator of (3.6), which from (1.14) is found to be

$$\mathcal{L} = x_2 \frac{\partial}{\partial x_1} - (\omega^2 x_1 + 2\zeta\omega x_2) \frac{\partial}{\partial x_2} + \frac{1}{2} \frac{\partial^2}{\partial x_2^2}. \tag{3.9}$$

From the three equations $\mathcal{L} x_1^2 = 0$, $\mathcal{L} x_1 x_2 = 0$, $\mathcal{L} x_2^2 = 0$, we obtain the stationary moments,

$$E\{x_1^2\}^{-1} = 4\zeta\omega^3 = \sigma^2_{\widehat{\omega}^2}\ , \quad E\{x_2^2\}^{-1} = 4\zeta\omega = \sigma^2_{2\widehat{\zeta\omega}}\ , \tag{3.10}$$

and thus the error variances. The coefficients of variation [7] are defined as the ratio of the standard deviation of the errors to the mean value. Thus, we have respectively,

$$\delta_{\widehat{\omega}^2} = \frac{\sqrt{4\zeta\omega^3}}{\omega^2} = 2\sqrt{\frac{\zeta}{\omega}} \quad ; \quad \delta_{\widehat{\zeta\omega}} = \frac{\sqrt{4\zeta\omega}}{2\zeta\omega} = \sqrt{\frac{1}{\zeta\omega}}\ . \tag{3.11}$$

But, since the damping ratio ζ is generally quite small we have that the relative error in the estimate of $\zeta\omega$ is much greater than the relative error in the estimate of ω^2. Indeed from (3.11) we see that the relative error for $\zeta\omega^2$ will be $1/\zeta$ times the relative error for ω^2.

Discussion

We have a very complete study of the properties of maximum likelihood estimates of parameters of systems, linear or non-linear, driven by white noise. However, many problems of interest remain. For example, we have not included observation noise in our theory. The white noise driven system is well understood probabilistically, but white noise is not physical noise. Therefore, one would like to have similar asymptotic results for the physical noise case for both linear as well as non-linear systems. Finally, structural parameter estimates have been studied from earthquake or wind excitations. These are non-stationary. The asymptotic properties of the estimates should be studied in this case as well. The objective of these studies would be to guarantee almost sure asymptotic convergence of the estimates to the true values, and to obtain the asymptotic distribution of the estimate in order to obtain some idea of the errors of estimation for finite times.

References

1. T.S.Lee, F.Kozin, Almost Sure Asymptotic Likelihood Theory for Diffusion Processes, J.Appl. Prob., Vol.14, pp.527-537 (1977).

2. F.Nakajima, F.Kozin, A Characterization of Consistent Estimators, IEEE Trans. Auto. Contr., Vol.AC-24, pp.758-765 (1979).

3. I.V.Girsanov, On Transforming a Certain Class of Stochastic Processes by Absolutely Continuous Substitution of Measures, Theory of Prob. and Appls., Vol.5, pp.285-301 (1960).

4. E.Wong, Stochastic Processes in Information and Dynamical Systems, McGraw-Hill,NY(1971).

5. R.Z.Khazminskii, Ergodic Properties of Recurrent Diffusion Processes and Stabilization of the Solution to the Cauchy Problems for Parabolic Equations, Theory of Prob. and Appls., Vol.5, pp.179-196 (1960).

6. W.M.Wonham, Liapunov Criteria for Weak Stochastic Stability, Jour. Diff. Eqs., Vol.2, pp.195-207 (1966).

7. W.Gersch, N.Nielsen, H.Akaike, Maximum Likelihood Estimation of Structural Parameters From Random Vibration Data, Journ. Sound and Vibr., Vol.31, pp.1-14 (1973).

ANALYSIS OF MECHANICAL SYSTEMS EXCITED BY RANDOM VIBRATIONS

A. Lingener [1)]

Summary: Experimental investigations of random excited mechanical systems are based on a digital processing of measured data. Different kinds of excitation are compared and advantages as well as disadvantages are discussed. A very efficient random excitation can be realised by pseudorandom signals. Some applications are presented, where a pseudorandom excitation is especially useful.

1. Introduction

This paper deals with applications of random signals in the field of experimental investigations of mechanical systems. Until some years ago, these investigations in most cases were carried out with harmonic or transient excitations.

The field of research consists firstly in experimental investigations, based on measurements of our own or on measurements of engineers from industry recorded on tapes.

The behaviour of a mechanical system caused by any given excitation is commonly described in frequency domain by transfer function H(f). In many cases of application it is sufficient, to know the transfer function. In other cases, for example, in cases of modelling, the transfer function is the origin of further investigations.

The connections between measured data and transfer function can be written in simple formulae as

$$\hat{X}_a(f) = H(f)\,\hat{X}_e(f) \qquad (1)$$

where $\hat{X}_e(f)$ and $\hat{X}_a(f)$ are the complex amplitude density spectra of input and output and where the excitation $x_e(t)$ is realised by changes in time of one (or several) mechanical quantities at the input(s) of the system.

Input and output are measurable in time domain. The transformation into frequency range is realised by Fourier transformation, usually digitally as Fast Fourier Transformation.

Multiplying this formula by the conjugate complex spectrum of the input $\hat{X}_e^*(f)$, we find a relation between power density spectra, from which

1) Prof.Dr.sc.techn. A. Lingener, Technische Hochschule "Otto von Guericke" Magdeburg, Sektion Maschinenbau, PSF 124, DDR 3010 Magdeburg

Table 1: Possibilities of system excitation

Kind of excitation	Advantages	Disadvantages
sinusoidal excitation	classical analog technique high signal to noise ratio high power feed high accuracy	high time expense danger of damage in resonance frequencies; exciting effort
impact excitation	very small excitation effort (hammer) little time needed (10^{-2}) wide field of application (light and middle-heavy structures)	low density of energy; unknown spectrum of excitation; averages required (only some)
step excitation	small excitation effort (static pretension) high density of energy at low frequencies small time effort especially for very heavy structures (and very light in weight)	unknown spectrum of excitation low density of energy at high frequencies
random excitation (natural random processes)	"natural" load of the system system excitable during operation elimination of noise by averages recognition of disturbances by coherence analysis	leakage exciting expense little density of energy many averages required

H(f) can be determined easier:

$$S_{ea}(f) = H(f)\, S_{ee}(f)\ . \tag{2}$$

Therefore, the complex cross power spectrum $S_{ea}(f)$ has to be divided by the real power spectrum S_{ee} of the input. Because of the symmetry in reference to the frequencies zero often the so-called one-sided power spectrum, specified here as G(f), is used:

$$G_{ea}(f) = 2\, S_{ea}(f)\ , \qquad f \geq 0\ . \tag{3}$$

All these mathematical relations have to be realised approximately on the basis of sampled time functions.

2. Possibilities for exciting mechanical systems

In every case an excitation of a system is necessary in order to find out its mechanical properties. The excitation acting at the input causes the system to vibrate. Exciting a mechanical system demands a supply of energy. Therefore we have some possibilities, differing in their efficiency and the effort required. The analysis of mechanical systems demands statements in a determined frequency range and consequently the exciting energy also has to be distributed in this frequency range.

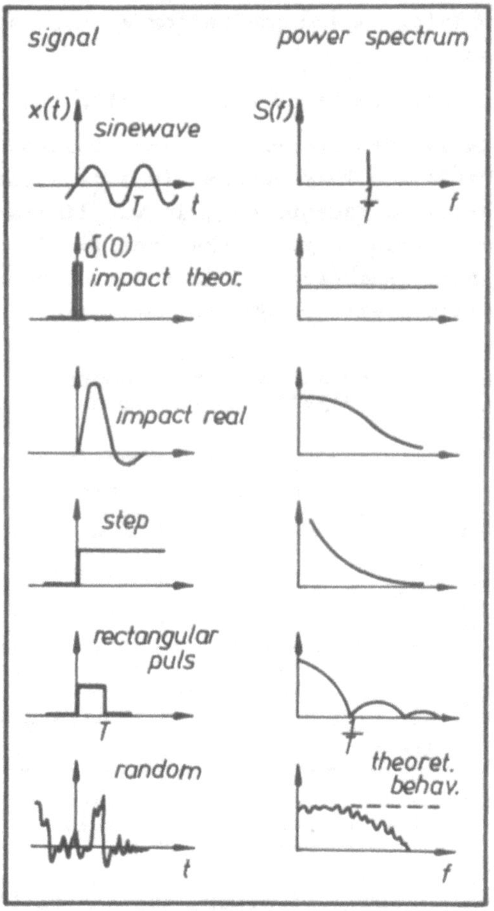

Fig. 1: Different types of exciting signals and their power spectra

The possibilities of the Table 1 are to be discussed briefly.

The best known and still often used method is an excitation with harmonical forces by a sine wave generator and a corresponding power amplifier. Great exciting power can be fed into the system at a single frequency. Great amplitudes can be generated. But always we need information on the system in a range of frequencies, and therefore, the system must be excited with different frequencies.

In this kind, frequency response can be found point for point by many measurements at a high expense of time.

When the system is excited by a mechanical vibrator with defined unbalance, an excessive excitation could destroy the system in its resonance frequencies. The other possibilities of excitation have common properties in so far as the system is excited simultaneously with all frequencies of the exciting signal.

A precondition for an application to analysis of mechanical systems is, that the excitation comprises all frequencies of the frequency range to be investigated. In Fig. 1 we see different excitations in time and frequency domain. It becomes evident that, when considering the upper frequency limit, given in each case, all shown signals are adapted for excitation of mechanical systems.

3. Special aspects of excitation with random signals

Pseudorandom

Applying random excitation we must take into consideration some special aspects.

The advantages of random excitation are connected with some disadvantages.

The sources of error in experimental investigations are very different (Table 2). They consist in a necessary limitation to a finite time interval in measurement, in the limitation of frequency range by the Nyquist frequency, in the quality of the used exciting signals, i.e. of random generator and in noise disturbances, which are of essential importance with regard to small exciting amplitudes.

Table 2: Sources of error, consequences and remedies in digital analysing of random processes

Source of error restriction	Consequences	Remedy
finite time intervals	leakage	window functions
randomness of finite time interval	great variance of estimates	averages
sampling frequency	reflection on Nyquist frequency	low pass filtering
noise disturbances	defective output results	coherence analysis

These sources of error, the finite time interval length and the randomness of this interval are the most serious disadvantages. After giving some examples pseudorandom will be introduced avoiding these disadvantages.

Leakage caused by finite time intervals

Starting from a sampled time series with a sampling frequency f_s we get a discrete spectrum only when the selected time interval T is continued periodically on the whole time axis. In this way we change the real time function and we have to consider the consequences.

The discrete values of spectrum appear in a distance of $f = \frac{1}{T}$, the bandwidth of the spectrum. As an example we consider the harmonic function (Fig. 2). When we select the finite time interval equal to the period T_o of the harmonic function then we find only a single line in frequency domain at the frequency $1/T_o$ in the usual part of the spectrum.

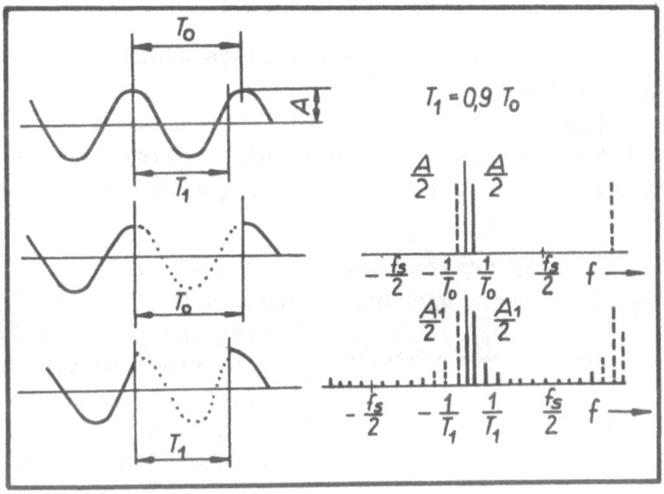

Fig. 2: Leakage in spectrum in consequence of a finite time interval

In the lower part of Fig. 2 the time interval does not coincide with the period T_o. Here the case $T_1 = 0.9\ T_o$ is considered and we find frequencies in the spectrum, that really do not occur in the time signal. In the case of any real signal, this behaviour is valid for all frequencies. So much about leakage.

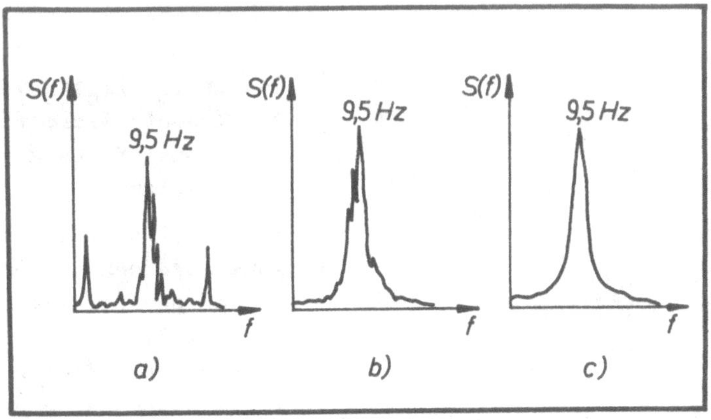

Fig. 3: Output power spectra of a single mass system excited by random signals
a) random generator without averages
b) 50 averages
c) pseudorandom signal without averages

Randomness of time interval

As an example for randomness and length of the time interval we take the excitation of a single mass system with the output signal of a random generator.

Assuming that the power spectrum is constant in the frequency range to be investigated, the power spectrum of the system output represents the transfer function of the system. The eigenfrequency of the considered test system was 9.5 Hz, the time interval of measurement had a length of 8 s. Fig. 3 shows the result: case a) without averages, case b) 50 averages. The time interval of 8 s is too short to provide a useful spectrum. But also 50 x 8 s do not provide a satisfactory result.

Pseudorandom signals

These disadvantages can be avoided by using pseudorandom signals as excitation. Pseudorandoms are marked as follows:

- They are periodical in a time window of length T
- The power spectrum consists in defined values at all discrete frequencies of a determined range (e.g., constant at every frequency of this range)

Pseudorandoms must be generated by inverse Fourier transformation in a digital computer. Starting from power spectrum, magnitudes of particular frequencies of amplitude spectrum are defined, and every magnitude is associated with a random phase angle. Inverse Fourier transformation then gives a time function with the spectrum given before in a time window of length $T_o=1/f_o$; f_o is the lowest discrete frequency of the given spectrum.

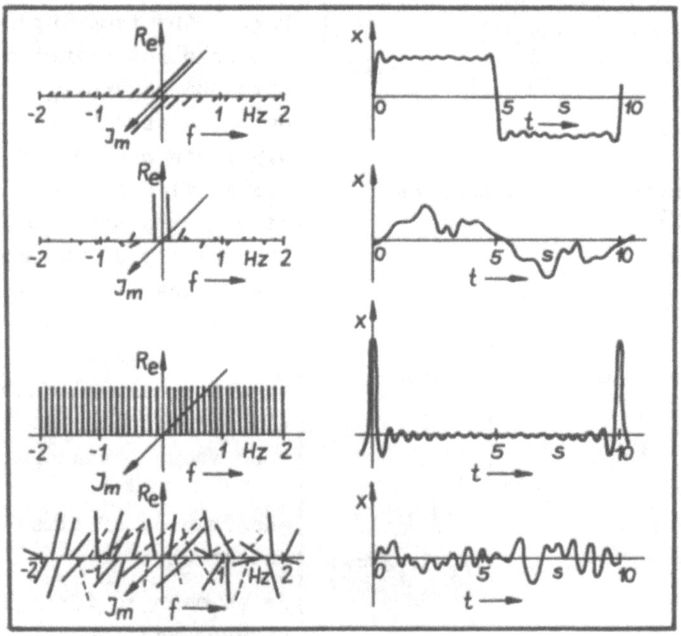

Fig. 4: Complex amplitude spectra with 20 lines each and associated time signals
a) rectangular wave
b) pseudorandom
c) impact
d) pseudorandom

The time function so generated can be continued periodically. In Fig. 3, case c) we see the efficiency of pseudorandoms. This transfer function results from a single measurement without averages. In

Fig. 4 are presented some simple examples for pseudorandom, calculated on a pocket calculator from 20 discrete frequency lines. This is to underline the discrete character of the results. Usually are calculated $1024 = 2^{10}$ values within one period. Fig. 4a) presents the well-known spectrum of a rectangular wave. In the complex discrete amplitude spectrum all values are imaginary, real parts are zero. From random phase angles in the amplitude spectrum (Fig. 4b)), i.e. a complex spectrum, we obtain periodic time signals not differing in shape from real random signals. In Fig. 4b) one period of such a signal is shown.The amplitude spectra only differ in phase angles; the power spectra of both signals are the same. Signals of similar kind occur as spectra of road excitations.

Fig. 4c) presents the spectrum of an impact and Fig. 4d) a random process similar to white noise. The real-valued shock spectrum and the spectrum of the pseudorandom process again only differ in the phase angles of the amplitude spectrum. Fig. 4 underlines the decisive influence of phase information on the shape of a time signal in spite of the same power spectrum.

Using pseudorandom for testing mechanical systems no leakage effects occur. Averages are not necessary. Errors from the randomness of the realised complex of phase angles may be abolished by joining several independent pseudorandom signals. These decisive advantages have made the pseudorandoms an indispensible device in system analyses based on random excitation.

4. Coherence analysis

The advantage of small amplitudes of random excitation contrasts with the disadvantage of sensitiveness to noise or other disturbances. Consequently, the power of the output involves the noise power and is greater than it can be expected without noise (Fig. 5).

The ratio of expected output power to measured output power is defined as coherence function and can be calculated from measured data as to the formula

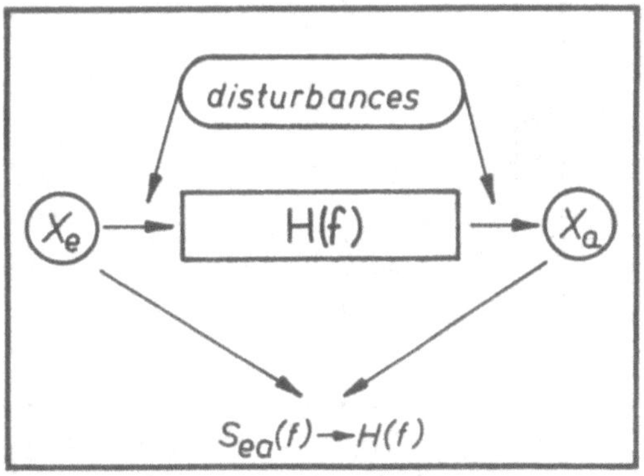

Fig. 5: Influence of disturbances in system analysis

$$\gamma^2(f) = \frac{|S_{ea}(f)|^2}{S_{ee}(f) \cdot S_{aa}(f)} . \tag{4}$$

If the coherence function equals one, this means, that the system is without noise; if it equals zero, it means, that there is no connection between excitation and output signal or that the output signal is caused only by noise. Fig. 6a) gives measured magnitudes of transfer function of a single mass system disturbed at its input. The coherence function between input and output identifies the peak at 24 Hz as the natural frequency of the system, while the vibrations lower than 16 Hz are identified as noise.

Another example of measurements at real industrial plants is given in Fig. 6b). Here are presented the power spectra of two functions measured at motor bearings and turbocompressor bearings of an industrial compressor. The coherence functions shows that two frequencies f_1 and f_2, the rotational speeds of motor and compressor, are reasonable connected with

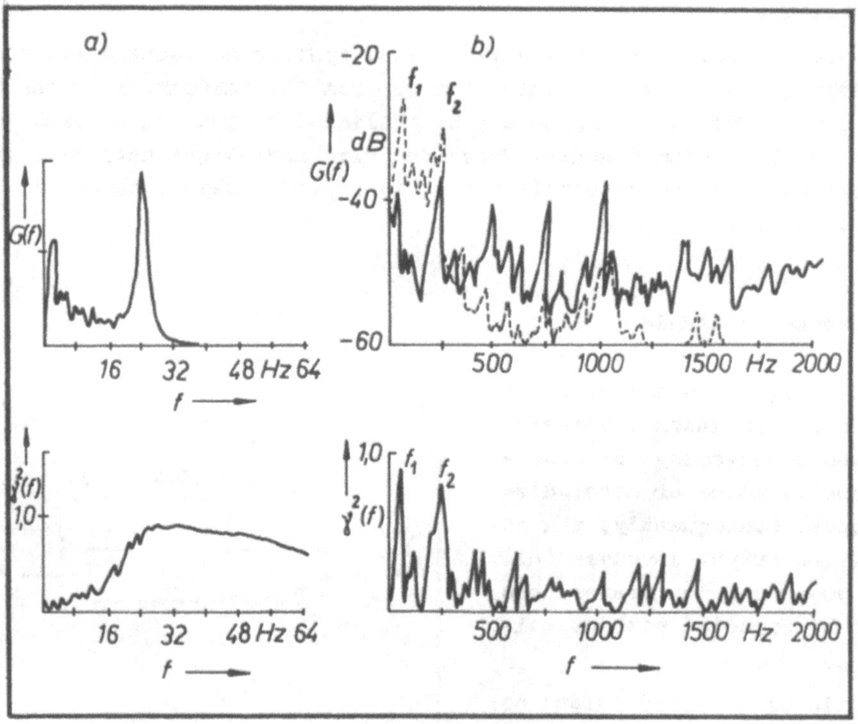

Fig. 6: Coherence functions of disturbed system
a) Output power spectrum disturbed by noise at the input of a single-mass-system
b) Spectra and coherence function of two points of measurement on an industrial turbocompressor

each other, and nearly all the other components of both spectra are caused by noise.

Especially in the compressor spectrum (the continuous curve) we see several peaks with equal intensities, while we find in the motor spectrum only the rotational speeds of motor and compressor as essential frequencies.

5. Applications

Three applications are presented, where it is especially advantageous to use pseudorandom signals.

First example

The first application relates to investigation of influences of vibrations to man. When selecting an adapted excitation, attention must be paid to the fact that physical loads remain small and that the "system man" does not adapt to excitation (e.g. by cramping).

Great physical loads would result from transient excitation (great amplitudes occur here), adapting may be the result of sinusoidal excitation. Exciting with pseudorandom signals, amplitudes of excitation remain small and only one short measurement is necessary. Pseudorandom is generated digitally in the computer, displayed by digital to analog converter and recorded on a tape. In order to carry out the measurement the pseudorandom signal is amplified, and an electromechanical exciter is admitted (Fig. 7). Excitation and measured time functions y_1 and y_2 at different test points are recorded again on a tape. The analyses take place with appropriate analysing programmes in a digital computer. In Fig. 7 are presented real and imaginary parts of two transfer functions of a sitting man. The time of mea-

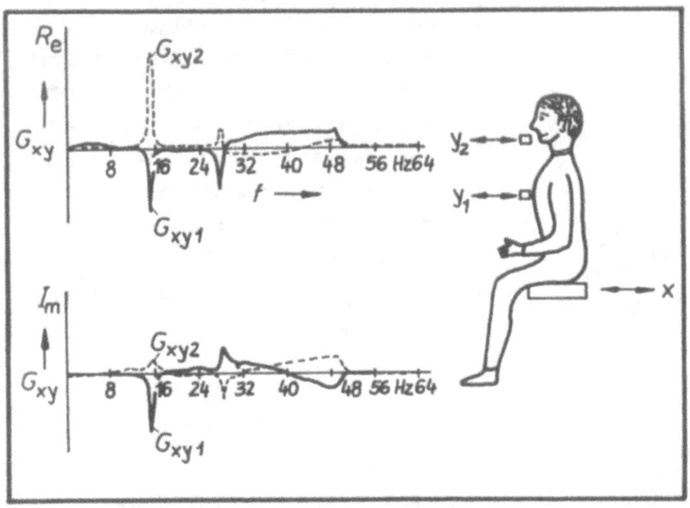

Fig. 7: Transfer function of a sitting man, calculated from measurements of Zentralinstitut für Arbeitsschutz, Dresden

surements was 4 s. From the transfer function can be calculated parameters of mechanical models.

Investigations of this kind shall help to find better results about allowed vibration load of man. Tests of reactivity of sense organs can be dealt with similarly. So, e.g., a light point on a display can be controlled by a pseudorandom signal, and a test person must follow up a second light point. From this test can be found the transfer function of the transmission channel eye-controller of the second light point.

Second example: Investigation of a nonlinear system

An equivalent linear system is often needed instead of a nonlinear system in order to describe approximately the behaviour of the original system. Admitting the system with sinusoidal or transient excitation we get distorted Nyquist-diagrams, from which the calculation of parameters of an equivalent linear system is impossible. When the Nyquist-diagram is determined by exciting with one pseudorandom signal, we also obtain no proper circular diagram. By repeating the measurement with other pseudorandoms, we have the possibility of averaging, and as a result of this averaging the wanted Nyquist-diagrams are obtained.

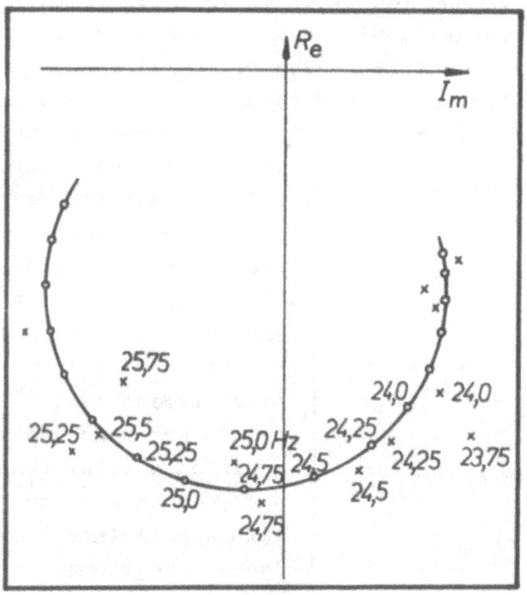

Fig. 8: Linearised Nyquist-diagram of a nonlinear system with a differential equation

$$u''(\tau)+2Du'(\tau)u(\tau)+0.325u^3(\tau)=f(t).$$

× one measurement, ∘ after 256 averages

In Fig. 8 we see such measurements at a strong nonlinear system. The crosses result from one single measurement. Realising 256 averages we find a very useful circular diagram, from which the parameters of the equivalent linear system can be determined. The application of a random generator in spite of averaging provided even worse results, as in the case if the linear system in Fig. 3.

Third example: Exciting signals available for fatigue tests

Real world fatigue testing is used in an increasing rate for tests of elements. Definite types of loads may be classified

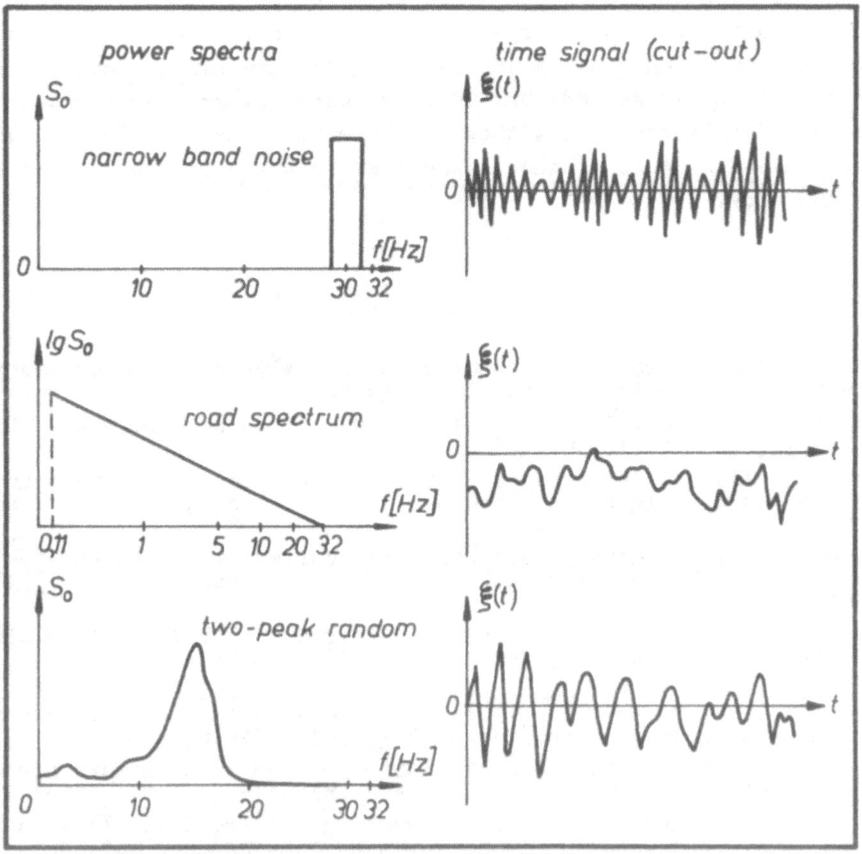

Fig. 9: Power spectra and time functions of pseudorandom signals for fatigue testing
a) narrow band noise b) road spectrum
c) load of an element with two peaks in the power spectrum

by frequency analyses of loads. Thereby we often find random loads with two or three peaks in power spectra.

To admit real structures with such signals real excitations might be measured, stored and be used for controlling servo-hydraulic equipment. But real random signals have the same disadvantages as signals from a random generator. Therefore, more and more simulated excitations with defined spectra are used. The principle of generation is the same as for constant power spectrum. Random phase angles are generated, associated to the defined magnitudes of the amplitude spectrum, and the inverse Fourier transformation is carried out.

In Fig. 9 some power spectra and associated time functions are shown.

In the upper part a narrow-band noise. In the middle we see dominant parts of low frequencies as it is characteristic for roads. At the bottom we have a two-peak power spectrum and an associated time function. Experimental problems consist in realising of time signals with a length of not only 10 or 20 seconds but 20 or 30 minutes. Another problem is to realise a definite spectrum without undue errors. Several long realisations of this kind were produced, and these signals are applied in automobile industry.

References

1 Bendat, J.S.; Piersol, A.G.: Random Data: Analyses and Measurement Procedures. John Wiley & Sons, Intersience 1971

2 Bilý, M.; Čačko, J.: Real time simulation of random processes. VDI-Z. 124(1982),Nr.5-März(I),186-188

3 Experimentelle Analyse stochastischer Vorgänge und Anwendungen auf Probleme des Maschinenbaus. Brochure of TH Magdeburg, Sektion Maschinenbau 1981

4 Holzweißig, F.; Hardtke, H.J.: Verfahren zur experimentellen Modellfindung in der Maschinendynamik. Maschinenbautechnik, Berlin 26(1977), 11,500-504

5 --- : Zur Entwicklung der Maschinendynamik in der DDR von 1969 bis 1978. Maschinenbautechnik, Berlin 28(1979),10,442-447

6 Lingener, A.; Wahl, F.: Einsatz von Klein- und Prozeßrechnern bei der experimentellen Untersuchung stochastischer Schwingungen und von mechanischen Systemen. Techn. Mechanik 1(1980)1,21-26

7 Styles, D.D.; Dodds, C.J.: Simulation of random environments for structural dynamic testing. Experimental Mechanics 1976, Nov., 416-424

8 Wahl, F.: Erzeugung, Analyse und Anwendung von Pseudozufallssignalen. Vortrag zum Problemseminar 'Stochastische Schwingungen und Identifikation', Ber. des Weiterbildungszentrums der TU Dresden, Heft 3/1978.

A COMPARATIVE STUDY OF THE ORDER DETERMINATION METHODS FOR THE AUTOREGRESSIVE-MOVING AVERAGE MODEL

Takayoshi NAKAMIZO* and Hitoshi KANEKO

Summary Much attention has been paid recently to the problem of identifying the mixed autoregressive-moving average (ARMA) model from a finite number of measurement data. The key to identification problem is a good technique for determining the order of ARMA model, because the structure of the model is not known a priori. In this paper, different tests suitable for the determination of the order of the ARMA model have been compared with the help of simulation study and real process experiment. A description is given of four tests based on : Graupe's criterion, the minimal realization, the pole-zero cancellation, the general information criterion.

1. INTRODUCTION

It is known that any gaussian stationary time series can be characterized as the output of a time-invariant linear filter driven by a white noise. From this point of view, the autoregressive (AR) model and the moving average (MA) model have been defined. Moreover, in order to obtain a parsimonious parametrization(use of the fewest number of parameters), it will be useful to include both AR term and MA term, so that

$$x_t+a_1x_{t-1}+ \dots +a_px_{t-p} = e_t+b_1e_{t-1}+ \dots +b_qe_{t-q} \tag{1}$$

where the input e_t is a discrete white noise with the variance σ_e^2. The model represented by eqn.(1) is referred to as an autoregressive-moving average model, and is denoted by ARMA(p,q) model. Using the backward shift operator z^{-1} defined as $z^{-k}x_t=x_{t-k}$, the time series model eqn.(1) can be written as

$$A(z^{-1})x_t = B(z^{-1})e_t \tag{2}$$

where

$$\begin{aligned} A(z^{-1}) &= 1+a_1z^{-1}+a_2z^{-2}+ \dots +a_pz^{-p} , \\ B(z^{-1}) &= 1+b_1z^{-1}+b_2z^{-2}+ \dots +b_qz^{-q} . \end{aligned} \tag{3}$$

It can be observed from eqn.(2) that the ARMA process x_t can be generated by passing a white noise through a time-invariant linear filter with a transfer function $B(z^{-1})/A(z^{-1})$. Detailed discussions on AR, MA and ARMA models are given in Box and Jenkins[1].

Given a finite number of the measurement data $\{x_t: t=1,2,\dots,N\}$, the time series analysis reduces to the model fitting problem. Further the problem of identifying an appropriate time series model is equivalent to the estimation of the transfer function $B(z^{-1})A^{-1}(z^{-1})$. There have been proposed many parameter estimation methods based on the statistical estimation theory. In many practical situations,however, the structure of the model is not known a priori. Thus the order of the transfer function is not available beforehand. Consequently simple and effective methods for determining the order are necessary. Although considerable interest has been paid to the methods for determining the order of ARMA model, comparative evaluation of relative usefulness for practical problems under realistic conditions have rarely been made in the literature.

Since a comparison of different methods by the theoretical analysis is very complicated, the most common way of studying and comparing different tests of the order no

* Department of the Mechanical Engineering, The National Defense Academy, Yokosuka, Kanagawa, Japan

doubt is simulation, which indeed is a natural and valuable tool. It should be noted, however, that most methods seem to work extremely well on simulated data, but not always that well on actual data. In this paper, different methods suitable for the determination of the order of ARMA model have been compared with the help of simulation study and real process experiments. A description is given of four tests based on:

(1) the behavior of AR fitting (Graupe's criterion)[2]

(2) the minimal realization concept[3]

(3) the pole-zero cáncellation effect

(4) the general information criterion (AIC) [4]

Although the results are not presented here, the other tests such as F-test[5], Chow's test[6], the behavior of the autocorrelation function[7] etc. are attempted. However the results are less impressive, so omitted here.

2. DESCRIPTION OF THE EXAMPLES IN THE SIMULATION STUDY

For the evaluation and comparison of the order determination methods, the following two simulated cases are used :

Case A : $x_t - 0.5x_{t-1} = e_t + 0.2e_{t-1}$;

Case B : $x_t - 1.8x_{t-1} + 1.3x_{t-2} - 0.4x_{t-3} = e_t + 1.1e_{t-1} + 0.28e_{t-2}$.

The model is excited by a white noise sequence e_t with zero mean and unit variance. The output signal x_t only is available for estimation , 1024 samples in this study. In Fig.1, the true pole-zero patterns of the test cases are given in the unit circle of the z-plane. The parameter estimation method used here is the standard maximum likelihood method with the Gauss-Newton algorithm[8].

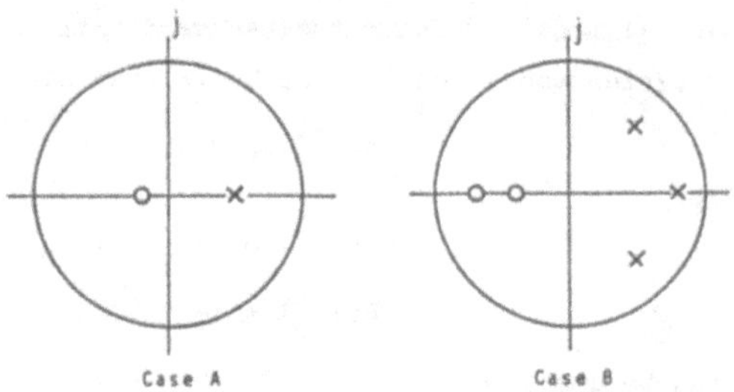

Fig.1 Pole-Zero pattern of simulated cases

3. BRIEF DESCRIPTION OF THE METHODS

3.1 Graupe's method The stable ARMA model described by eqn.(2) may be approximated with sufficient accuracy as an AR model of the form

$$C(z^{-1})x_t = e_t \tag{4}$$

where

$$C(z^{-1}) = 1 + c_1 z^{-1} + c_2 z^{-2} + \dots + c_m z^{-m} \tag{5}$$

provided that the order m is sufficiently large. This implies

$$A(z^{-1})B^{-1}(z^{-1}) = C(z^{-1}) . \tag{6}$$

Now we can easily obtain the estimation of the AR parameters c_i $(i=1,2,\dots,m)$ by the conventional least square method or other modified methods. One procedure to determine the order m of the AR model is the final prediction error (FPE) criterion developed by Akaike[9]. The FPE is defined as the one step ahead prediction error variance when the least square estimates of AR parameters are used for prediction of another process. In fact the order selection is carried out by fitting of successively increasing order computing the appropriate estimate of FPE for the model and adopting the one with the minimum of the FPE value. The FPE is expressed as

$$FPE = \frac{N+m-1}{N-m-1}\,\hat{\sigma}_e^{\,2}(m) \,. \tag{7}$$

In practical computation, $\hat{\sigma}_e^{\,2}(m)$ can be successively obtained by using the Levinson-Durbin (PACOR) algorithm[10]. It should be noted that the FPE criterion takes into account both the error due to model fitting, and the error due to the inaccuracies in estimating the parameters. As an alternative , the criterion autoregressive transfer function (CAT criterion) developed by Parzen[11] may also be useful.

Once the AR model is estimated, the order $(\hat{p},\hat{q})$ of ARMA model is assumed, and then the parameters a_i and b_i are estimated by the maximum likelihood method. Thus we have

$$\hat{A}(z^{-1})\hat{B}^{-1}(z^{-1}) = \hat{C}(z^{-1}) + \varepsilon(z^{-1}) \tag{8}$$

in which $\varepsilon(z^{-1})$ is a polynomial of $(\hat{q}+m)$th order as

$$\varepsilon(z^{-1}) = \sum_{i=1}^{\hat{q}+m} \varepsilon_i z^{-i} \tag{9}$$

whose coefficients ε_i can be calculated by equating the coefficients of same power in eqn.(8). If the assumed order $(\hat{p},\hat{q})$ is correct, then it is evident that $\varepsilon(z^{-1}) = 0$. From this fact, it is reasonable that the quantity

$$J(p,q) = \Sigma\varepsilon_i^{\,2} \tag{10}$$

can be taken as the criterion to select the order $(\hat{p},\hat{q})$ of ARMA model.

3.2 <u>Pole-zero cancellation</u> Suppose that $\hat{p}=\hat{q}=m$, where $m>\max(p,q)$. Then the ARMA model can be expressed as

$$A_m(z^{-1})x_t = B_m(z^{-1})e_t \tag{11}$$

where the suffix m denotes the assumed order. For the over-fitted model

$$\hat{G}_m(z^{-1}) = \hat{B}_m(z^{-1})/\hat{A}_m(z^{-1}) \tag{12}$$

the pole-zero cancellation will be seen. A useful, but primitive way to do the test of common factors in the identified model is to compute the zeroes of two polynomials $\hat{A}_m(z^{-1})$ and $\hat{B}_m(z^{-1})$, and compare them. In practical situations, however, we only have estimated values of parameters, leading to a less complete cancellation.

3.3 <u>The minimal realization method</u> The transfer function $B_m(z^{-1})/A_m(z^{-1})$ of ARMA model will be realized by an observable canonical form as

$$\begin{aligned} \underline{z}_{t+1} &= \underline{F}\,\underline{z}_t + \underline{g}\,e_t\,, \\ x_t &= \underline{h}^T\underline{z}_t + b_0 e_t \end{aligned} \tag{13}$$

where $\underline{z}_t$ is the m state vector and

$$\underline{F} = \left[\begin{array}{c|c} 0 \;\dots\; 0 & -a_m \\ \hline & -a_{m-1} \\ I_{m-1} & \vdots \\ & -a_1 \end{array}\right] \qquad \underline{g} = \begin{bmatrix} b_m - b_0 a_m \\ b_{m-1} - b_0 a_{m-1} \\ \vdots \\ b_1 - b_0 a_1 \end{bmatrix} \qquad \underline{h}^T = [0 \dots 0\ 1\,]$$

The minimal realization can be determined by a controllable subsystem. The controllability matrix $\underline{K}$ is defined by

$$\underline{K} = [\ \underline{g},\ \underline{Fg},\ \dots\ ,\underline{F}^{m-1}\underline{g}\ ].$$

Then the minimal realization is achieved by choosing[12]

$$\underline{\tilde{F}} = \underline{V}^T\underline{FV}, \quad \underline{\tilde{h}}^T = \underline{h}^T\underline{V}, \quad \underline{\tilde{g}} = \underline{Vg}$$

where $\underline{V} = [\underline{v}_1,\underline{v}_2,\dots,\underline{v}_r]$, $\underline{v}_i(i=1,2,\dots,r)$ are normal eigenvectors of $\underline{KK}^T$ corresponding to the non-zero eigenvalues. The transfer function of this process is given by

$$\hat{G}_r(z^{-1}) = \tilde{h}^T(zI-\tilde{F})^{-1}\tilde{g} + b_0 \tag{17}$$

from which the ARMA model of the correct order could be obtained. In practical situations, we only have estimated parameters, leading to a less ideal set of eigenvalues which are not exactly zeros.

3.4 General information criterion To select the order of AR model, the FPE criterion is introduced and is successfully applied to the spectral estimation. Akaike then extended this criterion to any maximum likelihood situation. This will be referred as the Akaike's information criterion (AIC), which is defined by

$$AIC = -2\ \log(\text{maximized likelihood}) + 2k \tag{18}$$

where k is the number of independent adjustable parameters within the model. The AIC is introduced as an estimate of twice the negentropy of the structure with respect to the fitted model and a low value of AIC means a good fit.

The AIC criterion is also effectively used to determine the order of the ARMA model. In this case AIC is expressed as

$$AIC(\hat{p},\hat{q}) = N \log \hat{\sigma}_e^{\,2}(\hat{p},\hat{q}) + 2(\hat{p}+\hat{q})\,. \tag{19}$$

This criterion was obtained with the aid of an information theoretical interpretation of the maximum likelihood method. The order $(\hat{p},\hat{q})$ giving the smallest value of $AIC(\hat{p},\hat{q})$ should be chosen.

4. COMPARISON OF METHODS

To the two simulated cases A and B described in section 2, four tests decribed in the previous section are applied. In Tables 1 and 2 the results obtained from 21 trials are

Table 1 Simulation results for the case A

MA \ AR	1	2	3	4
1	11	3	3	
2	3	1		
3				
4				

J criterion

MA \ AR	1	2	3	4
1	18			
2		3		
3				
4				

minimal realization

MA \ AR	1	2	3	4
1	15	1	1	
2	3			
3	1			
4				

AIC

Table 2 Simulation results for the case B

MA \ AR	1	2	3	4
1	7	2		2
2		1	0	
3				
4	5			

J criterion

MA \ AR	1	2	3	4
1				
2			20	
3				
4				1

minimal realization

MA \ AR	1	2	3	4
1				4
2			16	
3	1			
4				

AIC

presented. It should be noted for the simulated case B that compared with other methods, the behavior of Graupe's criterion yields less impressive results, and all other tests giv• reasonably good results. Generally speaking, the minimal realization test yields the most reliable results.

For the *case A*, the identification is achieved by assuming the ARMA model of the order (3,3), and we obtained

$$\hat{G}_3(z^{-1}) = \frac{1-0.69767z^{-1}+0.73509z^{-2}+0.26748z^{-3}}{1-1.45218z^{-1}+1.46716z^{-2}-0.48265z^{-3}} .$$

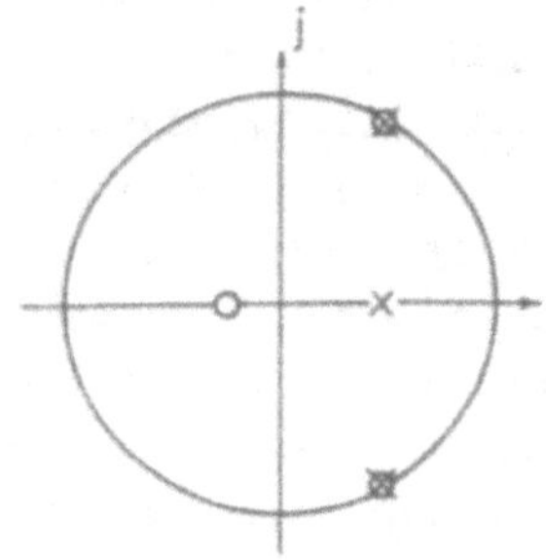

Fig.2 pole-zero pattern of model (case A)

The pole-zero cancellation effect is shown in Fig.2, from which the correct order (1,1) can be determined. Thus we are able to calculate the transfer function of the form

$$\hat{G}(z^{-1}) = \frac{1+0.2632z^{-1}}{1-0.49426z^{-1}} .$$

In a similar way, the identified model for the *case B* becomes

$$\hat{G}_4(z^{-1}) = \frac{1+0.48974z^{-1}-0.44104z^{-2}-0.21869z^{-3}+0.01070z^{-4}}{1-2.45778z^{-1}+2.45368z^{-2}-1.21854z^{-3}+0.25021z^{-4}}$$

where the assumed order is (4,4). The pole-zero pattern of $\hat{G}_4(z^{-1})$ is shown in Fig.3 in which only one pair of pole-zero is cancelled, leading to

$$\hat{G}(z^{-1}) = \frac{1+1.13232z^{-1}+0.29263z^{-2}-0.02208z^{-3}}{1-1.81518z^{-1}+1.29331z^{-2}-0.39682z^{-3}} .$$

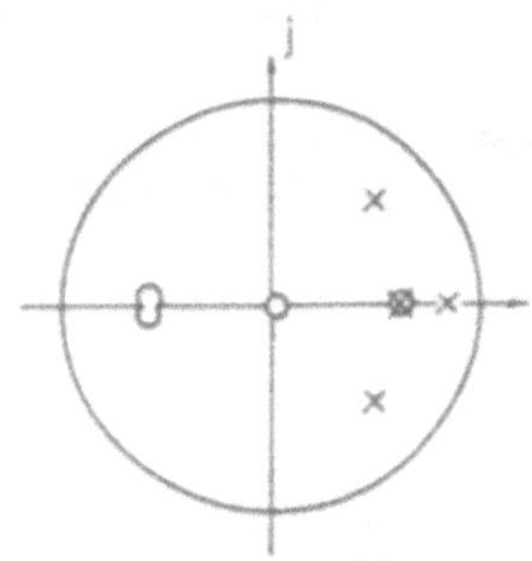

Fig.3 pole-zero pattern of model (case B)

Note that one zero is located in the neigbour of the origin. This indicates that the order can be determined as $\hat{p}=3$ and $\hat{q}=2$.

In Fig.4, the development of the pole-zero pattern is shown for the *case B*. It can be observed that three roots can be considered to be the characteristics of the

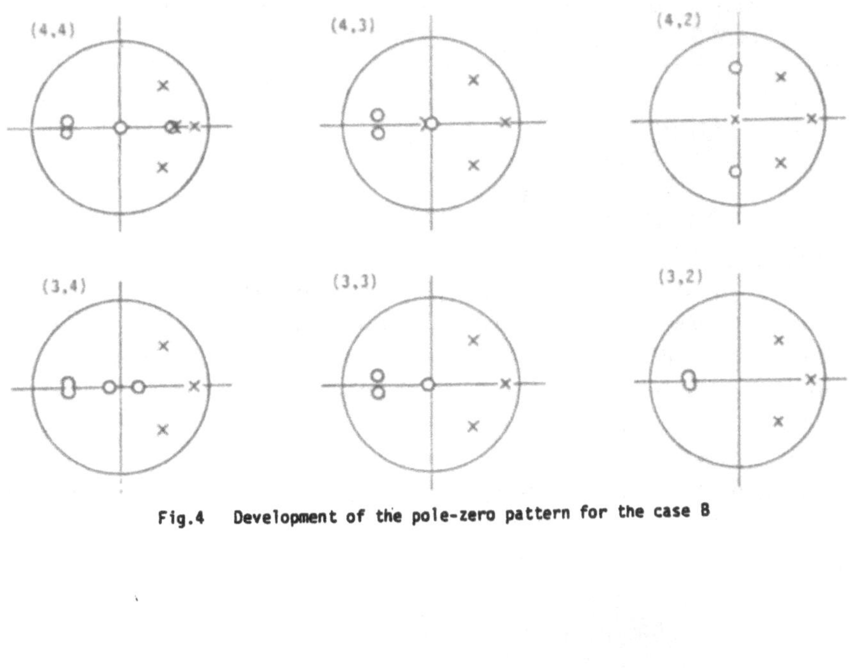

Fig.4 Development of the pole-zero pattern for the case B

investigated system.

In Table 3, eigenvalues of the matrix $\underline{KK}^T$ in the minimal realization test are given for the both identified models governed by $G_3(z^{-1})$ and $G_4(z^{-1})$. It can be safely stated for the *case A* that λ_2 and λ_3 are considered as zeros, leading to the correct order (1,1). For the *case B*, we will be able to choose the order (3,3) by taking λ_4=0 into consideration.

Table 3 Eigenvalues of $\underline{KK}^T$

	case A	case B*
λ_1	0.021499	254.5
λ_2	0.000081	2.5984
λ_3	0.000013	0.02797
λ_4	———	0.00005

Table 4 gives an example for order determination using AIC. It can be seen that the ARMA model of order (3,2) yields an adequate model. This is of course the true order of the model. For the same data, the loss function J(p,q) defined by eqn.(10) is tabulated in Table 5, indicating that the estimated order (4,1) is incorrect for this example. Note that the loss functions of J for the orders (3,2),(2,4) and (1,4) are also relatively small. This test principally depends on the accuracy in estimation of AR model. In this test, the AR model is firstly estimated. The estimated AR model is given by

$$\hat{C}(z^{-1}) = 1-2.823z^{-1}+3.933z^{-2}-3.767z^{-3}+2.988z^{-4}-2.247z^{-5}$$
$$+1.638z^{-6}-1.058z^{-7}+0.589z^{-8}-0.245z^{-9}-0.058z^{-10}$$

whose order was determined by using FPE criterion. The true model(*Case B*) can be

$$A(z^{-1})/B(z^{-1}) = 1-2.9z^{-1}+4.21z^{-2}-4.219z^{-3}+3.462z^{-4}-2.627z^{-5}$$
$$+1.920z^{-6}-1.377z^{-7}+0.977z^{-8}-0.689z^{-9}+0.484z^{-10}-\ldots$$

If we rewrite the estimated ARMA models of order (4,1),(3,2) and (1,4) as

$$\hat{A}_4(z^{-1})/\hat{B}_1(z^{-1}) = 1-2.845z^{-1}+3.974z^{-2}-3.823z^{-3}+3.056z^{-4}-2.841z^{-5}+\ldots,$$
$$\hat{A}_3(z^{-1})/\hat{B}_2(z^{-1}) = 1-2.842z^{-1}+4.002z^{-2}-4.468z^{-3}+3.745z^{-4}-2.341z^{-5}+\ldots,$$
$$\hat{A}_1(z^{-1})/\hat{B}_4(z^{-1}) = 1-2.821z^{-1}+3.950z^{-2}-3.847z^{-3}+4.092z^{-4}-4.344z^{-5}+\ldots$$

then it seems that each expression may be a good approximation of $\hat{C}(z^{-1})$.

Table 4 Values of AIC

AR \ MA	1	2	3	4
1	———	341.766	97.353	23.380
2	280.046	51.085	30.393	16.800
3	25.864	14.070	15.599	17.649
4	15.931	15.108	16.218	———

Table 5 Values of J criteion

AR \ MA	1	2	3	4
1	2.3760	0.2579	0.0725	0.0052
2	0.5767	0.0669	0.0163	0.0070
3	0.0207	0.0083	0.0158	0.0239
4	0.0040	0.0119	0.0430	0.1630

5. REAL DATA EXPERIMENT

In the simulation study, the true model to be identified is exactly described by a finite number of parameters. In practical situations, however, since any real data would be characterized by infinite number of parameters, it is often difficult to be conclusive about properties studied only by simulation. This needs the real data experiment to examine if the various tests work well.

As a test data, the acoustic data near the tool of a working lathe are used. The sound data measured through the microphone are applied to the analog highpass filter to cut the rotary sound produced by the lathe itself. And then the data are sampled at the rate of 20 KHz. Corresponding to four values of the width d of flanck wear, namely d=0, d=0.195,d=0.318 and d=0.374 mm, we obtain the time series after cutting off the low frequency components as shown in Fig.5. Assuming that the observed data are generated through ARMA model, our problem is to identify the proper ARMA model. The key to identification problem is a good technique for determining the order of ARMA model. Although the results are not presented here, the data tests on stationarity and normality are carried out by using the statistical run test and χ^2 test respectively.

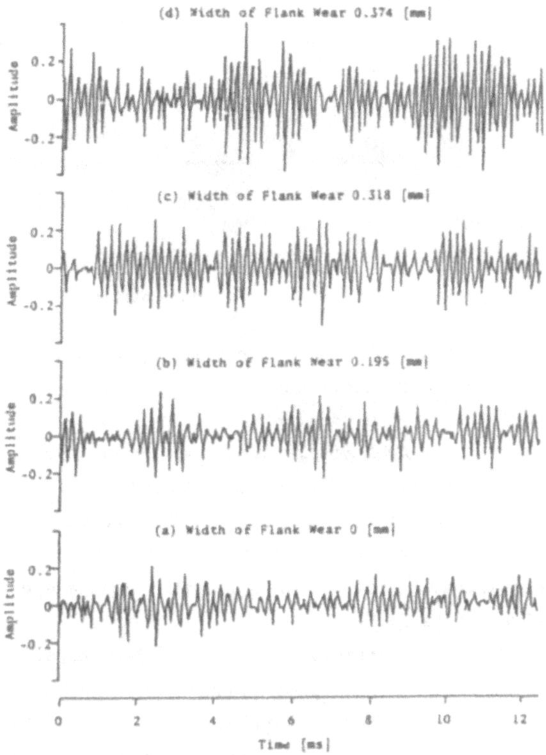

Fig.5 Wave forms of the real data after preprocessing

The identification is first achieved by assuming the ARMA model of the order (8,8). For three states of flanck wearing width, eigenvalues of the matrix $\underline{KK}^T$ can be calculated as shown in Table 6. It seems somewhat difficult to decide which of eigenvalues may be effectively regarded as zeros. As the help of decision, the pole-zero pattern of the identified model is shown in Fig.6. For the case of d=0.113 mm, four pairs of pole-zero in the left plane seem to be cancelled, leading to the ARMA model of the order (4,4). This indicates the eigenvalue λ_5 to be zero. For the case of d= 0.374 mm, however, the pole-zero cancellation effect is less pronounced. The only two pairs of pole-zero cancellation can be seen. Although the minimal realization test or pole-zero cancellation test appears to be the most effective method for the order determination in the simulation study as described in the previous sections; this method is somewhat subjective for the real data case under the actual conditions. The main advantage of this test is that only one estimation of ARMA model of the fixed order is needed, but the pole-zero cancellation becomes ambiguous, unless the

Table 6 Eigenvalues of KK^T (real data)

data	d=0.113	d=0.195	d=0.374
λ_1	92.749	13.515	21.590
λ_2	4.2716	3.5990	2.2266
λ_3	1.1976	0.2176	0.3599
λ_4	0.6231	0.1253	0.1647
λ_5	0.0115	0.0253	0.0257
λ_6	0.0056	0.0146	0.0191
λ_7	0.0037	0.0039	0.0049
λ_8	0.0000	0.0009	0.0006

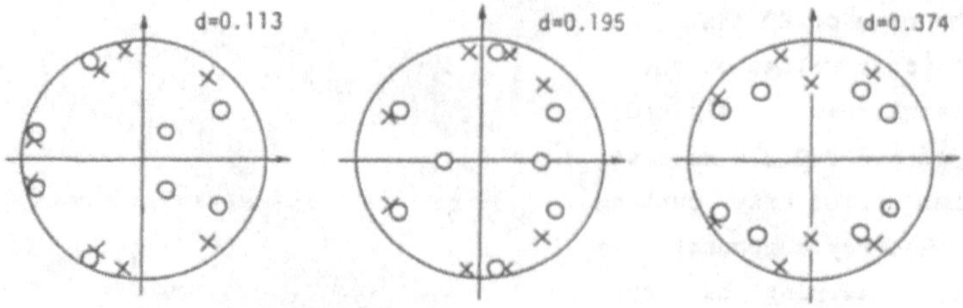

Fig.6 Pole-zero pattern of the identified models

parameter estimation is sufficiently accurate.

The order determination by the minimum AIC (or $J(\hat{p},\hat{q})$ criterion) method is carried out by fitting the ARMA models of successively increasing order $(\hat{p},\hat{q})$ and computing the AIC (or J) values for respective models. This requires a lot amount of computation to obtain a reasonable estimation of the order. The results are given in Table 7.

Table 7 Results obtained by minimum AIC test

DATA	AIC(ARMA)	FPE(AR)
d= 0	(5,6)	25
d=0.113	(5,3)	20
d=0.195	(4,4)	12
d=0.318	(4,6)	45
d=0.374	(6,7)	24

The main disadvantage of real data experiment is that the true model is not known. There is needed a diagnostic checking of the estimated model. There are several tests for determining if the identified model is adequate or not. In this paper, the whiteness test of the residuals and the comparison of estimated spectral densities are carried out.

5.1 Whiteness test of the residual This test is based on the fact that

$$\hat{e}_t = \hat{B}^{-1}(z^{-1})\hat{A}(z^{-1})x_t$$

is an estimate of the white noise $\hat{e}_t$, if the proper model is accepted. To test if the residual $\hat{e}_t$ is white noise, we compute the autocorrelation function of the residual. In Fig.7, the estimated auto-correlation functions of the residuals are shown for different values of d. It may be stated that the residuals are approximately white noises. This is confirmed with the help of statistical testing hypotheses. It is known [1] that the quantity

$$Q = N \sum_{k=1}^{K} \rho^2(k) \tag{20}$$

is approximately distributed as χ^2 with K-p-q degreesof freedom. In eqn.(20), $\rho(k)$ represents the normalized auto-correlation function of the residual $\hat{e}_t$. Note that the degree of freedom in χ^2 depends on the number of parameters in the model. For example in the case of d=0.374, the test quantity Q is given by

$$Q = 1024 \sum_{k=1}^{15} \rho^2(k) = 3.493.$$

From χ^2 table for K-p-q = 2 degree of freedom, we get $Q_{\alpha=0.05} = 5.99$ (1-α is the significant level in χ^2 test). This provides no evidence that the residual is correlated.

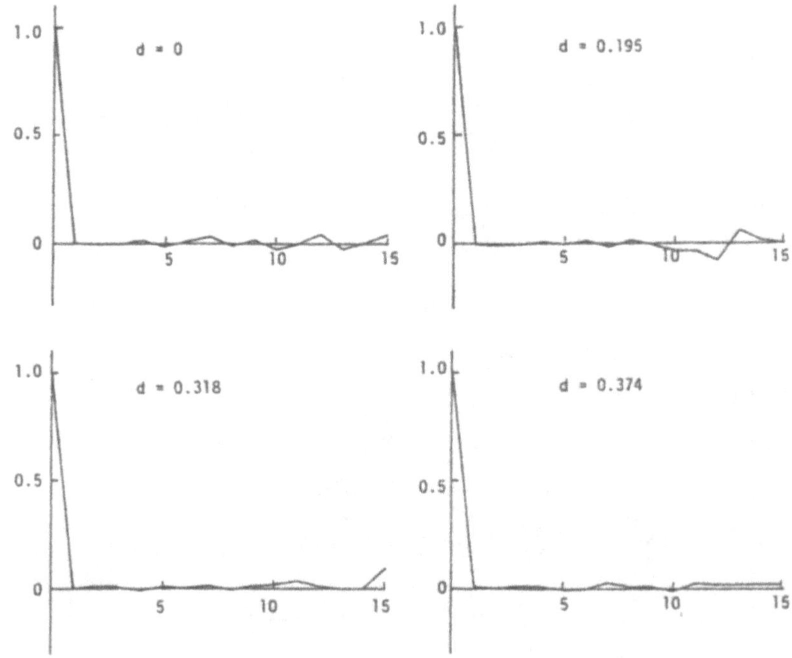

Fig.7 Estimated autocorrelation functions of the residuals

The values of Q criterion for other sets of data are given in Table 8 together with the valuesof 5% and 2.5% risk levels. Inspection of Table.8 shows however that only one suspiciously large value of Q occures for the case of d=0.195.

5.2 The estimation of spectral densities

Table 8 Values of Q criterion

d	Q	DOF	risk level 5%	2.5%
0	7.891	4	9.49	11.14
0.195	16.040	7	14.07	16.01
0.314	10.281	5	11.07	12.83
0.374	3.493	2	5.99	7.38

In Fig. 8 are shown the spectral densities calculated from the identified ARMA models for different values of d. The solid curve denotes the spectral estimation obtained by the minimum AIC method. The dotted curve represents the autoregressive spectral estimation obtained from FPE criterion. It can be observed that there seems to be in satisfactory agreement. Since the AR model is of the relatively high order, the estimated spectrum shows the fluctuating feature. The estimated ARMA spectrum passes through the region bounded by the oscillation of the estimated AR spectrum. Therefore the spectrum estimated by the ARMA model may be preferable in practice.

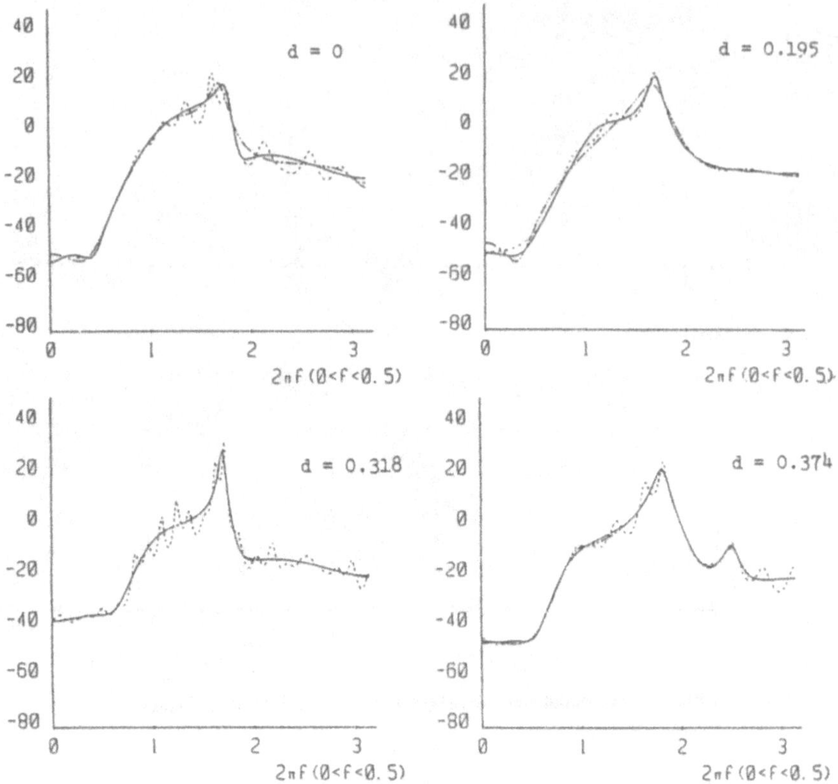

Fig.8 Estimation of the spectral densities

——— minimum AIC test (ARMA)
—·— minimum J criterion (ARMA)
------- minimum FPE test (AR)

References

[1] G.E.P.Box and G.M.Jenkins:Time Series Analysis,Forecasting and Control, Holden-Day 1970

[2] D.Graupe:Identification of Systems, Van Nostrand 1972

[3] K.Furuta:An application of realization theory to identification of multivariable process, IFAC symp. on Identification, the Hague/Delft (1973)939-942

[4] H.Akaike:A new look at the statistical model identification, IEEE Trans. on Autom. Control, AC-19 (1974) 716-723

[5] P.Eykhoff:System Identification, John Wiley 1974

[6] J.C.Chow:On the estimation of the order of autoregressive moving average with uncertain observation, IEEE Trans. on Autom. Control, AC-17 (1972) 707-709

[7] P.E.Krause:Prediction and filter identification, Ph.D thesis, Colorado Univ. 1972

[8] N.K.Gupta and R.K.Mehra:Computational aspect of maximum likelihood estimation and reduction in sensitivity function calculation, IEEE Trans. on Autom. Control, AC-19, (1974) 774-783

[9] H.Akaike:Fitting autoregressive models for prediction, Ann. Inst. Statist. Math., 21 (1969)

[10] J.Durbin:The fitting of time series models, Rev. Int. Inst. Statist.,(1960) 233-244

[11] E.Parzen:Some recent advances in time series analysis, IEEE Trans. on Autom. Control, AC-19 (1974) 716-723.

INFLUENCE OF STOCHASTIC PERTURBATIONS ON THE QUALITATIVE BEHAVIOUR OF NONLINEAR SYSTEMS

M.Peschel [1] and P.Rudolph[1]

Summary

The application of a STRUCTURE DESIGN PROCEDURE for the transformation of given ordinary differential equations into VOLTERRA EQUATIONS is illustrated for the case of stochastic disturbed systems.
The principal possibilities for the imbedding of deterministic VOLTERRA EQUATIONS into corresponding stochastic equations are described. The procedure will be demonstrated for simple mechanical systems.

Introduction

In /1/ we proposed a method for the transformation of a given set of ordinary differential equations into the VOLTERRA EQUATIONS, the basic equations for predator-prey competition in ecological systems:

$$F\, x_i = \sum_j G_{ij} x_j + \sum_s H_{is} y_s \quad \text{with} \quad F = d\ \ln/dt\,.$$

These equations were introduced by Verhulst /2/, Lotka /3/ and Volterra /4/. In /5/ a huge group of equivalence transformations for multinomial differential equations

$$F\, x_i = \sum_J A_{iJ} \prod_r x_r^{a_{Jr}}$$

was considered.
The socalled Riccati - representation for VOLTERRA EQUATIONS seems to be very useful for studying the qualitative properties of nonlinear systems.
The Riccati-representation has the form

$$F\, x_i = K_i \prod_r x_r^{k_{ir}}$$

This representation was analysed in /6/, /7/.
A lot of mechanical systems can be described by deterministic ordinary differential equations, therefore it is possible to transform these systems into VOLTERRA EQUATIONS or corresponding representations

[1] Academy of Sciences of the GDR, Central Institute for Cybernetics and Information Processes, 1086 Berlin, Kurstraße 30-34

(multinomial differential equations, Riccati-representation).
In all these cases the connection to stochastic perturbed equations is of interest. In this paper we give some insight into the stochastic perturbations by parameter fluctuations in autonomous systems.
The imbedding of the deterministic ordinary differential equations into stochastic concepts is considered in a lot of papers for example in /8/.

The STRUCTURE DESIGN CONCEPT in the deterministic case

We follow the image of structures in ecology consisting mainly of chains and cycles with a relatively small number of feedbacks within such a substructure element and links between them.
We are looking for a structural model for a given system described by a set of ordinary differential equations which can be roughly decomposed into chains and cycles.
In ecology, but also in economy, encymology etc., we often meet elementary functions determined by rate coupling, that means, two consecutive modules in a chain or a cycle are connected by the following local functional model

$$d\, x_i/dt = K_i x_i x_{i+1}$$

$F\, x_i = d \ln x_i/dt$ is the "growth-rate" of x_i.

The STRUCTURE DESIGN PRINCIPLE suggests a number of flexible rules for the design of the structure of complex systems:

1. We apply the basic operator $F = d \ln/dt$ successively to any (explicitly or implicitly) given signal $x(t)$ to try to build up a chain

   $$F\, x_i \sim x_{i+1}$$

   where $\sim$ stands for proportionality.
   In general $F\, x_i = A_i$ is an expression containing both familiar and new signals.

2. If A_i contains signals familiar to us because they have been already identified formerly, these are labeled according to the already existing structure.
   There are some degress of freedom in labelling the unknown signals, which are then expanded as in step 1.

3. This procedure stops when no new signals can be identified, thus yielding a finite structure.

4. Finally we introduce names x_i for the output signals of all basic modules F into the structure.

We demonstrate the formal applications of these rules for the example of a pendulum with moved top-point

$$d^2x/dt^2 + \beta dx/dt + (1 + A\cos\Omega t)\sin x = 0.$$

We put $x_o = x(t)$;

$$F x_o = \dot{x}/x = x_1; \quad F x_1 = \dot{x}^{\bullet} / \dot{x} - x_1, \quad x_2 = \dot{x}^{\bullet} / \dot{x},$$

$$x_2 = -\beta - (1 + A\cos\Omega t)\frac{\sin x}{\dot{x}},$$

$$x_2 = (1 + A\cos\Omega t)\sin x / \dot{x};$$

$$F x_2 = \frac{-A\Omega\sin\Omega t}{1 + A\cos\Omega t} + \frac{\cos x}{\sin x}\cdot\dot{x} - \tilde{x}_2,$$

$$x_3 = \frac{\sin\Omega t}{1 + A\cos\Omega t}, \quad x_4 = \frac{\cos x}{\sin x}\cdot\dot{x},$$

$$F x_3 = \Omega\frac{\cos\Omega t}{\sin\Omega t} + A\Omega x_3; \quad F x_4 = \frac{-\sin x.}{\cos x}\dot{x} - x_4 + \tilde{x}_2;$$

$$x_5 = \frac{\cos\Omega t}{\sin\Omega t}, \qquad x_6 = \frac{\sin x}{\cos x}\cdot\dot{x},$$

$$F x_5 = -\Omega\frac{\sin\Omega t}{\cos\Omega t} - \Omega x_5; \quad F x_6 = x_4 + x_6 + \tilde{x}_2;$$

$$x_7 = \frac{\sin\Omega t}{\cos\Omega t},$$

$$F x_7 = \Omega x_5 + \Omega x_7.$$

In this straight forward procedure we get the following Volterra equations

$$F x_o = x_1; \quad F x_1 = \tilde{x}_2 - x_1, \qquad \tilde{x}_2 = -\beta - x_2;$$

$$F x_2 = -A\Omega x_3 + x_4 - \tilde{x}_2; \qquad F x_3 = \Omega x_5 + A\Omega x_3;$$

$$F x_4 = -x_4 - x_6 + \tilde{x}_2; \qquad F x_6 = x_4 + x_6 + \tilde{x}_2;$$

$$F x_5 = -\Omega x_5 - \Omega x_7; \qquad F x_7 = \Omega x_7 + \Omega x_5.$$

This is not a useful Volterra representation because it leads always to a parasitary singularity even when a denominator assumes the value zero. This disadvantage can be easily avoided by passing to shifted variables. The consequence is an increase of the system order and a socalled regular Volterra representation with all $x_i \geqslant 0$.

Heuristic approach to the structure design under stochastic perturbations

We consider autocatalytic growth functions (for example growth of cracks in materials) following the deterministic differential equation

$$dx/dt = K\,x^k.$$

We assume, that the amplification K is superposed by a stochastic perturbation maybe by a stationary stochastic process

$$K = K_o + \varkappa(t).$$

Often people then decompose the stochastic process x(t) into an undisturbed component $x_o(t)$ changed by a resulting stochastic process $\xi(t)$

$$x(t) = x_o(t) + \xi(t)$$

where $\xi(t)$ is generated by the stochastic influence $\varkappa(t)$.

For our argument we assume that all trajectories x(t) are differentiable. A straight forward application of our STRUCTURE DESIGN RULES yields the following Volterra representation:

$$x = x_o + \xi,$$

$$F\,x_o = x_1; \quad F\,x_1 = (k-1)x_1;$$

$$F\,\xi = \xi_1 - \xi_2, \quad F\,\xi_1 = \dot{\varkappa}/(K_o + \varkappa) + (\xi_1 - \xi_2) + \xi_3,$$

$$F\,\xi_2 = x_1 + (\xi_2 - \xi_1),$$

$$F\,\xi_3 = \dot{\varkappa}/(K_o + \varkappa) + (k-1)\,\xi_3.$$

This is a wrong representation because the $\xi_i(t)$ are not neccessarily zero, if $\varkappa(t) \equiv 0$. The reason for this phenomenon is an incorrect "linear transfer assumption between $\varkappa(t)$ and $\xi(t)$".

But this error can be eliminated easily by not admitting the linear decomposition of x(t). Then we get the following regular Volterra representation

$$F\,x = x_1, \quad F\,x_1 = (k-1)\,x_1 + \dot{\varkappa}/(K_o + \varkappa).$$

As a second example let us use the same heuristic approach for the case of a parameter disturbed logistic growth function following the equation

$$dx/dt = K\,x\,(B - x) \quad \text{with}$$

$$K = K_o + \varkappa(t) \quad \text{and}$$

$$B = B_o + \beta(t)$$

and $\varkappa(t)$, $\beta(t)$ being stationary stochastic processes (with existing derivatives!).

We get the following Volterra equations (regular representation)

$$F\,x = x_1;\quad F\,x_1 = \dot{\mathscr{X}}/(K_0 + \mathscr{X}) + x_2 - x_3;$$

$$F\,x_2 = \ddot{\beta}/\dot{\beta} - x_2 + x_3;\quad F\,x_3 = \dot{\mathscr{X}}/(K_0 + \mathscr{X}) + x_1.$$

This stochastic model has the following properties:

The deterministic part is a model of a certain growth transition (EVOLON). The stochastic perturbation first leads to an increasing deviation from the deterministic trend, the variance increases at first monotoneously. In approaching the saturation stage the trend is accompanied by a stochastic component with decreasing variance, at the end we meet only the fluctuation $\beta(t)$ of the saturation level B.
Such a behaviour we expect general for growth transitions, we speak in such a case of the socalled 'sausage model of growth .

Imbedding of the deterministic Volterra representation into a more relevant set of stochastic differential equations

The need for the introduction of randomized evolutionary models stems not only from taking into account intrinsic and external randomness. It is wellknown, that randomized models can lead to qualitative different solutions compared with the deterministic ones. Arnold /9/ gave an example of an unstable system, which is stabilized by randomizing the parameters. Similar effects are known from control experience, the socalled vibration linearization /10/ has its stochastic analogon, that means a step-function can be substituted by a power-function of first (or higher) order, if we superpose the input an equally distributed (or some independent equally distributed) stochastic variables /11/. Sometimes in general randomness can have a smoothing effect on growth phenomena, for example a sequence of different EVOLONs can be smoothed in such a way that we get with high accuracy an envelope in the form of only one EVOLON. In the neighbourhood of unstable equilibrium points, of bifurcation points etc. even a very small stochastic influence can switch the behaviour of the whole system essentially. In such a case randomness can not be neglected.

Bottom-up approach: From the step-transition level to the continuous deterministic differential equation

Let us consider a homogenous Markov-process with values only in a discrete state space $\mathfrak{X}$.
Such processes are characterized by the infinitesimal probabilities (/12/, /13/)

$$q(x,B) = \lim_{t \downarrow 0} \frac{1}{t}\,(P(x,t,B) - P(x,0,B)) \quad \text{with} \quad P(x,0,B) = \chi_B(x)$$

(transition from B to x in time t).
It is assumed that the limit exists uniformly and is continuous.
From wellknown properties of semigroups of operators (/14/) it is enough in many cases to solve the Kolmogoroff forward and backward equation to get the transition probabilities. For a bounded continuous function f(u) the corresponding equations are

$$\frac{d}{dt} E_x f(x_t) = \int_{\mathcal{X}} E_z f(x_t)\, q(x,dz) \quad \text{and}$$

$$\frac{d}{dt} P^u(t,B) = \int_{\mathcal{X}} q(x,B)\, P^u(t,dx)$$

with

$P^u (t,B) = EP(u,t,B)$ for a random variable u.

For the special cases $f(x) = \chi_{\{y\}}(x)$ and $P(u = x) = 1$ we get the wellknown Kolmogoroff equations.

For Markoff processes in general the Bartlett-Moyal equation describes the process as well as the infinitesimal operator (/15/, /16/). This equation can be defined with the help of the function

$$F(x,t,\lambda) = \lim_{h \searrow 0} \frac{1}{h} \left(E_x \left(\exp \left(\lambda^T (x_{t+h} - x\right) / x_t = x \right)\right.$$

as a differential equation for the characteristic function

$$\varphi(x,t,\lambda) = E_x \exp(\lambda^T x_t),$$

$$\partial\varphi / \partial t = E_x (\exp(\lambda^T x_t) F x_t, t, \lambda)) .$$

It is important to mention, that the Bartlett-Moyal equation leads to a stochastic analogon of the deterministic differential equation

$$\frac{d}{dt} E_x x_t = E_x \frac{d}{d\lambda} F(x_t, t, 0) = \lim_{t \searrow 0} \frac{1}{t} E_x (x_t - x) .$$

We assume, that the state space consists of N states and maybe the process $x_{t,N}$ a density dependent jump process with

$$q(x, x+h) = N \bar{q} \left(\frac{x}{N}, h\right).$$

Then under some restrictions on $\bar{q}$ (/17/) $x_{t,N}$ converges for N approaching infinity stochastically and uniformly in any finite time interval to a solution of the deterministic differential equation

$$dx/dt = \sum_h h\, \bar{q} (x, x+h).$$

From the above presented considerations follows, that the homogenous Markov process with a discrete state space, for which the Lotka-Volterra equation in the mean must be fulfilled, must have an infinitesimal first moment of the form

$$\sum_h h\, \bar{q}_i\,(x,\, x+h) = x_i \left(\sum_j a_{ij}x_j + \varepsilon_i\right).$$

Depending on the signs a_{ij} or ε_j, resp., we can interpret these parameters as birth- or death-rates, respectively

Top-down approach: From a global stochastic equation to the continuous deterministic equation

The stochastic influence on a process, globally described by the Lotka-Volterra equation is in literature often taken into account by augmenting the deterministic equation with stochastical terms.
Therefore let us assume that some parameters in the Lotka-Volterra equations are influenced by 'real noise' in the form of stationary stochastic processes. If we further assume, that the noise is stochastically continuous we get the following random version of the Lotka-Volterra-equation

$$F\, x_i = \sum a_{ij}x_j + \varepsilon_i + \sigma_i(x)\,\Lambda_i.$$

Here σ_i (x) is a matrix and Λ_i a random vector with components being stationary stochastic processes.
The equality is understood with probability 1, samplewise or in the p-mean sense (/18/).
If the realistic noise tends to white noise the solution of this equation tends to the Ito-process as solution of a stochastic differential equation of the form

$$dx = (\, F(x) + \alpha D(x)\,)\; dt + G(x)\; dW.$$

Here W is a vectorial Wiener process with in general correlated components and $0 \leq \alpha \leq 1$.

For $\alpha = 1/2$ it is the Stratonovich version of stochastic integrals.
By Ito's differentiation rule the function D can be obtained.
The relevance of this representation for the process depends on the closedness of the real noise to the white noise. Nikitin a.o. /19/ gave a mathematical foundation of this fact.
In general we get the following stochastic versions of the differential equation. Let the right side of the differential equations be a function of the process itself and a noise process:

$$F\,(x) + G\,(x)\,\xi$$

Then the following conclusions can be drawn (Arnold /8/):

. If ξ is a real noise (any ordinary stochastic process) with piecewise continuous sample paths, x must not neccessarily be Markovian. Then

$$dx/dt = F(x) + G(x)\xi$$

is a meaningful stochastic version of the given deterministic equation with different solutions depending on the interpretation of the equation to be fulfilled in probability 1, samplewise or in quadratic mean (/18/).

If ξ is a white noise, x is a diffusion process which is not differentiable.

Then the Ito differential equation

$$dx = (F(x) + \alpha D(x))\, dt + G(x)\, dW$$

is a meaningful stochastic differentiation rule.

From the Bartlett-Moyal equation follows

$$\frac{d}{dt} E_x x_t = E_x\, (F(x_t) + \alpha D(x_t)).$$

Such processes can be described by the corresponding infinitesimal operator L:

$$L = \sum (F + \alpha D)_i\, \partial / \partial x_i + \frac{1}{2} \sum_{i,j} (GG^T)_{ij}\, \partial^2 / \partial x_i\, \partial x_j .$$

Then via the result from halfgroup theory

$$\frac{d}{dt} E_x\, f(x_t) = L\, E_x\, f(x_t) = E_x L\, f(x_t)$$

the Kolmogoroff equations are obtained in the form

$$\frac{\partial}{\partial t} E_x f(x_t) = \sum_i (F + \alpha D)_i\, \frac{\partial}{\partial x_i} E_x f(x_t) + \frac{1}{2} \sum_{i,j} (GG^T)_{ij}\, \frac{\partial^2}{\partial x_i\, \partial x_j} E_x f(x_t),$$

$$\frac{\partial}{\partial t} p^u(t,x) = - \sum \frac{\partial}{\partial x_i} ((F + \alpha D)_i\, p^u) + \frac{1}{2} \sum_{i,j} \frac{\partial^2}{\partial x_i\, \partial x_j} ((GG^T)_{ij} p^u) .$$

Discussion of another example

1. Regular representation of a harmonic oscillation

$x_o(t) = A\,(\sin wt + a)$ with $a > 1$ as regularisation parameter.

We introduce the following auxiliary variables

$$x_1 = \frac{b + \cos wt}{a + \sin wt},\ x_2 = \frac{1}{a + \sin wt},\ x_3 = \frac{a + \sin wt}{b + \cos wt},\ x_4 = \frac{1}{b + \cos wt}$$

with another regularisation parameter $b > 1$.
The result are the following regular Volterra equations

$$F x_0 = w(x_1 - bx_2), \quad F x_1 = w (a x_4 - x_3) - F x_0,$$

$$F x_2 = - F x_0 , \quad F x_3 = - F x_1, \quad F x_4 = w (x_3 - a x_4).$$

We assume, that A and w are stochasticly disturbed by stationary stochastic processes (may be with bounded amplitudes)

$$A = A_0 + \alpha (t), \qquad w = w_0 + \Lambda (t).$$

One might think that it is enough to take into account these stochastic disturbances in the deterministic Volterra equations we derived, that means to propose the following randomized variant

$$F x_i = w_0 \sum \tilde{G}_{ij} x_j + \Lambda (t) \sum \tilde{G}_{ij} x_j .$$

But this result is wrong, because A has no influence at all.

Relevant stochastic equations we get by the proposed heuristic approach by applying our Structure Design Principle immediately on the stochastic trajectories. We assume bounds of $\alpha(t)$ and $\Lambda (t)$ in such a way, that always holds $A(t) \geqslant 0$ and $w(t) \geqslant 0$. After such a procedure we get the following randomized Volterra equations.

$$F x_0 = F A + x_1 - b x_2, \quad F x_1 = F(w + \mathring{w} t) + (ax_4 - x_3) - F x_0,$$

$$F x_2 = F(w + \mathring{w}t) - F x_0, \quad F x_3 = F(w + \mathring{w}t) + F x_0 - (ax_4 - x_3),$$

$$F x_4 = F(w + \mathring{w}t) + (x_3 - ax_4).$$

The x_i used here differ from the formerly used by the factor w.

References

/1/ Peschel M., Mende W.: Strukturentwurf für instationäre und nichtlineare Systeme, msr 10 (1981), 581-583, VEB Verlag Technik, Berlin

/2/ Verhulst P.F.: Nuov. Mem. Acad. Roy. Bruxelles 18,1 and 20,1,1847

/3/ Lotka A.J.: Proc. Natl. Acad. Scie. U.S., 6, 410, 1920

/4/ Volterra V.: Leçons sur la théorie mathématique de la lutte pour la vie, Gauthier - Villars, Paris 1931

/5/ Peschel M., Mende W.: Volterra-Modelle für nichtlineare und instationäre Systeme und deren Äquivalenztransformationen, msr 2 (1982), 102-104, VEB Verlag Technik, Berlin

/6/ Peschel M., Mende W.: Die Riccati-Darstellung und mögliche meßtechnische Konsequenzen, msr 11 (1982), VEB Verlag Technik, Berlin

/7/ Peschel M., Mende W.: Leben wir in einer Volterra-Welt? - Ein ökologischer Zugang zur angewandten Systemanalyse, Akademie-Verlag, Berlin 1983

/8/ Arnold L.: Qualitative theory of stochastic nonlinear systems, Series on Stochastic Nonlinear Systems, Springer-Verlag, Berlin-Heidelber-New York 1981

/9/ Arnold L.: A new example of an unstable system being stabilized by random parameter noise, Inf. Comm. of Math. Chem. 7, 1979, 133-140

/10/ Göldner K.: Mathematische Methoden der Regelungstechnik, VEB Fachbuch-Verlag, Leipzig, 4. Auflg. 1980

/11/ Peschel M., Köhler L.: Integration und Differentiation statischer Kennlinien mit statistischen Hilfsmitteln, msr 2 (1970), 64-65

/12/ Koch G.: Stochastic models in biology, to be published in EIK 1982

/13/ Prochorov J.W., Rosanov J.L.: Theorie der Wahrscheinlichkeit (russ.), Nauka 1973

/14/ Lamperti J.: Stochastic Processes: Survey of the mathem.theory, Springer, Berlin-Heidelber-New York 1977

/15/ Mc Garty T.P.: Stochastic Systems and State Estimation, Wiley 1974

/16/ Bailey N.T.J.: The Elements of Stoch. Processes with Appl. to the Natural Sciences, Wiley 1964

/17/ Kurtz T.G.: Approximation of discontinuous processes by continuous processes. Series Stoch. Nonlinear Systems, Springer, Berlin-Heidelberg-New York 1981

/18/ Bunke H.: Gewöhnliche Differentialgleichungen mit zufälligen Parametern, Akademie-Verlag 1972

/19/ Nikitin N.N., Rasewitsch W.D.: Über die Berechnung des Übertragungskoeffizienten eines Markowschen Prozesses nach der stochastischen Gleichung eines dynamischen Systems (russ.). Avtomatika i Telemechanika, 6, 1976

/20/ Ricciardi L.M.: Diffusion Processes and Related Topics in Biology. Lecture Notes in Biomath. 18, Springer, Berlin-Heidelberg-New York 1977.

ON TIME-DOMAIN IDENTIFICATION OF MULTIBODY SYSTEMS BY INSTRUMENTAL VARIABLE TECHNIQUE

R. G. Schwarz [1], P. C. Müller [2]

A time-domain identification method based on the instrumental variable technique is formulated to identify the physical parameters and the number of degrees of freedom of mechanical multibody systems using measurements of a complete set of displacements only (but no velocities nor accelerations). Explicit and recursive parameter-estimation formulas are presented. Particular attention is given to the influence of input noise and measurement noise; both being allowed to be coloured. The new identification algorithm has been tested so far for a system with 4 degrees of freedom and 36 parameters yielding very good results for this practical application.

1 Introduction and Problem Statement

Time-domain identification methods have a number of advantages compared with frequency-domain methods. But so far only very few time-domain methods especially suited for mechanical systems have been published and even fewer have been tested successfully. The present paper proposes a time-domain identification method based on the instrumental variable technique, especially adapted to linear time-invariant mechanical multibody systems

$$M\,\ddot{y}(t) + P\,\dot{y}(t) + Q\,y(t) = S\,(u(t) + w(t)). \tag{1}$$

Here, $y(t)$ means the f-dimensional displacement vector, $u(t)$ is the p-dimensional vector of exciting testsignals while $w(t)$ represents a p-dimensional vector of input noise. The matrices M, P, Q, S have to be of appropriate dimensions and represent the influences of inertia, damping and gyroscopic, stiffness and circulatory forces as well as the excitation forces, respectively.

Information about the dynamical behaviour of the mechanical system is obtained by noisy measurements of the complete set of displacements

---

1) Dr.-Ing. R. G. Schwarz, Rheinstr. 60,D 6805 Heddesheim, F.R.G.

2) Prof. Dr. P. C. Müller, Sicherheitstechnische Regelungs- und Meßtechnik, Universität -GH- Wuppertal, Gaußstr. 20,D5600 Wuppertal 1, F.R.G.

$$z(t) = H\,y(t) + v(t)\ , \tag{2}$$

where the $f \times f$ - matrix H is assumed to be regular and known, and v(t) represents the vector of measurement noise.

The identification problem consists in estimating the physical parameters of the mechanical system, i.e. the elements of M, P, Q, S by using suitable testsignals u(t) and processing the measured input and output data. More specific it is the aim to find an identification algorithm which works in the time domain, uses as little a priori knowledge as possible (for example it is not assumed that P, Q are symmetric), and defines the experimental conditions (testsignal u(t), stochastic behaviour of v(t), w(t)) which must be given to get consistent estimates of the unknown parameters. Furthermore, the algorithm must also allow the consistent estimation of the number f of degrees of freedom if it is not a priori known.

This problem is discussed in a comprehensive investigation of the first author /1/. The critical survey on the present state of identification techniques leads to the conclusion to adapt the instrumental variable technique from state space systems to mechanical systems. To do so, a new treatment of the mechanical identification problem is essential to obtain a linear estimation problem suited for the instrumental variable method. With this special representation and based on the structural properties of multibody systems the general conditions for the identifiability of the parameters can be shown, the experimental requirements for consistent estimates can be defined, and an efficient identification algorithm can be formulated. The present paper is based on /1/ showing the most important results for the practical application in the identification of mechanical systems.

2 Results on Identifiability

Identifiability considers the existence, uniqueness and consistency of the estimated parameter matrices M, P, Q and S in the assumed model structure when processing known signals y(t), u(t) in the case of vanishing noise w(t), v(t). A first result is the following /1,2/.

Theorem 1: The system matrices M, P, Q, S are uniquely identifiable independent of the special choice of measurements z(t) = y(t) if and only if

(i) system (1) is completely controllable,

(ii) one of the matrices M, P, Q, S is known and regular.

Both conditions are obviously meaningful. Condition (i) guarantees that each eigenmode of the system will be excited; condition (ii) requires the knowledge of certain physical parameters in the space of all parameters to define the physical representation of (1). One should notice that Theorem 1 is independent of the special identification technique.

The second result sums up all the experimental conditions one has to care for when using the proposed instrumental variable technique.

Theorem 2: Assuming Theorem 1 and measurements $u(k\tau)$, $z(k\tau)$, $k = 0,1,2...$, at equally spaced time-instants, then M, P, Q, S are consistently identifiable by the proposed instrumental variable technique if and only if

(i) the system (1) (with $u(t) = 0$, $w(t) = 0$) is asymptotically stable,

(ii) the exciting testsignal $u(t)$ is persistent and piecewise constant in the time-intervals $(k-1)\tau \leq t < k\tau$,

(iii) the sampling time τ satisfies

$$\tau < \frac{\pi}{2\omega_{max}} , \tag{3}$$

$$\lambda_i \neq \lambda_j \Rightarrow e^{\lambda_i \tau} \neq e^{\lambda_j \tau} , \tag{4}$$

where ω_{max} is the largest eigenfrequency of (1) and λ_k are the eigenvalues of (1),

(iv) the input and output noise processes $w(t)$, $v(t)$ are
- unbiased,
- uncorrelated with $u(t)$,
- correlated only on a finite time interval.

Particularly, it should be noted that the input and output noise processes have not to be white but can be coloured. Furthermore, the condition (ii) of Theorem 2 will be satisfied using pseudo-random binary sequences (PRBS) of sufficient length.

3 Towards a Linear Estimation Problem

The identification problem described in section 1 represents a nonlinear estimation problem because in (1) the unknown elements of P are multiplied by the unkown element $\dot{y}$ of the state vector $\begin{bmatrix} y \\ \dot{y} \end{bmatrix}$. To overcome this difficulty it is shown in /1/ that the nonlinear continuous-time estimation problem (1,2) can be transformed into a linear discrete-time estimation problem:

$$z_{k+2} + P_d\, z_{k+1} + Q_d\, z_k = S_{1d}\, u_{k+1} + S_{2d}\, u_k + e_{k+2} \tag{5}$$

where

$$z_k = z\,(k\tau)\ ,\ u_k = u\,(k\tau) \tag{6}$$

and

$$e_{k+2} = v_{k+2} + P_d\, v_{k+1} + Q_d\, v_k + S_{1d}\, w_{k+1} + S_{2d}\, w_k , \tag{7}$$

$$v_k = v(k\tau)\ ,\ w_k = w(k\tau) . \tag{8}$$

Theorem 3: Assuming measurements (6) for $k = 0,1,2,...$ at equally spaced time-instants where the testsignal $u(t)$ as well as the noise input $w(t)$ are piecewise constant on the time-intervals $(k-1)\tau \leq t < k\tau$, then the measurements $z(k\tau)$ of (1,2) are equivalently represented by the discrete -time difference equation (5) with certain matrices P_d, Q_d, S_{1d}, S_{2d} depending on P, Q, S_1, S_2 and sampling time τ .

Remark 1: The parameter identification of P_d, Q_d, S_{1d}, S_{2d} by measurements (6) represents a linear estimation problem.

Remark 2: If the noise input $w(t)$ is not piecewise constant, which is the case in real problems, then the stochastic process e_{k+2} of the error of the difference equation (5) has some modified statistical properties which do not influence the identification problem if the instrumental variable technique is used.

In the next section the linear estimation problem (5,6) will be solved yielding estimates $\hat{P}_d$, $\hat{Q}_d$, $\hat{S}_{1d}$, $\hat{S}_{2d}$. Once these estimates are given the question arises how to obtain the interesting parameter matrices M, P, Q, S of the continuous-time system (1). To get these, a transformation back to the continuous-time representation is necessary. In a first step a state space description is used,

$$F = \begin{bmatrix} 0 & I_f \\ -\hat{Q}_d & -\hat{P}_d \end{bmatrix} \quad , \quad G = \begin{bmatrix} \hat{S}_{1d} \\ \hat{S}_{2d} - \hat{P}_d\, \hat{S}_{1d} \end{bmatrix} . \tag{9}$$

The next step is the transformation back to continuous-time. Here the matrix logarithmic function is used expressed by the areatangens function, which has better convergence properties when expanding into a matrix power series for numerical evaluation. This power series converges if the sampling time τ fulfils condition (3):

$$A = \begin{bmatrix} A_{11} & A_{12} \\ A_{21} & A_{22} \end{bmatrix} = \frac{1}{\tau} \ln F = \frac{2}{\tau} \text{Arth } (F+I_{2f})^{-1} (F-I_{2f}), \tag{10}$$

$$B = \begin{bmatrix} B_1 \\ B_2 \end{bmatrix} = (F-I_{2f})^{-1} A G . \tag{11}$$

Using the submatrices of A and B the estimates of the parameter matrices of (1) can be determined.

Theorem 4: Assuming condition (ii) of Theorem 1 then the estimates of the physical parameters of (1) are uniquely defined by (9-11) and e.g. by

(i) M regular, known:

$$\begin{aligned} \hat{P} &= - M H^{-1} (A_{11} + A_{12} A_{22} A_{12}^{-1}) H , \\ \hat{Q} &= \quad M H^{-1} (- A_{12} A_{21} + A_{12} A_{22} A_{12}^{-1} A_{11}) H , \\ \hat{S} &= \quad M H^{-1} (A_{12} B_2 - A_{12} A_{22} A_{12}^{-1} B_1) ; \end{aligned} \tag{12}$$

(ii) Q regular, known:

$$\begin{aligned} \hat{M} &= Q H^{-1} [- A_{12} A_{21} + A_{12} A_{22} A_{12}^{-1} A_{11}]^{-1} H , \\ \hat{P} &= - Q H^{-1} [- A_{12} A_{21}+A_{12} A_{22} A_{12}^{-1} A_{11}]^{-1} (A_{11}+A_{12}A_{22}A_{12}^{-1})H, \\ \hat{S} &= Q H^{-1} [-A_{12} A_{21}+A_{12}A_{22}A_{12}^{-1}A_{11}]^{-1} (A_{12}B_2-A_{12}A_{22}A_{12}^{-1} B_1) . \end{aligned} \tag{13}$$

In the case when P or S are regular and known then similar results are obtained.

Remark 3: The submatrix B_1 has to be the zero matrix. Therefore, the numerical smallness of B_1 can be used to check the validity of the results when applying the proposed identification technique. When performing these transformations no numerical difficulties were observed, neither was any noticeable error in the original parameters caused.

As result of this section the original identification problem (1,2) is transformed into the linear estimation problem (5,6).

4 Identification Algorithm

To choose an appropriate estimator for (5,6), one has to appraise the statistical properties of the equation error e_{k+2} , cf. Theorem 2 (iv) and Remark 2. Using the instrumental variable technique only the weak restrictions (iv) of Theorem 2 for the input and output noise processes have to be considered. That leads to the following estimator, cf. /1/.

An estimation of the unknown parameters combined in one matrix

$$X = [P_d \; Q_d \; S_{1d} \; S_{2d}] \tag{14}$$

based on (N+1) sampling points is given by

$$\hat{X}_N = Z_N H_N^T \; (D_N H_N^T)^{-1} \; . \tag{15}$$

Here, D_N and Z_N are matrices summarizing all measured input and output data:

$$D_N = [d_1 \; d_2 \; \dots \; d_N] \; , \; d_k = \begin{bmatrix} -z_k \\ -z_{k-1} \\ u_k \\ u_{k-1} \end{bmatrix} \; , \tag{16}$$

$$Z_N = [z_2 \; z_3 \dots \; z_{N+1}] \; . \tag{17}$$

The matrix H_N contains data of the instrumental variables y_H ,

$$H_N = [h_1 \; h_2 \; .. \; h_N] \; , \quad h_k = \begin{bmatrix} -y_{H,k} \\ -y_{H,k-1} \\ u_k \\ u_{k-1} \end{bmatrix} \; , \tag{18}$$

generated by an auxiliary model.
This auxiliary model has the same structure as (5) and is driven by the testsignal:

$$y_{H,k+2} + P_H \, y_{H,k+1} + Q_H \, y_{H,k} = S_{1H} \, u_{k+1} + S_{2H} \, u_k \; . \tag{19}$$

It must be asymptotically stable and completely controllable. From the theoretical point of view the matrices P_H, Q_H, S_{1H}, S_{2H} can be chosen quite arbitrarily, but to get an efficient estimator these matrices should approximate the matrices P_d, Q_d, S_{1d}, S_{2d} as good as possible. Therefore, in the following recursive version of the estimator (15-19) an adaptive auxiliary model (19) is used.

<u>Theorem 5</u>: The recursive instrumental variable estimator is given by

$$\hat{X}_{N+1} = \hat{X}_N + (z_{N+2} - \hat{X}_N \, d_{N+1}) k_{N+1}^T \; , \tag{20}$$

$$k_{N+1}^T = \frac{1}{1 + h_{N+1}^T \, \Phi_N \, d_{N+1}} \; h_{N+1}^T \, \Phi_N \; , \tag{21}$$

$$\Phi_{N+1} = \Phi_N - \Phi_N\, d_{N+1}\, k^T_{N+1} \ . \tag{22}$$

It looks very much like a Kalman filter, but it differs in the details. The estimation is updated by a weighted prediction error. One should note that Φ_N is not the covariance matrix of the estimation error but $\Phi_N = (D_N\, H^T_N)^{-1}$.

As already mentioned, to increase efficieny it is useful to have an adaptive auxiliary model (19) realized for example by a low pass filter driven by the estimation with some time delay ν :

$$x_{H,N} = (1 - \delta)x_{H,N-1} + \delta\, \hat{x}_{N-\nu} \quad , \quad 0 \leq \delta < 1 \ . \tag{23}$$

The initial values of the recursive estimator may be generated by an explicit estimator (16-19) or may be chosen arbitrarily (but meaningfully).

The proposed estimator consistently identifies the parameters of the discrete-time system (5) leading to consistent estimates of the physical parameters of the continuous-time system (1) according to (12-13). Particularly, the estimates are unbiased under weak conditions.

This identification algorithm has been tested for a centrifuge system with 4 degrees of freedom and 36 parameters /1/. Processing of simulated data verifies the theoretical considerations. The influence of input and measurement noise proves to be as predicted by the theory. Then, hardware measurements, recorded during a laboratory scale experiment with a real centrifuge, were used to identify the physical parameters of this system. To give an impression from this practical application the realized testsignal u(t) is shown in Fig. 1. Although the signal is not piecewise constant as assumed by the theory, the proposed identification algorithm yields very

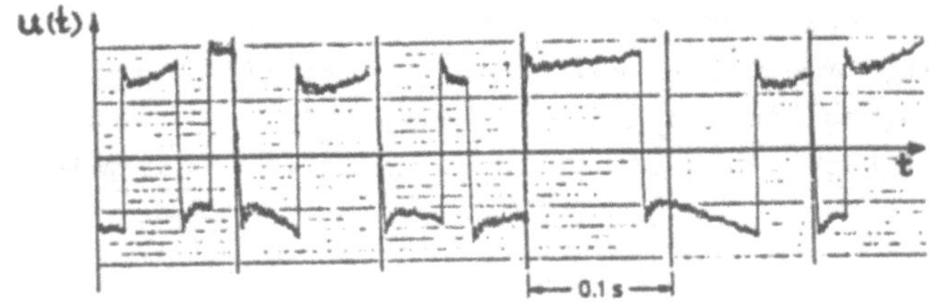

Fig. 1. Testsignal u(t) in the experiment

good results. A typical result is shown in Fig. 2 . In this picture the absolute error of one parameter (here P_{d22}) is plotted over the number N of sampling points.

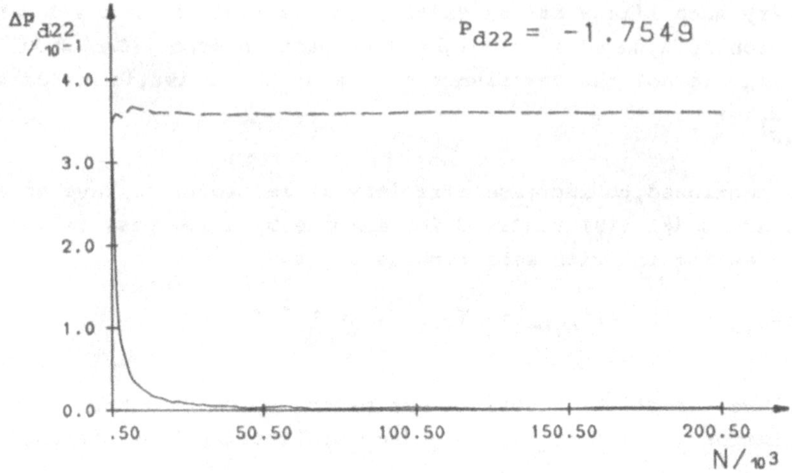

Fig. 2 . Absolute error ΔP_{d22} of the estimate of P_{d22} over a number N of sampling points using an instrumental variable estimator (full line) or a least squares estimator (dotted line)

This plot shows the convergence of the instrumental variable estimator as well as the different consistency properties of the estimator(full line) and of a least squares estimator (dotted line). The last one shows a large constant bias while in the case of the proposed estimator (20-22) the error disappears very quickly. It should be noted that the recursive estimator (20-22) was started in this case by the least squares estimation based on 500 sampling points.

In /1/ further points of interest are considered in this application. The investigation of the convergence behaviour dependent on input and output noise levels demonstrates that the number of needed sampling points may increase with growing levels but even for high noise levels the estimator converges and gives small errors in the parameters. Also the advantages of an adaptive auxiliary model (19, 23) could be clarified showing that such a model smoothes curves like in Fig. 2 considerably and leads to much faster convergence. Altogether the proposed identification method leads to very good results.

5 StructureIdentification Method

The parameter identification algorithm discussed in section 4 needs the knowledge of the number f of degrees of freedom. In the case of the above-mentioned centrifuge problem the number f of that multibody system is easily determined. But there are many mechanical systems where the number of degrees of freedom is not known in advance, e.g. think of mechanical systems with elastic parts which are modeled approximately by a system (1). In those cases it is necessary to estimate the number of degrees of freedom before applying the identification algorithm of section 4 to determine the parameters.

This problem of structure identification can be solved by the extended information matrix $D_N H_N^T$, that was already used in the parameter estimation formula (15), cf. /1/. It is assumed that the number $\hat{f}$ of measurements $z(t)$ (2) is not less than the unknown number f of degrees of freedom. Then from the upper left block matrix of $D_N H_N^T$ the matrix

$$\Psi_N = \frac{1}{N} \sum_{k=o}^{N-1} z_{k+1} \, y_{H,k+1}^T \tag{24}$$

is defined. It has the property that - assuming Theorem 2 - Ψ_N converges in probability to a matrix Ψ

$$\operatorname*{plim}_{N \to \infty} \Psi_N = \Psi , \tag{25}$$

with

$$\operatorname{rank} \Psi = \begin{cases} f & \text{if} \quad f \leq f_H = \hat{f} , \\ f_H & \text{if} \quad f > f_H = \hat{f} , \end{cases} \tag{26}$$

where f_H is the number of degrees of freedom of the auxiliary model. One has to notice that based on the number $\hat{f}$ of measurements the number f_H of instrumental variables y_H has to coincide with $\hat{f}$.

Because (26) must not hold for Ψ_N in the case of a finite number N of samples, we change to the check of the ratios of principal minors. Defining the principal minors of Ψ_N ,

$$D_i = \det [\Psi_{N,kj}] , \quad k,j = 1,..,i, \tag{27}$$

the ratios

$$DV_i = \frac{D_i}{D_{i+1}} , \quad i = 1,..,\hat{f} - 1 , \tag{28}$$

are used to determine the number f. The ratios DV_i change only a little if the new row of D_{i+1} compared with D_i is linearly independent of the preceding rows, but they vary essentially if the new row depends approximately on the other rows. In this later case the dependent row (it is the (i+1)th row of Ψ_N) and the corresponding column of Ψ_N has to be eliminated. After eliminating a row, the procedure of calculating the ratios (28) continues with the reduced matrix Ψ_N'. In this way all of the $\hat{f}$ coordinates of the measurement vectors are discussed. The number of the remaining coordinates determines the number f of degrees of freedom of the mechanical system.

<u>Theorem 6</u>: The number f of degrees of freedom is equal to the number of remaining measurement coordinates using the selection procedure of the ratios of principal minors (28).

By the aid of this number f and of the corresponding reduced f-dimensional measurement vector of the remaining coordinates the parameter ieentification method of section 4 can be put into operation.

Applying this result to the above-mentioned centrifuge problem of 4 degrees of freedom, it is for illustration assumed to start with 5 measurements. Then a typical plot based on hardware measurements of the ratios (28) looks like Fig. 3 from which the real number of degrees of freedom is identified to be 4. How large the ratio DV_4 increases depends on the number N of samplings and on the noise levels. The influences of these two quantities are shown in Figs.4 und 5.

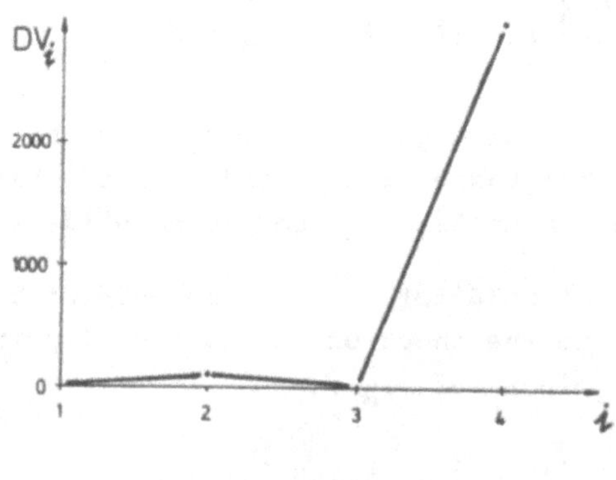

Fig. 3. Ratios DV_i

The experimental results show that the number f of degrees of freedom can be identified quite well by the instrumental variable technique.

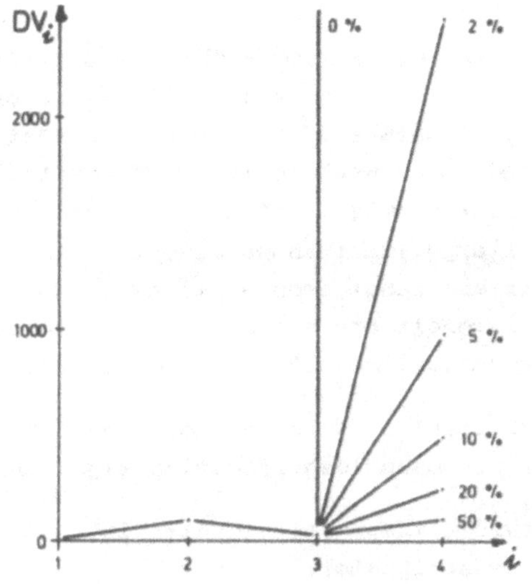

Fig. 4. Influence of the measurement noise level on the ratios DV_i in the case of 10 000 samples

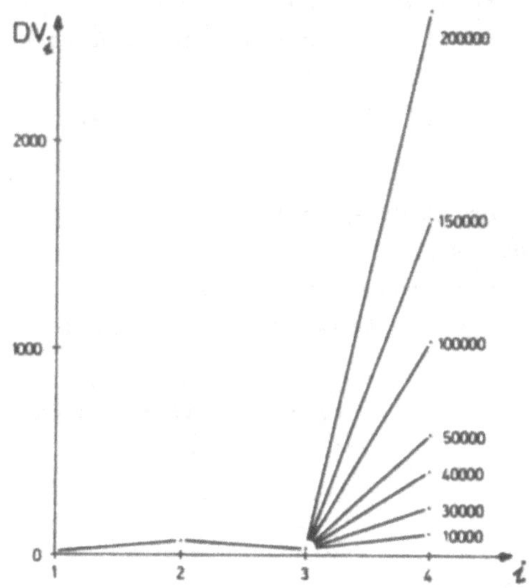

Fig. 5. Influence of the number N of samples on the ratios DV_i in the case of a measurement noise level of 5o %

6 Outlook

The proposed instrumental variable identification algorithm is a very powerful method to estimate the parameters as well as the order of mechanical systems. One essential advantage of the method is that consistent estimations can be guaranteed under weak restrictions on the input and output noise processes. Explicit knowledge of noise covariances are not necessary. The experimental experience shows also that the testsignal is allowed to differ in a certain range from a piecewise constant signal. Therefore, the experimental conditions are realistic. The practical application verifies the theoretical results in an outstanding manner.

Despite the successful results work is going on for further refinements and extensions of the proposed identification algorithm:

(i) piecewise linear testsignals (instead of piecewise constant testsignals);

(ii) more general type of measurements: f linear independent measurements of displacements, velocities or accelerations:

$$z(t) = H_1\, y(t) + H_2\, \dot{y}(t) + H_3\, \ddot{y}(t) + v\,(t)\ ,$$

$$\text{rank}\ [H_1\ H_2\ H_3] = f\ ;$$

(iii) refinement of the identification of the number f of degrees of freedom in the case of elastic structures and hybrid multibody systems with respect to an interesting frequency range of the elastic modes.

Other open questions refer to the generalization of the proposed identification algorithm to general dynamical systems when only subsets of the state variables are measured. Then again a nonlinear estimation problem is given and the question arises whether it is possible to find a suitable transformation into a linear estimation problem analog to the methods of section 3.

The authors would appreciate it very much if these open questions will stimulate further research.

References

/1/ Schwarz, R.G.: Identifikation mechanischer Mehrkörpersysteme. Fortschr.-Ber. VDI-Z. Reihe 8, Nr. 30, Düsseldorf 1980.

/2/ Schwarz, R.G.: Parameteridentifizierbarkeit linearer mechanischer Mehrkörpgersysteme. Z. Angew. Math. Mech. 60 (1980) T 79 - T 81.

FAST ALGORITHMS IN THE PARAMETER IDENTIFICATION OF DYNAMIC SYSTEMS

Walter Wedig[1]

In this paper we discuss second order moment properties of structural response processes for determining the stiffness and damping parameters of elastic systems. In the special case of stationary white noise excitations and simply damped structures, the symmetric stiffness matrix is already measurable by the mean squares of the displacement and velocity processes without any knowledge on the intensities or dependences of the excitations or where they are applied. To extend such estimations to general linear systems driven by non-white noise, we generate additional state processes the squared means of which allow to identify the system parameters as well as the excitation parameters.

1. Introduction

For the development of ambient response monitoring systems we have to utilize the natural excitations of elastic systems coming up from atmospheric turbulent flows or randomly profiled roads and rails. In a first step the excitations are assumed to be zero-mean broad-band processes with unknown intensities and statistical dependences. Since it is also unknown where such excitations are applied, we are looking for identification procedures which need only the measuring of the response processes in order to estimate the stiffness and damping parameters of the elastic structure.

The basic idea of such identification procedures can be reduced to the investigation of the linear oscillator under white noise $\dot{W}_t$ described by

$$\ddot{Z}_t + 2D\omega_1 \dot{Z}_t + \omega_1^2 Z_t = \frac{\sigma}{m} \dot{W}_t. \tag{1}$$

Herein the scalar processes Z_t and $\dot{Z}_t$ denote the displacement and the velocity, respectively. The parameter ω_1 is the natural frequency of the oscillator. D is its damping ratio and σ represents an excitation intensity related to the mass m of the system. In the stationary case the mean squares of both response processes satisfy the well known relation

$$E(\dot{Z}_t^2)/E(Z_t^2) = \omega_1^2. \tag{2}$$

Hence, the oscillator frequency ω_1 can be calculated from mean square measurements of the response without any knowledge on the damping ratio or on the intensity of the zero-mean white noise excitation [1].

---

[1] Prof. Dr.-Ing. W. Wedig, University of Karlsruhe, D-7500 Karlsruhe 1, F.R.G.

It is worth noting that numerical evaluations of the squared means of measured signals are simply and rapidly performable according to the ergodic theorem of stationary random processes. First applications of a moment technique for system parameter identification have been introduced by F. Kozin and T.S. Lee [2,3] in an extended measuring procedure including the acceleration and excitation processes. In contrast to [3], we are looking for identification equations restricted to the measurement of only the state processes. Particularly, we extend such second order estimations by means of so-called generators in order to identify the excitation intensities and their frequency distributions, as well.

2. Basic frequency relations

To introduce the basic frequency relations of mechanical structures subjected to stationary white noise $\dot{W}_t$ with the intensity σ, we first consider the special dynamic system of the form

$$\ddot{\underline{X}}_t + \frac{b}{m}\dot{\underline{X}}_t + \frac{1}{m} C \underline{X}_t = \underline{r}\frac{\sigma}{m}\dot{W}_t, \qquad E(\dot{W}_t\dot{W}_s) = \delta(t-s). \tag{3}$$

Herein, $\underline{X}_t$ is the n-dimensional displacement vector, $\underline{r}$ is the corresponding excitation vector and C denotes the stiffness matrix with nxn symmetric elements. The scalar parameters m and b characterize a homogeneous distribution of the system mass and damping, respectively. Notice that (3) is a special important case in structural mechanics.

Next, we perform a modal analysis in order to transform the structural equation of motion (3) into the uncoupled form of the principal coordinates. For this purpose, we set up the associated eigenvalue problem. Solving it, we obtain n eigenfrequencies ω_k and eigenvectors x_k which are then compiled to the diagonal frequency matrix Ω and the modal matrix X, respectively.

$$X = (x_{ik}), \qquad \Omega = \mathrm{diag}(\omega_k^2) = \frac{1}{m} X^{-1} C X. \tag{4}$$

The modal matrix X allows the principal transformation of the form

$$\underline{X}_t = X \underline{Z}_t, \qquad \ddot{\underline{Z}}_t + \frac{b}{m}\dot{\underline{Z}}_t + \Omega \underline{Z}_t = \underline{a}\frac{\sigma}{m}\dot{W}_t \tag{5}$$

where $\underline{Z}_t$ is the n-dimensional vector of the principal coordinates and the modal excitation vector is given by $\underline{a} = X^{-1}\underline{r}$.

The diagonal form (5) is well posed for the covariance analysis. As already shown in [1], we calculate the stationary mean squares of all processes $Z_t = (Z_{1t}, Z_{2t}, \dots Z_{nt})$ to

$$E(Z_{it}Z_{jt}) = \frac{\sigma^2}{2bm} 2\Delta_{ij}, \qquad E(\dot{Z}_{it}\dot{Z}_{jt}) = \frac{\sigma^2}{2bm} \Delta_{ij}(\omega_i^2 + \omega_j^2),$$
$$\Delta_{ij} = a_i a_j [\omega_i^2 + \omega_j^2 + \varkappa(\omega_i^2 - \omega_j^2)^2]^{-1}, \qquad \varkappa = m^2/2b^2. \tag{6}$$

From this result it follows the basic frequency relation

$$E(\dot{Z}_{it}\dot{Z}_{jt}) = \frac{1}{2} (\omega_i^2 + \omega_j^2) E(Z_{it}Z_{jt}), \qquad (i,j = 1,\dots n) \tag{7}$$

which is an extended version of (2). Consequently, for the special oscillating system (3), it seems to be also possible to calculate the natural frequencies ω_i by the formula (7) provided all covariances (6) are known by any suitable measuring of the response of the elastic structure.

To perform such measurements, it is suitable to go over to the matrix form which is equivalent to (7)

$$E(\dot{\underline{Z}}_t\dot{\underline{Z}}_t^T) = \frac{1}{2}[\Omega E(\underline{Z}_t\underline{Z}_t^T) + E(\underline{Z}_t\underline{Z}_t^T)\,\Omega^T] \tag{8}$$

and multiply (8) by the modal matrix X from the left side and by its transposed version X^T from the right-hand side. Subsequently, we use the transformation $\underline{X}_t = X\,\underline{Z}_t$ and $X^{-1}\,C\,X = m\,\Omega$, already noted in (4), in arriving finally at the following basic identification equation:

$$2m\, E(\dot{\underline{X}}_t\dot{\underline{X}}_t^T) = C\, E(\underline{X}_t\underline{X}_t^T) + E(\underline{X}_t\underline{X}_t^T)\, C^T. \tag{9}$$

Note that (9) has the form of a Ljapunow matrix equation determining the unknown stiffness matrix $C = C^T$ provided that (9) is solvable and the response processes $\underline{X}_t$ and $\dot{\underline{X}}_t$ are stationary. In contrast to (7), we now have the possibility to measure the covariance coefficients of the matrix equation (9) since they are representing the mean squares of natural response processes. In particular, such estimations are performable without any knowledge of the special damping b in (3) and the excitation distribution in $\underline{r}$.

3. Measuring of stiffness matrices

Naturally, it is possible to generalize the basic equation (9) to a more general version of (3) given by

$$M\ddot{\underset{\sim}{X}}_t + D\dot{\underset{\sim}{X}}_t + C\underset{\sim}{X}_t = \underset{\sim}{\sigma}\dot{W}_t, \qquad \underset{\sim}{\sigma}\underset{\sim}{\sigma}^T = (\sigma_{ij}), \quad (i,j = 1,..n). \tag{10}$$

Herein, M is a positiv definit mass matrix, D and C are damping and stiffness matrices, respectively, with the symmetric restrictions $D = D^T$ and $C = C^T$. As known, an excitation by n statistically dependent white noise components can be reduced to the n-dimensional intensity vector $\underset{\sim}{\sigma}$ multiplied by scalar white noise $\dot{W}_t$ such that the matrix (σ_{ij}) gives the auto- and cross-spectra of all white noise excitations.

We rewrite (10) into the form of a first order system and apply the Ljapunow equation technique [4] in order to set up the associated four covariance equations. Under the stationary condition $\dot{E}(.) = 0$, we then obtain algebraic equations from which we finally derive the following two matrix equations [1]:

$$\begin{aligned} C\,E(\underset{\sim}{X}_t\underset{\sim}{X}_t^T) + E(\underset{\sim}{X}_t\underset{\sim}{X}_t^T)\,C^T + D\,E(\dot{\underset{\sim}{X}}_t\underset{\sim}{X}_t^T) - E(\dot{\underset{\sim}{X}}_t\underset{\sim}{X}_t^T)\,D^T &= \\ = M\,E(\dot{\underset{\sim}{X}}_t\dot{\underset{\sim}{X}}_t^T) + E(\dot{\underset{\sim}{X}}_t\dot{\underset{\sim}{X}}_t^T)\,M^T&, \end{aligned} \tag{11}$$

$$M\,E(\dot{\underset{\sim}{X}}_t\underset{\sim}{X}_t^T)\,C^T - C\,E(\dot{\underset{\sim}{X}}_t\underset{\sim}{X}_t^T)\,M^T + D\,E(\dot{\underset{\sim}{X}}_t\dot{\underset{\sim}{X}}_t^T)\,M^T + M\,E(\dot{\underset{\sim}{X}}_t\dot{\underset{\sim}{X}}_t^T)\,D^T = \underset{\sim}{\sigma}\underset{\sim}{\sigma}^T.$$

Instead of (9), there are now two coupled matrix equations of the Ljapunow type which are sufficient to calculate the stiffness matrix C and the damping matrix D, as well provided that the mass matrix M is given and both equations in (11) are solvable. Furthermore, we have to require that all displacement and velocity processes of (10) are stationary so that their covariance matrices in (11) are measurable.

The identification procedure (11) has two significant disadvantages. First, it is restricted to symmetric matrices C and D. Second, we have to measure the spectral excitation matrix (σ_{ij}) in addition. However, this last disadvantage can be avoided provided that the structure is uniformly damped by $D = b\,I$ where I denotes the identity matrix and b is a scalar damping coefficient. Obviously, this simplification leads to the matrix equation

$$C\,E(\underset{\sim}{X}_t\underset{\sim}{X}_t^T) + E(\underset{\sim}{X}_t\underset{\sim}{X}_t^T)\,C^T = M\,E(\dot{\underset{\sim}{X}}_t\dot{\underset{\sim}{X}}_t^T) + E(\dot{\underset{\sim}{X}}_t\dot{\underset{\sim}{X}}_t^T)\,M^T. \tag{12}$$

This is a slightly generalized version of (9) which has been previously derived by means of the real-valued modal analysis.

An example of those cases shows figure 1 representing a one-dimensional vibration chain of two degrees of freedom with a uniform mass and damping distribution and two different random excitations.

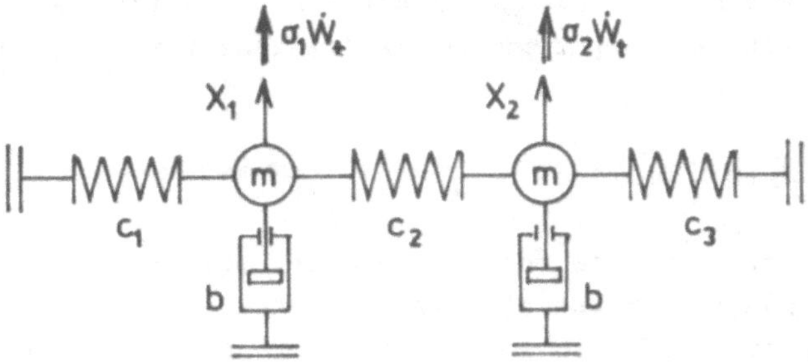

Fig. 1: One-dimensional vibration chain

The corresponding identification equation (12) is evaluated in the following vector form:

$$\begin{pmatrix} 2\,k_{11} & 2\,k_{12} & 0 \\ k_{12} & k_{11}+k_{22} & k_{12} \\ 0 & 2\,k_{12} & 2\,k_{22} \end{pmatrix} \begin{pmatrix} c_{11} \\ c_{12} \\ c_{22} \end{pmatrix} = 2\,m \begin{pmatrix} \dot{k}_{11} \\ \dot{k}_{12} \\ \dot{k}_{22} \end{pmatrix} . \qquad (13)$$

Taking into account that the covariance matrices $(k_{ij}) = E(\underline{X}_t\underline{X}_t^T)$ and $(\dot{k}_{ij}) = E(\dot{\underline{X}}_t\dot{\underline{X}}_t^T)$ are symmetric, the fourth equation of (12) can not be applied because of its linear dependence. Therefore, the evaluation of (12) is restricted to the symmetric case $c_{ij} = c_{ji}$ for $i,j = 1,2,..,n$.

The vector equation (13) is solvable if its determinant Δ is not vanishing. From the theory of stationary gaussian processes [5], we further know that the covariance matrix (k_{ij}) is positiv definit in the stationary case. Consequently, the conditions

$$\Delta = 4(k_{11}+k_{22})(k_{11}k_{22}-k_{12}^2) \neq 0, \qquad k_{11}k_{22}-k_{12}^2 > 0 \qquad (14)$$

have to be satisfied in order to measure the stiffness coefficients c_{ij} of the two-degree-of-freedom system represented in Fig. 1.

In [1] we noted some measurements in case of three degrees of freedom using electric noise generators and analogue computers for the experimental simulation. Furthermore, we showed in [1] that the parameter identification can efficiently be improved by application of least squares estimations provided that there is a band structure in the stiffness matrix C or more specially, if the elastic system possesses a uniform stiffness distribution. Finally, we gave three significant examples of non-measurable systems and the associated physical interpretations that those structures have coinciding eigenfrequencies, a special node excitation or a loss of asymptotic stability arising from an unsufficient fixing of the structure.

4. Measuring of system matrices

It is now important to extend the second order moment identification to general damping and stiffness matrices without the restrictions of (12). To show the basic idea of such an extension, we first consider the linear oscillator excited by white noise $\dot{W}_t$. Its analogue simulation is represented on the left side of figure 2. On the right-hand side we have indicated the digital mean square evaluation of the measured signals X_t and $\dot{X}_t$. In order to obtain now further informations from X_t, we apply a low-pass filter which generates a new state process Y_t.

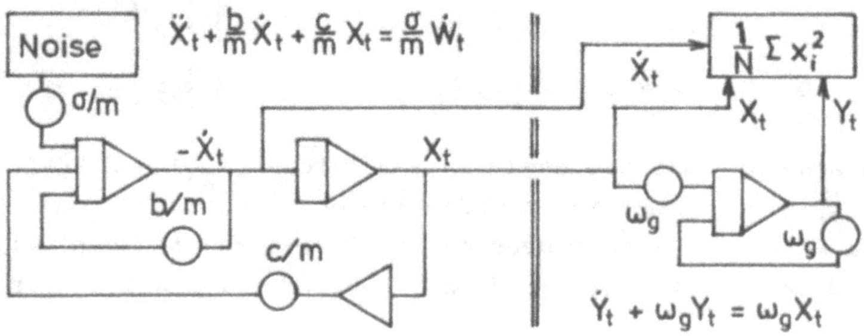

Fig. 2: Linear oscillator with post-filtering of X_t

We emphasize that the only objective of this post-filtering is to obtain an arbitrary third state process Y_t allowing to evaluate further second order moments. Therefore, the limiting frequency ω_g of Y_t can arbitrarily be chosen in the stable range $0 < \omega_g < \infty$. Calculating the squared means of all processes in Fig. 2, we obtain

$$E(X_t^2) = \frac{\sigma^2}{2bc}, \qquad E(Y_t^2) = E(X_tY_t) = \frac{\sigma^2}{2bc}\frac{\omega_g}{\Delta}(\omega_g + \frac{b}{m}), \tag{15}$$

$$E(\dot{X}_t^2) = \frac{\sigma^2}{2bm}, \qquad E(\dot{X}_tY_t) = -\frac{\sigma^2}{2bm}\frac{\omega_g}{\Delta}, \quad \Delta = \frac{c}{m} + \omega_g(\omega_g + \frac{b}{m}).$$

This results obviously in the parameter estimation

$$\frac{c}{m} = \frac{E(\dot{X}_t^2)}{E(X_t^2)}, \qquad \frac{b}{m} = -\omega_g - \frac{E(\dot{X}_t^2)\ E(X_tY_t)}{E(X_t^2)\ E(\dot{X}_tY_t)} \tag{16}$$

by which we are able to calculate the stiffness and the damping coefficient of the oscillator from the measured data of the response processes without any knowledge on the intensity σ of the noise excitation.

To get a general extension of the estimation prescription (16), we next consider a dynamic system of the first order form

$$\dot{\underline{X}}_t = A\ \underline{X}_t + \underline{\sigma}\ \dot{W}_t, \qquad \dot{\underline{Y}}_t = -\omega_g\ \underline{Y}_t + \omega_g\ \underline{X}_t. \tag{17}$$

Herein, $\underline{X}_t$ is an n-dimensional state vector of the system, A denotes an nxn matrix with constant elements and the vector $\underline{\sigma}$ contains the intensities of the n-dimensional excitation by scalar white noise $\dot{W}_t$. Assuming that all n state processes X_{it} are stationary and measurable, we can apply to each component of $\underline{X}_t$ a low-pass filter with the same limiting frequency ω_g generating n new state processes Y_{it} for i = 1,2,..,n. Now, the second order moments of both, the system processes and the filter processes, are sufficient to estimate all elements of the unknown matrix A as well as the intensity matrix (σ_{ij}) of the white noise excitation. To derive the corresponding identification procedure, we interpret the equations (17) in their Itô forms [6] and set up the increment $d(\underline{X}_t\underline{Y}_t^T)$ of both vector processes $\underline{X}_t$ and $\underline{Y}_t$:

$$d(\underline{X}_t\underline{Y}_t^T) = (A\underline{X}_t dt + \underline{\sigma}\, dW_t)\ \underline{Y}_t^T + \underline{X}_t(-\omega_g\underline{Y}_t^T + \omega_g\underline{X}_t^T)dt. \tag{18}$$

Then, we take the expectation in (18) under consideration that the increment dW_t of the zero-mean Wiener process W_t is independent on all state processes in (17):

$$\dot{E}(\underline{X}_t\underline{Y}_t^T) = A\ E(\underline{X}_t\underline{Y}_t^T) - \omega_g E(\underline{X}_t\underline{Y}_t^T) + \omega_g E(\underline{X}_t\underline{X}_t^T). \tag{19}$$

In conclusion, we go over to the stationary case $\dot{E}(\underline{X}_t\underline{Y}_t^T) = 0$ leading to

$$A = \omega_g[I - E(\underline{X}_t\underline{X}_t^T)\ E(\underline{X}_t\underline{Y}_t^T)^{-1}]. \tag{20}$$

This is a second basic result which gives us the possibility to calculate all elements of the system matrix A by measuring the covariance matrices $E(\underline{X}_t\underline{X}_t^T)$ and $E(\underline{X}_t\underline{Y}_t^T)$ of the stationary response and filter processes in (17). In comparison with the first result (12), the identification equation (20) has no restrictions with respect to a special damping of the elastic structure or with respect to any special symmetric conditions in the system matrices. Knowing the matrix A from (20), we finally can calculate the excitation intensities $\underset{\sim}{\sigma}\underset{\sim}{\sigma}^T = (\sigma_{ij})$ by means of

$$\underset{\sim}{\sigma}\underset{\sim}{\sigma}^T = - A\, E(X_t X_t^T) - E(X_t X_t^T)\, A^T. \tag{21}$$

This matrix equation follows from the expectation of the increment $d(\underset{\sim}{X}_t \underset{\sim}{X}_t^T)$ in an analogous manner as previously shown.

It should be noted that there are several modifications or extensions of the basic estimation equation (20). Instead of the low-pass filtering of $\underset{\sim}{X}_t$ in (17), we can also apply a higher dimensional filter or the following chain of two low-pass filter.

$$\dot{\underset{\sim}{Y}}_t + \omega_g \underset{\sim}{Y}_t = \omega_g \underset{\sim}{X}_t, \qquad \dot{\underset{\sim}{U}}_t + \omega_g \underset{\sim}{U}_t = \omega_g \underset{\sim}{Y}_t. \tag{22}$$

Setting up the increments $d(\underset{\sim}{X}_t \underset{\sim}{Y}_t^T)$ and $d(\underset{\sim}{X}_t \underset{\sim}{U}_t^T)$ and taking their expectations, we go over to the stationary case $\dot{E}(.) = 0$ what results in

$$\begin{aligned} 0 &= A\, E(\underset{\sim}{X}_t \underset{\sim}{Y}_t^T) - \omega_g [E(\underset{\sim}{X}_t \underset{\sim}{Y}_t^T) - E(\underset{\sim}{X}_t \underset{\sim}{X}_t^T)], \\ 0 &= A\, E(\underset{\sim}{X}_t \underset{\sim}{U}_t^T) - \omega_g [E(\underset{\sim}{X}_t \underset{\sim}{U}_t^T) - E(\underset{\sim}{X}_t \underset{\sim}{Y}_t^T)]. \end{aligned} \tag{23}$$

Obviously, there are now two independent matrix equations for the determination of the system matrix A which both can be used to improve the covariance measurements applying a least square procedure. Second, we consider a mechanical system of the form (10)

$$M\, \ddot{\underset{\sim}{X}}_t + D\, \dot{\underset{\sim}{X}}_t + C\, \underset{\sim}{X}_t = \underset{\sim}{\sigma}\, \dot{W}_t, \qquad \dot{\underset{\sim}{Y}}_t + \omega_g \underset{\sim}{Y}_t = \omega_g \underset{\sim}{X}_t. \tag{24}$$

In this special case, only the displacement processes have to be filtered in order to identify the stiffness matrix C and the damping matrix D by means of the following two coupled matrix equations:

$$\begin{aligned} C\, E(\underset{\sim}{X}_t \underset{\sim}{Y}_t^T) + D\, E(\dot{\underset{\sim}{X}}_t \underset{\sim}{Y}_t^T) &= \omega_g M\, [E(\dot{\underset{\sim}{X}}_t \underset{\sim}{X}_t^T) - E(\dot{\underset{\sim}{X}}_t \underset{\sim}{Y}_t^T)], \\ C\, E(\underset{\sim}{X}_t \underset{\sim}{X}_t^T) + D\, E(\dot{\underset{\sim}{X}}_t \underset{\sim}{X}_t^T) &= M\, E(\dot{\underset{\sim}{X}}_t \dot{\underset{\sim}{X}}_t^T). \end{aligned} \tag{25}$$

In comparison with (20), the modification (25) reduces the number of the applied filter equations. In case of the linear oscillator, the equations (25) are reducible to the special result (16). Furthermore, if the structural system contains nonlinear terms, the identification procedure can be extended to higher order moments. As an example let us consider a nonlinear first order system with only one state process X_t.

$$\dot{X}_t + \nu_g X_t + \varepsilon X_t^3 = \sigma \dot{W}_t, \qquad \dot{Y}_t + \omega_g Y_t = \omega_g X_t. \tag{26}$$

Associated to the increments $d(X_tY_t)$ and $d(X_tY_t^3)$, there are stationary moments equations as follows:

$$\begin{aligned} \nu_g\, E(X_tY_t) + \varepsilon\, E(X_t^3Y_t) &= \omega_g\, [\, E(X_t^2) - E(X_tY_t)], \\ \nu_g\, E(X_tY_t^3) + \varepsilon\, E(X_t^3Y_t^3) &= 3\omega_g [E(X_t^2Y_t^2) - E(X_tY_t^3)]. \end{aligned} \tag{27}$$

Both are linear equations in ν_g and ε. They are sufficient to estimate the linear and nonlinear parameter of the system (26). For $\varepsilon = 0$, they degenerate to linearly dependent equations that can be verified by means of the gaussian moments properties.

Some more extensions of the post-filtering technique are given in [7] wherein it is shown that the response spectra, too, can be estimated by the mean squares of a simple second order filter. Thus, there are several possibilities for an optimal filtering. Applications in problems of rotor dynamics have been recently mentioned by H.I. Weber [8]. Finally, it has to be noted that the post-filtering can be avoided by means of maximum likelihood techniques [9] or by utilizing focussing effects in homogeneously distributed systems [10].

5. Excitation identification

As already mentioned, the mean square identification gives estimations for both, the system parameters and the excitation intensities. The same is true in the non-white noise case, too. To show this extension, we formulate a stationary n-dimensional excitation process $\underset{\sim}{Z}_t$ generated from scalar white noise $\dot{W}_t$ by means of shaping filters with an unknown nxn frequency matrix Ω and an unknown n-dimensional intensity vector $\underset{\sim}{\sigma}$:

$$\begin{aligned} \dot{\underset{\sim}{Z}}_t &= \Omega\, \underset{\sim}{Z}_t + \underset{\sim}{\sigma}\, \dot{W}_t, & \dot{\underset{\sim}{U}}_t &= -\alpha \underset{\sim}{U}_t + \alpha\, \underset{\sim}{X}_t, \\ \dot{\underset{\sim}{X}}_t &= A\, \underset{\sim}{X}_t + \underset{\sim}{Z}_t, & \dot{\underset{\sim}{V}}_t &= -\beta\, \underset{\sim}{V}_t + \beta\, \underset{\sim}{U}_t. \end{aligned} \tag{28}$$

The excited dynamic system possesses the n-dimensional state vector process $\underset{\sim}{X}_t$ and the unknown nxn-matrix A. $\underset{\sim}{U}_t$ and $\underset{\sim}{V}_t$ are n-dimensional processes of the post-filtering with any limiting frequencies α and β suitably chosen in the range $0 < \alpha, \beta < \infty$.

We derive the identification equations of interest by setting up the increments $d(\underset{\sim}{X}_t\underset{\sim}{U}_t^T)$, $d(\underset{\sim}{X}_t\underset{\sim}{V}_t^T)$, $d(\underset{\sim}{X}_t\underset{\sim}{X}_t^T)$, $d(\underset{\sim}{Z}_t\underset{\sim}{U}_t^T)$, $d(\underset{\sim}{Z}_t\underset{\sim}{V}_t^T)$ and the corresponding covariance equations. Then, we go over to the stationary case $\dot{E}(.) = 0$ and eliminate the non-measurable covariance matrices $E(\underset{\sim}{Z}_t\underset{\sim}{U}_t^T)$, $E(\underset{\sim}{Z}_t\underset{\sim}{V}_t^T)$, $E(\underset{\sim}{Z}_t\underset{\sim}{X}_t^T)$

in arriving to the following two equations:

$$(\Omega - \beta I)[(\beta I - A)\, E(\underset{\sim}{X}_t \underset{\sim}{V}_t^T) - \beta\, E(\underset{\sim}{X}_t \underset{\sim}{U}_t^T)] + \\ + \beta[(\alpha I - A)\, E(\underset{\sim}{X}_t \underset{\sim}{U}_t^T) - \alpha\, E(\underset{\sim}{X}_t \underset{\sim}{X}_t^T)] = 0, \tag{29}$$

$$\alpha[(A + \Omega - \alpha I)\, E(\underset{\sim}{X}_t \underset{\sim}{X}_t^T) + E(\underset{\sim}{X}_t \underset{\sim}{X}_t^T)\,(A^T + \Omega^T - \alpha I)] + \\ + (\alpha I - \Omega)(\alpha I - A)\, E(\underset{\sim}{X}_t \underset{\sim}{U}_t^T) + E(\underset{\sim}{U}_t \underset{\sim}{X}_t^T)\,(\alpha I - A^T)(\alpha I - \Omega^T) = 0. \tag{30}$$

Now, these are non-linear matrix equations of the Ljapunow type. Since all coefficients of (29) and (30) are measurable in the stationary case, they are sufficient to determine the frequency matrix Ω of the excitation as well as the parameter matrix A of the dynamic system. Upon their calculation we obtain from the increment $d(\underset{\sim}{X}_t \underset{\sim}{Z}_t^T)$

$$2\alpha\, E(\underset{\sim}{Z}_t \underset{\sim}{Z}_t^T) = A\, E(\underset{\sim}{U}_t \underset{\sim}{Z}_t^T)\,(\alpha I - \Omega^T) + (\alpha I - \Omega)\, E(\underset{\sim}{Z}_t \underset{\sim}{U}_t^T)\, A^T + \\ + \Omega\,(\alpha I - \Omega)\, E(\underset{\sim}{Z}_t \underset{\sim}{U}_t^T) + E(\underset{\sim}{U}_t \underset{\sim}{Z}_t^T)\,(\alpha I - \Omega^T)\,\Omega^T \tag{31}$$

which allows us finally to calculate the excitation intensities $E(\underset{\sim}{Z}_t \underset{\sim}{Z}_t^T)$ and where they are applied.

In conclusion, we give some first measurement results in case of the linear oscillator (1) under white noise and of a filtering of type (17).

$$\dot{Y}_{1t} = -\omega_g Y_{1t} + \omega_g X_t, \qquad \dot{Y}_{2t} = -\omega_g Y_{2t} + \omega_g \dot{X}_t. \tag{32}$$

In correspondence to (20), the identification equations read as follows:

$$\frac{c}{m} = \frac{E(\dot{X}_t^2)}{E(X_t^2)}, \qquad \frac{b}{m} = \omega_g \left[\frac{E(Y_{1t}^2)}{E(Y_{2t}^2)}\,\frac{E(\dot{X}_t^2)}{E(X_t^2)} - 1\right]. \tag{33}$$

As shown in Fig. 2, we apply an analogue computer to simulate the oscillator and the low-pass filtering with the frequencies $\omega_1 = \omega_g = 10$ [1/s]. The measured signals are successively and separately evaluated by means of a Fourier analyzer with a scan-rate of 0.6 s in performing N = 1000 and N = 2000 averages. The following table contains the measured mean squares and estimates for the scaled parameters of the oscillator with the true values $\overline{c/m} = 1$ and $\overline{b/m} = 0.2$, respectively.

N \ M	$E(X_t^2)$	$E(\dot{X}_t^2)$	$E(Y_{1t}^2)$	$E(Y_{2t}^2)$	$\overline{c/m}$	$\overline{b/m}$
1000	2.2832	2.4158	1.2573	1.0501	1.0581	0.2669
2000	2.3345	2.3468	1.3030	1.0612	1.0053	0.2344

1000	2.5303	2.4656	1.3789	1.0966	0.9744	0.2253
2000	2.3983	2.3451	1.3544	1.0952	0.9778	0.2092

Tab. 1: Frequency and damping estimates of the oscillator

Naturally, such measurements can considerably be improved by a directly and simultaneously performed evaluation of the mean squares if several A/D - channels are available. The applied micro-computer with two channels performed simultaneously 500 averages with a given scan-rate of 0.1 s such that the total estimation duration was T = 50 s . The following table 2 contains the results of all together 14 trials under same condition for $\overline{c/m} = E(\dot{X}_t^2)/E(X_t^2)$ with the true value $\overline{c/m} = 1$.

1.0006	0.9836	0.9792	0.9926	0.9909	0.9926	0.9628
0.9996	0.9983	0.9818	0.9807	1.0150	0.9951	0.9800

Tab. 2: Results of 14 trials with N = 500 and T = 50 s

References

[1] Wedig, W.: Stochastic identification of stiffness and damping matrices, to appear in CISM courses on "Structural Identification and Parameter Estimation", Springer, Wien 1983

[2] Kozin, F., Lee, T.S.: Almost sure asymptotic likelihood theory for diffusion processes, J. Appl. Prob., Vol. 14, 527-537 (1977)

[3] Kozin, F.: Some statistical methods of structural identification and parameter estimation, to appear in CISM courses on "Structural Identification and Parameter Estimation", Springer, Wien 1983

[4] Bucy, R.S., Joseph, P.D.: Filtering for Stochastic Processes with Application to Guidance, Wiley, New York 1968

[5] Snyders, J.: Stationary probability distributions for linear time-invariant systems, SIAM J. Control and Optimization, 15, 3, 428-437 (1977)

[6] Itô, K.: On stochastic differential equations, Amer. Math. Soc. (1951), (Memoirs. Amer. Math. Soc. No. 4)

[7] Wedig, W.: Anwendung einer Spektralfiltertechnik zur Identifikation linearer und nichtlinearer Systeme, Ing.-Archiv 49, 413-425 (1980)

[8] Weber, H.I.: Partnerschaftsvorhaben Rotordynamik, non-published lecture of Oberwolfach-Tagung "Schwingungstheorie" (1982)

[9] Kozin, F.: Estimation of parameters for systems driven by white noise excitations, to appear in Proceedings of the IUTAM-Symposium on "Random Vibrations and Reliability", Akademie-Verlag, Berlin 1983, pp.163-172

[10] Wedig, W.: Bettungs- und Dämpfungsanalyse homogen verteilter Systeme, VDI-Berichte Nr. 456, Düsseldorf, 201-207 (1982).

SECTION R A N D O M V I B R A T I O N S

WIDE BAND RANDOM EXCITATION OF SQUARE PLATES

Stephen H. Crandall[1] and Wei Qiu Zhu[2]

Summary. The local mean-square velocity response of a thin rectangular plate due to point excitation by a stationary wide-band random force depends on the source location, the response location, and the number N of resonant modes excited. As N increases the spatial distribution of response tends to become more uniform with the residual deviations in response concentrated in narrowing lanes of intensified or reduced response. The intensification ratios remain essentially independent of N for plates with non-repeated natural frequencies and negligible modal overlap. In a square plate multiple frequencies are scattered irregularly throughout the spectrum. Their effects are estimated by using results from the theory of numbers. It is shown that the spatial response variance is increased for a square plate as compared to a rectangular plate of equal area and that local intensification ratios for a square plate can increase without limit as N increases.

Introduction. When a uniform thin elastic plate of retangular shape and light damping is excited by a point force, normal to the plate, with a time history that is a stationary random process, the bending response of the plate is a stationary space-time random process. As the bandwidth of the excitation increases, the number of resonant modes responding increases, and the bandwidth of the response at each point also increases but the spatial distribution of mean-square response tends toward a simple pattern. Crandall and Wittig [1] showed that when the response contains several hundred resonant modes the mean-square velocity is nearly uniform over the surface of the plate except for narrow strips along the edges and for a tic-tac-toe pattern of lanes with one intersection at the driving point. The response along the edges is intensified or reduced depending on whether the boundary is free or supported. The response along the lanes making up the tic-tac-toe pattern is intensified independently of the boundary conditions. Similar patterns of intensified response have been observed [2] for plates with various symmetries; e.g., square, circle, or equilateral triangle. In the present paper the analysis for a rectangular plate is reviewed with emphasis on the spatial variation in response and on the maximum local response, and how these depend on the number of resonant modes participating in the response. The special case of a square plate is then considered in detail. The symmetry of the square is responsible for an irregular sequence of multiple frequencies which result in qualitative and quantitative differences in the response of a square plate as compared to that of a rectangular plate of the same area subjected to the same excitation. The major qualitative difference (the presence of diagonal lanes of intensification) was described in [2] but important quantitative differences were overlooked. We will show that the spatial variance of response for the square plate is greater than that for the rectangular plate and that the maximum local intensification ratio for a square plate can increase without limit as

[1]Ford Professor of Engineering, Massachusetts Institute of Technology, Cambridge, Mass 02139, U.S.A.
[2]Visiting Scientist, M.I.T. Permanent address: Zhejiang University, Hangzhou, P.R.C.

the number of resonant modes increases whereas the corresponding ratio for a rectangular plate is essentially independent of the number of resonant modes.

Rectangular plate. Consider a uniform simply supported elastic plate of thickness h and rectangular shape with sides of length a and b excited by a source of normal force located at the driving point P_s with coordinates $\xi, \eta (0 < \xi < a,\ 0 < \eta < b)$. If the time history of the force is a stationary random process with mean-square spectral density $S(\omega)$ the mean-square velocity response at the receiving point P_r with coordinates $x, y (0 < x < a,\ 0 < y < b)$ is given by the modal sum [2]

$$E[v^2(P_r)] = \sum_{m=1}^{\infty} \sum_{n=1}^{\infty} \psi_m(P_r)\psi_n(P_r)\psi_m(P_s)\psi_n(P_s) \int_{-\infty}^{\infty} H_m(-\omega)S(\omega)H_n(\omega)d\omega \qquad (1)$$

where the natural modes of the plate are indicated by either the single subscript notation $\psi_n(P_r)$ or the double subscript notation $\psi_{jp}(x,y)$ as is convenient. For the simply supported plate we take

$$\psi_n(P_r) = \psi_{jp}(x,y) = 2 \sin\frac{j\pi x}{a} \sin\frac{p\pi y}{b}. \qquad (2)$$

The corresponding natural frequencies are

$$\omega_n = \omega_{jp} = \left(\frac{D}{\rho h}\right)^{\frac{1}{2}} \left[\left(\frac{j\pi}{a}\right)^2 + \left(\frac{p\pi}{b}\right)^2\right] \qquad (3)$$

where D is the bending modulus and ρh is the mass per unit area of the plate. In (1) the $H_n(\omega)$ are the modal frequency response functions for the velocity

$$H_n(\omega) = \frac{i\omega}{m(\omega_n^2 - \omega^2 + i\omega\beta)} \qquad (4)$$

where $m = \rho hab$ is the plate mass and β is the modal bandwidth, assumed to be constant due to damping forces proportional to the plate velocity. The modes (2) satisfy the orthonormality conditions

$$\int_A \psi_m(P_r)\psi_n(P_r)\,dA_r = \delta_{mn},$$
$$\int_0^a dx \int_0^b \psi_{jp}(x,y)\psi_{kq}(x,y)dy = \delta_{jk}\delta_{pq}. \qquad (5)$$

The single subscript natural frequencies ω_n are assumed to be arranged in order of increasing magnitude. The sequence of double subscripts in the corresponding ω_{jp} advances in an irregular manner. The situation may be visualized by considering the integer lattice points in the first quadrant of a Cartesian j, p plane to represent the natural modes. For each frequency ω_n eq. (3) defines an elliptical locus in the j, p plane which passes through one of the integer lattice points j, p. The spacing between ellipses is irregular and there is no simple rule for predicting the j, p values associated with ω_{n+1} given the j, p values associated with ω_n. This feature of the theory of vibrations of uniform two-dimensional structures is perhaps the major difficulty encountered in extending the theory of uniform one-dimensional structures where the natural frequencies are distributed systematically. Although the individual modal spacings

$$\Delta\omega_n = \omega_{n+1} - \omega_n \tag{6}$$

for the rectangular plate are erratic it is known [3] that asymptotically the number of modes N with frequencies ω_n satisfying $\omega_n < \omega_c$ is linearly proportional to ω_c. To see this consider the first quadrant of the ellipse in the j, p plane determined by eq. (3) for the large frequency ω_c. The number of modes N with natural frequencies less than ω_c is simply the number of lattice points within the ellipse. By associating with each lattice point a unit area we see that N is asymptotically equal to the area bounded by the ellipse. This area is $\pi JP/4$ where J and P are the semi-major and semi-minor axes of the ellipse

$$J = \frac{a}{\pi}\left(\frac{\rho h\omega_c^2}{D}\right)^{\frac{1}{4}} \quad P = \frac{b}{\pi}\left(\frac{\rho h\omega_c^2}{D}\right)^{\frac{1}{4}} . \tag{7}$$

Thus we obtain

$$N \sim \frac{ab}{4\pi}\left(\frac{\rho h}{D}\right)^{\frac{1}{2}} \omega_c \tag{8}$$

which implies that there is an asymptotic average modal spacing $\overline{\Delta\omega}$ defined by

$$\frac{\omega_c}{N} \sim \frac{4\pi}{ab}\left(\frac{D}{\rho h}\right)^{\frac{1}{2}} = \overline{\Delta\omega} . \tag{9}$$

In the following it is assumed that the damping is light in the sense that the modal overlap ratio $\beta/\overline{\Delta\omega}$ is small compared to unity. This means that modal overlap (the occurrence of successive natural frequencies with spacing (6) less than β) is infrequent for most rectangles in which a and b are incommensurate (which rules out repeated frequencies) and whose ratio is not too close to an integer fraction with small integers. The subsequent discussion of the rectangular plate is based on the assumption that modal overlap effects are negligible. For the square plate, however, repeated frequencies are the rule rather than the exception and the modal overlap of modes with rep eated frequencies must be taken into account.

Response to band-limited white noise. We take the excitation spectral density $S(\omega)$ to have uniform density S_o out to a cut-off frequency ω_c; i.e.,

$$S(\omega) = \begin{cases} S_o, & |\omega| < \omega_c \\ 0, & \omega_c < |\omega| \end{cases} . \qquad (10)$$

The natural modes ψ_n with $\omega_n < \omega_c$ are called resonant modes while the modes with $\omega_n > \omega_c$ are called nonresonant. The local mean-square velocity (1) due to the spectrum (10) depends on the source locations P_s, or (ξ,η), and the receiver location P_r, or (x,y), and on the cut-off frequency ω_c or, on what is asymptotically equivalent, the number N of resonant modes. To indicate this we introduce the notation

$$E[v^2(P_r)] = \phi(P_r,P_s,N). \qquad (11)$$

A rough indication of the spatial distribution of ϕ is obtained by examining the spatial mean and the spatial variance of ϕ as functions of N. We also examine the maximum value of ϕ.

Consider the spatial mean m_ϕ of the response $\phi(P_r,P_s,N)$, averaging over the source location P_s and receiver location P_r. Using (1) and the orthonormality conditions (5) we have

$$m_\phi = \langle \phi \rangle_{r,s} = \frac{1}{A^2}\int_A dA_r \int_A \phi(P_r,P_s,N)\, dA_s = \sum_{n=1}^{\infty} \int_{-\omega_c}^{\omega_c} S_o |H_n(\omega)|^2\, d\omega . \qquad (12)$$

The sum can be divided into resonant contributions for $1 \le n \le N$ and nonresonant contributions for $N + 1 \le n < \infty$. In the resonant terms very little error is made if the limits of integration are extended to infinity provided that the resonant frequency ω_n is not too close to the cut-off frequency ω_c. For these values of n the contribution of each term to the sum (12) is very nearly equal to

$$\int_{-\infty}^{\infty} \frac{S_o\omega^2 d\omega}{m^2\left[\left(\omega_n^2 - \omega^2\right)^2 + \beta^2\omega^2\right]} = \frac{\pi S_o}{m^2\beta} = V_1^2 . \qquad (13)$$

If the resonant bandwidth β is small compared with the average modal spacing $\overline{\Delta\omega}$ the approximation (13) may be used for ω_n values up to $\omega_c - \overline{\Delta\omega}$. In the nonresonant terms very little error is made if the damping term in the frequency response is neglected provided again that the resonant frequency ω_n is not too close to the cut-off frequency ω_c. For these values of n the contribution of each term to the sum (12) is approximately

$$\int_{-\omega_c}^{\omega_c} \frac{S_o\omega^2 d\omega}{m^2\left(\omega_n^2 - \omega^2\right)^2} = \frac{\beta}{\pi} V_1^2 \left[\frac{\omega_c}{\omega_n^2 - \omega_c^2} - \frac{1}{2\omega_n} \ln \frac{\omega_n + \omega_c}{\omega_n - \omega_c}\right] . \qquad (14)$$

The approximation (14) may be used for ω_n values greater than $\omega_c + \overline{\Delta\omega}$. In the neighborhood of $\omega_n = \omega_c$ there is a rapid change in the magnitude of the contribution to the sum (12) from each resonant frequency ω_n, falling from essentially the value V_1^2 for $\omega_n = \omega_c - \overline{\Delta\omega}$, to about half this value for $\omega_n = \omega_c$, and continuing down until for $\omega_n = \omega_c + \overline{\Delta\omega}$ the contribution has dropped to about

$$V_1^2 \frac{\beta}{\overline{\Delta\omega}} \frac{1}{2\pi} \left(1 - \frac{\ln N}{N}\right) \tag{15}$$

which is an asymptotic estimate of (14) evaluated at $\omega_n = \omega_c + \overline{\Delta\omega}$. To estimate the sum (12) for the case where ω_N is the largest resonant frequency smaller than the cut-off frequency ω_c, we take exactly N of the individual modal contributions (13) as the total resonant contribution, arguing that any decrease in contribution from a resonance just below cut-off is on balance made up by the contribution from a resonance just above cut-off. For the total nonresonant contribution we sum the individual modal contributions (14) over the infinite set of modes with resonant frequencies $\omega_n > \omega_c + \overline{\Delta\omega}$. The resulting estimate for the spatial mean of the local mean-square response is

$$m_\phi = V_1^2 \left[N + \frac{1}{2\pi} \frac{\beta}{\overline{\Delta\omega}} \ln(2N)\right]. \tag{16}$$

Although the nonresonant contribution in (16) is unbounded as $N \to \infty$ it grows slowly and is always a negligible fraction of the resonant contribution. For example, if the modal overlap ratio $\beta/\overline{\Delta\omega} = 1/3$ and $N = 100$ the resonant contribution is 100 V_1^2 while the nonresonant contribution is 0.28 V_1^2; if N is increased to 1000, the resonant contribution increases to 1000 V_1^2 while the nonresonant contribution only increases to 0.40 V_1^2. In the following we omit nonresonant contributions to our estimates.

Spatial variance of response. We consider next the spatial variance σ_ϕ^2 of the response ϕ, averaging over source location and receiver location

$$\sigma_\phi^2 = \langle \phi^2 \rangle_{r,s} - \left(\langle \phi \rangle_{r,s}\right)^2. \tag{17}$$

To evaluate the spatial mean square $\langle \phi^2 \rangle_{r,s}$ we assume that by neglecting modal overlap effects the double summation (1) can be reducted to a single summation of its main-diagonal terms for which m = n. If in addition the summation is limited to the resonant terms and the approximation (13) is employed for the frequency integrals we have the following approximation for (1)

$$\phi(P_r,P_s,N) = V_1^2 \sum_{n=1}^{N} \psi_n^2(P_r)\psi_n^2(P_s). \tag{18}$$

In double subscript notation the spatial mean square is

$$\langle\phi^2\rangle_{r,s} = V_1^4 \sum_{jp} \sum_{kq} \frac{1}{A^2} \int_A \int \psi_{jp}^2(x,y)\psi_{kq}^2(x,y)\,dxdy \int_A \int \psi_{jp}^2(\xi,\eta)\psi_{kq}^2(\xi,\eta)\,d\xi d\eta. \tag{19}$$

Carrying out the integrations for the modes (2) we find

$$\langle\phi^2\rangle_{r,s} = V_1^4 \sum_{jp}\sum_{kq} \begin{cases} 1 & \text{for } j \neq k,\ p \neq q \\ \left(\frac{3}{2}\right)^2 & \text{for } \begin{cases} j \neq k,\ p = q, \text{ or} \\ j = k,\ p \neq q \end{cases} \\ \left(\frac{3}{2}\right)^4 & \text{for } j = k,\ p = q \end{cases} . \tag{20}$$

Now the summation here over j and p (and similarly over k and q) must include all combinations of positive integer j and p such that the corresponding natural frequencies ω_{jp} given by (3) are less than the cut-off frequency ω_c. These combinations constitute the integer lattice points in the first quadrant of the j, p plane bounded by the ellipse

$$\frac{j^2}{J^2} + \frac{p^2}{P^2} = 1 \tag{21}$$

where the semi-major and semi-minor axes (7) have the asymptotic values

$$J \sim \left(\frac{4Na}{\pi b}\right)^{\frac{1}{2}}, \quad P \sim \left(\frac{4Nb}{\pi a}\right)^{\frac{1}{2}}. \tag{22}$$

To estimate the contribution to (20) arising from $j \neq k$, $p \neq q$ consider that for each lattice point j, p the number of permitted k, q combinations is the total number N minus the number for which either $k = j$ or $q = p$. The sum over all j, p combinations is approximated by an integral over the quadrant of the ellipse. Thus

$$\begin{aligned} \sum_{jp}^{j\neq k}\sum_{kq}^{p\neq q} 1 &\simeq \int_0^J dj \int_0^{P(1-j^2/J^2)^{\frac{1}{2}}} dp\left[N - J(1 - p^2/P^2)^{\frac{1}{2}} - P(1 - j^2/J^2)^{\frac{1}{2}} + 1\right] \\ &= (N + 1)\pi JP/4 - \frac{2}{3} J^2P - \frac{2}{3} JP^2 \\ &\sim N^2 + N - \frac{2}{3}\left(\frac{4N}{\pi}\right)^{3/2}\left[\left(\frac{a}{b}\right)^{\frac{1}{2}} + \left(\frac{b}{a}\right)^{\frac{1}{2}}\right]. \end{aligned} \tag{23}$$

By a similar argument

$$\sum_{jp}^{j=k,}\sum_{kq}^{p\neq q}\left(\frac{3}{2}\right)^2 \simeq \int_0^J dj \int_0^{P(1-j^2/J^2)^{\frac{1}{2}}} dp\left[P(1 - j^2/J^2)^{\frac{1}{2}} - 1\right]\left(\frac{3}{2}\right)^2 = \left(\frac{2}{3} J^2P - \pi JP/4\right)\left(\frac{3}{2}\right)^2 \tag{24}$$

and

$$\sum_{jp}^{j\neq k,}\sum_{kq}^{p=q}\left(\frac{3}{2}\right)^2 \simeq \left(\frac{2}{3} JP^2 - \pi JP/4\right)\left(\frac{3}{2}\right)^2. \tag{25}$$

Adding to these

$$\sum_{jp}^{j=k}\sum_{kq}^{p=q}\left(\frac{3}{2}\right)^4 = N\left(\frac{3}{2}\right)^4 \tag{26}$$

we obtain

$$\langle\phi^2\rangle_{r,s} = V_1^4\left\{N^2 + \frac{5}{6}\left(\frac{4N}{\pi}\right)^{3/2}\left[\left(\frac{a}{b}\right)^{\frac{1}{2}} + \left(\frac{b}{a}\right)^{\frac{1}{2}}\right] + \left(\frac{5}{4}\right)^2 N\right\} \tag{27}$$

for the spatial mean square. Taking the resonant mode estimate $m_\phi = NV_1^2$ for the spatial mean, we have for the spatial variance (17)

$$\sigma_\phi^2 = V_1^4 \left\{ \frac{5}{6} \left(\frac{4N}{\pi}\right)^{3/2} \left[\left(\frac{a}{b}\right)^{\frac{1}{2}} + \left(\frac{b}{a}\right)^{\frac{1}{2}}\right] + \left(\frac{5}{4}\right)^2 N \right\} . \tag{28}$$

For $N \to \infty$ the asymptotic value of the relative spatial deviation is

$$\frac{\sigma_\phi}{m_\phi} \sim \left\{ \frac{5}{6} \left(\frac{4}{\pi}\right)^{3/2} \left[\left(\frac{a}{b}\right)^{\frac{1}{2}} + \left(\frac{b}{a}\right)^{\frac{1}{2}}\right] \right\}^{\frac{1}{2}} N^{-\frac{1}{4}} . \tag{29}$$

This implies an approach to relative uniformity of response as the number of resonant modes increases without limit.

The spatial variance (28) does not result from a uniform distribution of small deviations but is in fact mostly concentrated in large deviations in the neighborhood of the edges and along the narrow lanes of intensified response which form the characteristic tic-tac-toe pattern. The intensification ratio

$$I(P_r) = \frac{\phi(P_r)}{m_\phi} \tag{30}$$

along these lanes estimated [2] by calculating the spatial mean of $\phi(P_r)$ when P_r is restricted to remain on a lane is 3/2.

<u>Driving-point intensification factor</u>. The maximum intensification factor I_{max} occurs at the driving point ($P_r = P_s$) and its three image points. A spatial average driving-point intensification ratio can be calculated by placing in the numerator of (30) the average of ϕ_{max} with respect to the source location P_s. The result is

$$\langle I_{max} \rangle_s = \frac{m_{\phi_{max}}}{m_\phi} = \frac{16}{N\,ab} \int_0^a d\xi \int_0^b d\eta \sum_{jp} \sin^4 \frac{j\pi\xi}{a} \sin^4 \frac{p\pi\eta}{b} = \left(\frac{3}{2}\right)^2 . \tag{31}$$

The spatial variance of ϕ_{max} can also be estimated in a manner similar to that outlined above for σ_ϕ^2 except that ϕ_{max}^2 is averaged with respect to source location only. The asymptotic value of the relative spatial deviation of ϕ_{max} is

$$\frac{\sigma_{\phi_{max}}}{m_{\phi_{max}}} \sim \left\{ \frac{35}{27} \left(\frac{4}{\pi}\right)^{3/2} \left[\left(\frac{a}{b}\right)^{\frac{1}{2}} + \left(\frac{b}{a}\right)^{\frac{1}{2}}\right] \right\} N^{-1/4} \tag{32}$$

which implies that when N is large the deviations from the average driving-point intensification factor (31) must be small for almost all driving-point locations. It is shown in [2] that there are special driving-point locations which have distinctly different driving-point intensification factors: if P_s lies on one midline of the rectangle but not the other, the driving-point intensification factor is 3; if P_s lies at the center of the rectangle the driving-point intensification factor is 4. For driving-point locations not too close to the edges or to these special locations, the driving-point intensification asymptotically approaches (31) as N increases.

In reviewing the preceding results for a rectangular plate, $a \neq b$, we note that the spatial mean of the response $m_\phi = NV_1^2$ is independent of the aspect ratio a/b but the spatial variance σ_ϕ^2 of (28) does vary with a/b. The average driving-point intensification factor (31) is independent of the number N of resonant modes in the response as well as being independent of the aspect ratio. These results depend for the most part on the assumption of non-overlapping modes. They cannot be immediately extended to a square plate by simply setting b = a because the spectrum of natural frequencies for a square consists primarily of repeated frequencies of various multiplicities. The several modes sharing a common natural frequency have total overlap with each other and thereby generate important differences in the response distribution.

Square plate. In place of the rectangular plate with sides a and b, we consider a simply supported square plate of side L. The natural modes are given by (2) with a = L and b = L. The natural frequencies (3) become

$$\omega_n = \omega_{jp} = \frac{\pi^2}{L^2}\left(\frac{D}{\rho h}\right)^{\frac{1}{2}} \nu_{jp} \tag{33}$$

where

$$\nu_{jp} = j^2 + p^2 . \tag{34}$$

Repeated frequencies are common. If $j \neq p$ then the modes ψ_{jp} and ψ_{pj}, which are mutually orthogonal, have the same natural frequency since $\nu_{jp} = \nu_{pj}$. In [2] the response of a square plate is predicted taking into account repeated frequencies of multiplicity two. It was noted there that multiplicities of higher order could occur and examples of multiplicity 3, 4, and 6 were given but it was (incorrectly) assumed that such higher order multiplicities were of relatively rare occurence. For larger values of ν higher order multiplicities become increasingly common. Table 1 shows some results from a

TABLE 1

Multiplicities of Modes of a Square Plate

Range of ν - values	Number of Modes	1	2	3	4	5	6	7	8
1 - 100	69	6	26	1	2	0	0	0	0
10,001 - 10,100	85	1	12	0	12	0	2	0	0
20,001 - 20,100	74	0	11	0	10	0	2	0	0
30,001 - 30,100	86	0	11	0	6	0	4	0	2

computer study over the range $1 < \nu < 32{,}000$. The number of modes and their multiplicities within a ν-interval of 100 are listed for intervals beginning at 1, 10,001, 20,001, and 30,001. Note that modes with odd multiplicity are relatively rare. In the total of 24,944 modes encompassed in the range $1 \leq \nu \leq 32{,}000$ there are only 67 unrepeated frequencies, 56 with multiplicity three, 4 with multiplicity five, one with multiplicity seven, and none with higher-order odd multiplicity. In this same range there are 30 distinct frequencies with multiplicity 12.

In order to account for these multiple frequencies we adopt some notation from the theory of numbers. Let $M(\nu)$ be the number of distinct representations of

$$\nu = j^2 + p^2 \tag{35}$$

for positive integer values of j and p; i.e., $M(\nu)$ is the multiplicity of the frequency (33). The argument ν is taken to be the sequence of positive integers. It is known that the odd and even properties of j and p are related to the magnitude of ν modulo 4; i.e., for any integer value of n

if $\nu = 4n$ then both j and p are even,

if $\nu = 4n + 1$ then either j is odd and p is even or vice versa,

if $\nu = 4n + 2$ then both j and p are odd,

if $\nu = 4n + 3$ there are no integer values of j and p which satisfy (35).

The sum of the multiplicities over all the integers from $\nu = 1$ to $\nu = \nu_c$

$$N = \sum_{\nu=1}^{\nu_c} M(\nu) \tag{36}$$

gives the total number of distinct modes with natural frequencies (33) corresponding to ν_{jp} values which satisfy $\nu_{jp} \leq \nu_c$. If ν_c is the largest integer which satisfies

$$\nu_c \leq \frac{L^2}{\pi^2} \left(\frac{\rho h}{D}\right)^{\frac{1}{2}} \omega_c \tag{37}$$

where ω_c is the cut-off frequency of (10) then N, as given by (36), is the number of resonant modes in the response of the square plate excited by the spectrum (10). The asymptotic estimate of N obtained by assigning unit area to each integer lattice point in the first quadrant of the j, p plane bounded by the circle of radius $(\nu_c)^{\frac{1}{2}}$ is

$$N \sim \frac{\pi}{4} \nu_c . \tag{38}$$

See also Eq. (A-6) of the Appendix. We shall also require an asymptotic estimate of the sum of the squares of the multiplicities. It can be shown (see Appendix) that

$$\sum_{\nu=1}^{\nu_c} M^2(\nu) \sim \frac{\nu_c}{4} \ln \nu_c + 0.477\, \nu_c . \tag{39}$$

It is interesting to compare the asymptotic estimate for the number of resonant modes as given by (38) for $\nu_c = 32{,}000$, namely 25,133 with the actual number of 24,944 obtained by computer. The estimate (38) also predicts that in a ν-interval which spans a range of 100 the asymptotic number of modes is 78.5. The actual numbers of modes in four such intervals are tabulated in Table 1.

<u>Response of a square plate to band-limited white noise</u>. The excitation is taken to be a force acting at the driving-point P_s (with coordinates ξ, η) whose time history is a

stationary random process with the spectrum (10). The integrals in the mean-square response (1) have the form

$$\int_{-\omega_c}^{\omega_c} H_m(-\omega)S_oH_n(\omega)d\omega. \tag{40}$$

If $\omega_m = \omega_n$ we approximate this by $V_1^2 = \pi S_o/m^2\beta$ as in (13). In order to compare results for a square plate with those of a rectangular plate it is convenient for both plates to have the same value of V_1^2(i.e., the same mass) which implies that for equal plate thickness and density the plate dimensions should satisfy $L^2 = ab$. If $\omega_m \neq \omega_n$ we approximate (40) by zero; i.e., we neglect modal overlap. In the case of a square plate there can never be nearly overlapping modes. There is either total overlap with a multiple frequency or there is at least a certain minimum spacing $(\Delta\omega)_{min}$ between adjacent resonances where $\Delta\omega_{min}$ corresponds to a unit change in ν. From (33) it follows that

$$(\Delta\omega)_{min} = \frac{\pi^2}{L^2}\left(\frac{D}{\rho h}\right)^{\frac{1}{2}} \tag{41}$$

which is of the same order of magnitude as the average modal spacing $\overline{\Delta\omega}$ defined by (37) and (38)

$$\frac{\omega_c}{N} \sim \frac{4\pi}{L^2}\left(\frac{D}{\rho h}\right)^{\frac{1}{2}} = \overline{\Delta\omega}\,. \tag{42}$$

Note that (9) and (42) become equal when $ab = L^2$.

The mean square response (1) for a square plate with these approximations and with the neglect of the nonresonant contribution is

$$\phi(P_r,P_s,N) = V_1^2\left\{\sum_{n=1}^{N} \psi_n^2(P_r)\psi_n^2(P_s) + \sum_{\substack{m\\ \omega_m=\omega_n,m\neq n}}^{N}\sum_{n}^{N} \psi_m(P_r)\psi_n(P_r)\psi_m(P_s)\psi_n(P_s)\right\}\,. \tag{43}$$

The first summation on the right is equivalent to (18). The second summation represents the additional contribution due to the repeated roots. These terms have an effect on the variance of the response and on local intensification ratios but they do not affect the spatial mean m_ϕ. Because of the orthogonality conditions (5) we have only the contribution from the first summation

$$m_\phi = \langle\phi(P_r,P_s,N\rangle_{r,s} \sim NV_1^2 \tag{44}$$

where the same argument as for the rectangular plate has been used to estimate the contributions from the resonances just below and just above the cut-off frequency.

Spatial variance of the response of a square plate. To evaluate the spatial variance (17) it is necessary to calculate the spatial mean square of (43) averaging with respect to source location and receiver location. When (43) is squared three terms result. The first is a double sum identical in form to (19). Its space average is given by (27) with $a = b$. The second term is a triple sum crossproduct whose spatial average can be shown to vanish. The third term is a quadruple sum representing the square of the double sum in (43). Because of orthogonality its spatial average is

$$\frac{1}{L^4}\int_o^L dx\int_o^L dy\int_o^L d\xi\int_o^L d\eta \sum_{\substack{jp\\ j\neq k,\ p\neq q\\ \omega_{jp}=\omega_{kq}}}\sum_{kq}(16)^2 \sin^2\frac{j\pi x}{L}\sin^2\frac{k\pi x}{L}\sin^2\frac{p\pi y}{L}\sin^2\frac{q\pi y}{L}\cdot \sin^2\frac{j\pi\xi}{L}\sin^2\frac{k\pi\xi}{L}\sin^2\frac{p\pi\eta}{L}\sin^2\frac{q\pi\eta}{L} \tag{45}$$

$$= \sum_{\substack{jp\\ j\neq k,\ p\neq q\\ \omega_{jp}=\omega_{kq}}}\sum_{kq} 1 = \sum_{\nu=1}^{\nu_c}\left[M^2(\nu) - M(\nu)\right].$$

The spatial variance for the square plate therefore takes the form

$$\sigma_\phi^2 = V_1^4\left\{\frac{5}{3}\,\nu_c^{\,3/2} + \sum_{\nu=1}^{\nu_c}\left[M^2(\nu) - M(\nu)\right] + \left(\frac{5}{4}\right)^2 N\right\} \tag{46}$$

which is the result approached by the rectangular plate (28) when $a \to b$ plus the contribution (45) due to the repeated frequencies. Inserting the asymptotic approximations (38) and (39) we have

$$\sigma_\phi^2 = V_1^4\left\{\frac{5}{3}\,\nu_c^{\,3/2} + \frac{1}{4}\,\nu_c \ln \nu_c + 0.035\,\nu_c\right\}. \tag{47}$$

The effect of the repeated frequencies is to increase (slightly) the spatial variance of the response without changing the spatial mean. Since the leading term in (47) is unaffected by the multiplicities of the frequencies the increased variance for a square plate becomes negligible for very large ν_c (or $N \sim \pi\nu_c/4$). Even for small ν_c the effect is modest; e.g., the repeated frequencies contribute an additional 4.2% to the variance for $\nu_c = 200$ ($N \sim 157$), which drops to an additional 2.1% for $\nu_c = 2000$ ($N \sim 1571$).

Driving-point intensification factor for a square plate. The multiple frequencies of a square plate have a more pronounced effect on the maximum intensification ratio I_{max}. To illustrate this we consider the case where the driving point P_s is located at the center of the plate P_o with coordinates $\xi = L/2$ and $\eta = L/2$. The mean-square velocity response at the driving point is given by (43) with $P_r = P_s$

$$\phi(P_o,P_o,N) = V_1^2\left[\sum_{n=1}^{N}\psi_n^4(P_o) + \sum_{\substack{m\\ \omega_m=\omega_n,\ m\neq n}}^{N}\sum_{n}^{N}\psi_m^2(P_o)\psi_n^2(P_o)\right]. \tag{48}$$

Now $\psi_{jp}^2(P_o)$ equals 4 when j and p are both odd and vanishes otherwise which implies that

$$\phi(P_o,P_o,N) = 16\, V_1^2 \sum_{jp} \sum_{kq} 1 \qquad (49)$$
$$j,p,k,q \text{ odd}$$
$$\omega_{jp} = \omega_{kq}$$

For a repeated frequency with multiplicity M the number of units contributed to the sum (49) is simply M^2. Since the only values of ν, for which the j and p values which satisfy (35) are both odd, are of the form 4m + 2 the sum can be written as follows

$$\phi(P_o,P_o,N) = 16\, V_1^2 \sum_{n=1}^{(\nu_c+2)/4} M^2(4n-2) \qquad (50)$$

where the connection between ν_c and N is defined by (36). The corresponding intensification ratio is

$$I_o = \frac{\phi(P_o,P_o,N)}{m_\phi} = 16\, \frac{\sum_{n=1}^{(\nu_c+2)/4} M^2(4n-2)}{\sum_{\nu=1}^{\nu_c} M(\nu)} \; . \qquad (51)$$

An asymptotic approximation to I_o is obtained by assuming that the summation in the numerator over 1/4 of lattice points (those with odd j and odd p) is simply 1/4 of the sum over all the lattice points, and then using the asymptotic approximations (38) and (39). In this way we find

$$I_o \sim \frac{4}{\pi}(\ln \nu_c + 1.908) \; . \qquad (52)$$

The estimates provided by (52) are compared with computer evaluations of (51) in Table 2. Note that these intensification factors increase logarithmically with the number of resonant modes whereas the estimate of $I_o = 8$ given in [2] based on neglecting all repeated frequencies of multiplicity greater than two is independent of the number of resonant modes in the response.

TABLE 2

Driving-point intensification factors at the center of a square plate

ν_c	Approx. number resonant modes N	Intensification factor I_o from (51)	Asymptotic approximation (52)
125	98	9.04	8.58
250	196	11.02	9.46
500	393	10.98	10.34
1000	785	12.00	11.22
2000	1571	12.52	12.11
4000	3142	13.43	12.99
8000	6283	14.28	13.87
16000	12566	15.01	14.75
32000	25133	15.89	15.64

APPENDIX

The authors wish to thank Professor N.C. Ankeny for several interesting discussions concerning the theory of numbers and in particular for his guidance in obtaining the asymptotic estimate (39) for the function

$$f(\nu_c) = \sum_{\nu=1}^{\nu_c} M^2(\nu) . \tag{A-1}$$

The procedure followed is typical of the methods employed in number theory although the particular function $f(\nu_c)$ does not appear to have been studied previously. The generating function [4] for M^2 is

$$\psi(s) = \sum_{\nu=1}^{d} M^2(\nu)\ \nu^{-s} . \tag{A-2}$$

It can be shown using the theorem on p. 249 of [4] that ψ has a double pole at $s = 1$; i.e.,

$$\psi(s) = \frac{c_1}{(s-1)^2} + \frac{c_2}{s-1} + c_3 + \dots \tag{A-3}$$

where $c_1 = 1/4$. Then using the theorem on p. 239 of [5] we write

$$f(\nu_c) = \frac{1}{2\pi i}\int_{1+\delta-i\sqrt{\nu_c}}^{1+\delta+i\sqrt{\nu_c}} \frac{\nu_c^s}{s}\left[\frac{c_1}{(s-1)^2} + \frac{c_2}{s-1}\right] ds \tag{A-4}$$

and deform the contour to reduce (A-4) to the residual at $s = 1$ plus an integral of order $(\nu_c)^{\frac{1}{2}}$. In this way we obtain

$$f(\nu_c) \sim c_1\nu_c \ln \nu_c + (c_2 - c_1)\nu_c . \tag{A-5}$$

In principle the coefficient c_2 can be obtained analytically. In our case the coefficient in (39) was obtained by fitting an expression of the form (A-5) to the computer calculated results of $f(\nu_c)$ for $\nu_c = 2000\,k$ $(k = 1,\dots,16)$. In this range the deviations between (39) and the computed values were of the order of 2%. A similar procedure applied to $N(\nu_c)$ as defined by (36) leads to the asymptotic estimate

$$N = \sum_{\nu=1}^{\nu_c} M(\nu) \sim \frac{\pi}{4}\, \nu_c - 0.995\sqrt{\nu_c} . \tag{A-6}$$

REFERENCES

1. S.H. Crandall and L.E. Wittig, Chladnis patterns for random vibrations of a plate, pp. 55-71 in Dynamic Response of Structures (G. Herrmann and N. Perrone, eds.), Pergamon Press, N.Y., 1972.
2. S.H. Crandall, Random Vibration of One-and Two-Dimensional Structures, pp. 1-82, in Developments in Statistics, Vol. 2, (P.R. Krishnaiah, ed.), Academic Press, N.Y., 1979.
3. R. Courant and D. Hilbert, Methods of Mathematical Physics, Wiley (Interscience) N.Y., 1953.
4. G.H. Hardy and E.M. Wright. An Introduction to the Theory of Numbers, Oxford University Press, London, 1960.
5. B.M. Wilson, Proofs of some formulae enunciated by Ramanujan, Proc. London Math. Soc. 21, 235-255 (1923).

RESPONSE OF SYSTEMS WITH RANDOMLY VARYING PARAMETERS TO EXTERNAL EXCITATION

M. F. Dimentberg[1)]

This report consists of two parts. In the first part a method of moments is outlined, which provides deterministic equations for first and second order moments of the response of linear systems with randomly varying parameters to external excitation. Two examples are solved analytically with an additional use of the averaging over the period. The second part is concerned with stationary probability density of the response. Several analytical solutions to Fokker-Planck-Kolmogorov equations are obtained, which illustrate certain general features of probability densities in systems with randomly varying parameters.

1. The problem of the response of linear systems with randomly varying parameters to external excitation may be considered as relatively simple when only first and second order moments of the response are desired. Under certain restrictions a method of moments can be applied, which provides a closed system of deterministic equations for these moments [1,2] (and which has been extensively used for studies in stochastic stability). This method is based on the fact, that the vector solution to the system of Ito stochastic differential equations (SDE) is uncorrelated with the input white-noise vector (if the values of both vectors at the same time moments are considered); for simplicity the initial conditions are considered here to be either deterministic or uncorrelated with random inputs. Thus consider the "physical" or Stratonovich SDE

$$\dot{x}_i = \sum_{j=1}^{n} b_{ij}x_j + \sum_{j=1}^{n}\sum_{r=1}^{n^2}\sigma_{ijr}x_j\xi_r(t) + \sum_{r=1}^{n}\lambda_{ir}\xi_r(t) + \sum_{r=1}^{n}\gamma_{ir}\zeta_r(t) + f_i(t),$$

$$i = 1,\dots,n, \qquad (1.1)$$

where $\xi_r(t)$, $\zeta_r(t)$ are independent zero-mean Gaussian white noises with unit intensities, whereas the coefficients b_{ij}, σ_{ijr}, γ_{ir}, λ_{ir} may be deterministic functions of time as well as $f_i(t)$. Transforming (1.1) into the corresponding Ito SDE according to well-known rules [3] and taking the mathematical expectation (this operation will be denoted by angular brackets), we obtain the desired deterministic equations for the first-order moments $m_i = \langle x_i \rangle$:

$$\dot{m}_i = \sum_{j=1}^{n} b_{ij}m_j + \frac{1}{2}\sum_{j=1}^{n}\sum_{l=1}^{n}\sum_{r=1}^{n^2}\sigma_{ilr}\sigma_{ljr}m_j + \frac{1}{2}\sum_{j=1}^{n}\sum_{r=1}^{n}\sigma_{ijr}\lambda_{jr} + f_i(t). \qquad (1.2)$$

Now we want derive Ito's SDE for the variables $u_{ik} = x_i x_k$. This can

1) Senior scientist, Institute for Problems of Mechanics USSR Academy of Sciences, Prospekt Vernadskogo 101, 117526 Moscow, USSR

be done either by "direct" calculations of the derivatives $\dot{u}_{ik}=x_i\dot{x}_k+\dot{x}_ix_k$ from Stratonovich's SDE (1.1) with subsequent transition to Ito's SDE, or by applying Ito's differentiation rule [3] to Ito's SDE, which corresponds to (1.1). Both ways lead, of course, to the same result, and after applying the mathematical expectation operator it is possible to obtain the following deterministic equations for second order moments $K_{ik} = \langle u_{ik}\rangle = \langle x_i x_k\rangle$:

$$\dot{K}_{ik} = \sum_{j=1}^{n}(b_{ij}K_{jk}+b_{kj}K_{ij}) + f_i m_k + f_k m_i +$$
$$\sum_{l=1}^{n}\sum_{j=1}^{n}\sum_{r=1}^{n^2}[\sigma_{kjr}\sigma_{ilr}K_{jl} + \tfrac{1}{2}(\sigma_{ilr}\sigma_{ljr}K_{jk}+\sigma_{ljr}\sigma_{klr}K_{ij})] +$$
$$\sum_{j=1}^{n}\sum_{r=1}^{n^2}[\tfrac{1}{2}(\sigma_{ijr}m_k+\sigma_{kjr}m_j)\lambda_{jr} + \sigma_{kjr}m_j\lambda_{ir} + \sigma_{ijr}m_j\lambda_{kr} +$$
$$\sum_{r=1}^{n}\gamma_{ir}\gamma_{kr} + \sum_{r=1}^{n}\lambda_{ir}\lambda_{kr}. \qquad (1.3)$$

It should be noted that if the right hand sides of SDE (1.1) are proportional to a small parameter, then the restrictions on $\xi_r(t)$ may be somewhat relaxed (although the outlined scheme would then only be approximate rather than exact). Namely, the $\xi_r(t)$ may be considered simply as broadband random processes, not necessarily as Gausssian ones. Moreover, in this case it is possible to apply averaging (in time) over the period for the deterministic functions σ_{ijr}, γ_{ir}, λ_{ir}, b_{ij}, f_i (if they are periodic) according to the well-known Stratonovich-Hasminsky theorem (SHT) [4].

As a first example consider a system

$$\ddot{x} + 2\alpha\dot{x} + \Omega^2 x[1+\xi(t)] = y(t), \qquad \alpha \ll \Omega, \qquad (1.4)$$

where y(t) is a narrow-band random process, defined by the following second order shaping filter SDE

$$\ddot{y} + 2\beta\dot{y} + \nu^2 y = \zeta(t), \qquad \beta \ll \nu, \qquad \alpha \ll \Omega. \qquad (1.5)$$

Here $\xi(t)$, $\zeta(t)$ are broadband zero-mean random processes with spectral densities $\Phi_{\xi\xi}(\omega)$, $\Phi_{\zeta\zeta}(\omega)$, respectively; the latter are assumed to have small values of the same order as α, β and $|\Delta|$, where $\Delta = \nu-\Omega$. Then by change of variables

$$\begin{aligned} x &= x_c\cos\nu t + x_s\sin\nu t, & \dot{x} &= (-x_c\sin\nu t + x_s\cos\nu t),\\ y &= y_c\cos\nu t + y_s\sin\nu t, & \dot{y} &= (-y_c\sin\nu t + y_s\cos\nu t) \end{aligned} \qquad (1.6)$$

system (1.4),(1.5) can be transformed to the following "standard form" (where for brevity we have already applied time averaging over the period for the terms without $\xi(t)$, $\zeta(t)$):

$$\dot{x}_c = -\alpha x_c - \Delta x_s + \tfrac{1}{2}\Omega[x_c\sin 2\nu t + x_s(1-\cos 2\nu t)]\xi(t) - y_s/2\nu,$$
$$\dot{x}_s = -\alpha x_s + \Delta x_c - \tfrac{1}{2}\Omega[x_c(1+\cos 2\nu t) + x_s\sin 2\nu t]\xi(t) + y_c/2\nu, \qquad (1.7)$$

$$\dot{y}_c = -\beta y_c - \nu^{-1}\zeta(t)\sin\nu t, \qquad \dot{y}_s = -\beta y_s + \nu^{-1}\zeta(t)\cos\nu t \ .$$

Applying SHT we obtain four first order Ito SDE for "slow" variables $x_c(t)$, $x_s(t)$, $y_c(t)$, $y_s(t)$. Subsequent application of the method of moments according to the above outlined scheme leads to a system of ten equations for second order moments (full details can be found in [2]). For the case of constant $\alpha, \beta, \nu, \Omega, \Phi$ this system has an analytical steady-state solution [2], from which we obtain the following formula for the mean-square amplitude $\langle A^2\rangle = \langle x_c^2\rangle + \langle x_s^2\rangle$:

$$A^2 = \frac{(D_\zeta/2\beta\nu^4)(\alpha+\beta+q)}{[(\alpha+\beta+q)^2 + r^2](2\alpha - \frac{1}{2}\Omega^2 D_2)} \ . \tag{1.8}$$

Here

$$D_\zeta = 2\pi\Phi_{\zeta\zeta}(\nu), \quad q = \frac{1}{8}\Omega^2(D_0 - D_2), \quad D_0 = 2\pi\Phi_{\xi\xi}(0), \quad D_2 = 2\pi\Phi_{\xi\xi}(2\nu), \tag{1.9}$$

$$r = \Delta - \frac{1}{16}\Omega^2 E, \quad E = 2\int_0^\infty K(\tau)\sin 2\nu\tau \, d\tau$$

and $K(\tau)$ is a correlation function of $\xi(t)$.

It can be seen from (1.8) that if $\Omega^2 D_2 \ll 4\alpha$ and $\Phi_{\xi\xi}(0) > 3\Phi_{\xi\xi}(2\nu)(1+2\beta/3\alpha)$, then $\max\limits_{\nu}\langle A^2\rangle = \langle A^2\rangle|_{r=0}$ is less than $\langle A^2\rangle_0$, where $\langle A^2\rangle_0$ is the value of $\langle A^2\rangle$ when $\xi(t) \equiv 0$. This is the condition for reducing resonant response to external narrow-band random or periodic excitation by means of artificially induced fluctuations of natural frequency according to the idea due to Frolov [5] (reduction is understood here in the mean square sense , and formulae for the case of periodic external excitation were obtained in [2]). However, increasing the intensity of $\xi(t)$ may lead to decreasing of $\max\limits_{\nu}\langle A^2\rangle$ only to a certain limit: afterwards $\max\limits_{\nu}\langle A^2\rangle$ begins to increase because of the second cofactor in the denominator of (1.8) and becomes unbounded with approaching mean-square stability limit (when $\Omega^2 D_2 = 4\alpha$). It is also seen from (1.8), that the above effect of reducing resonant response becomes less pronounced with increasing the bandwidth of the input $y(t)$ (i.e. with increasing β). Note also, that this effect may be predicted only in the framework of approximate asymptotic solution: it does not exist in the special case of white-noise $\xi(t)$, for which it is possible to obtain an exact solution.

As a second example consider a pair of second order SDE

$$\begin{aligned} \ddot{x}_1 + 2\alpha_1\dot{x}_1 + \Omega_1^2 x_1 &= \lambda_{12}x_2\xi(t) + \zeta_1(t), \\ \ddot{x}_2 + 2\alpha_2\dot{x}_2 + \Omega_2^2 x_2 &= \lambda_{21}x_1\xi(t) + \zeta_2(t) \end{aligned} \tag{1.10}$$

with $\xi(t)$, $\zeta_i(t)$ once again being zero-mean broadband stationary random processes. Defining

$$x_i = A_i \cos \Theta_i, \quad \dot{x}_i = -\Omega_i A_i \sin \Theta_i, \quad \Theta_i = \Omega_i t + \varphi_i, \quad V_i = A_i^2 \tag{1.11}$$

where $i = 1,2$; making use of SHT it is possible to obtain the following pair of Stratonovich SDE for $V_1(t)$, $V_2(t)$ [2]:

$$\begin{aligned} \dot{V}_i = {} & [-2\alpha_i + \tfrac{1}{2}\gamma(D_+ - D_-)]V_i + \tfrac{1}{2}\gamma_i^2(D_+ + D_-)V_j + 2D_i + \\ & \gamma_i[(2D_+V_1V_2)^{1/2}\eta_+(t) - (2D_-V_1V_2)^{1/2}\eta_-(t)] + (8D_iV_i)^{1/2}\eta_i(t), \\ D_i = {} & (2\pi/\Omega_i^2)\Phi_{\zeta\zeta i}(\Omega_i),\ D_\pm = 2\pi\Phi_{\xi\xi}(\Omega_\pm),\ \Omega_\pm = \Omega_2 \pm \Omega_1,\ \gamma_i = \lambda_{ij}/\Omega_i, \\ \gamma = {} & \gamma_1\gamma_2,\ i = 1,2;\ j \neq i. \end{aligned} \tag{1.12}$$

Here $\eta_\pm(t)$, $\eta_i(t)$ are equivalent unit noises. Although the diffusion coefficients in SDE (1.12) are nonlinear, this special root-square-type nonlinearity does permit to apply method of moments. For steady-state mean values $m_i = \langle V_i \rangle$ the following formulae are obtained [2]:

$$\begin{aligned} m_i &= 2Q^{-1}\{D_i[2\alpha_j - \gamma(D_+ - D_-)] + D_j\gamma_i(D_+ + D_-)\}, \\ Q &= 2\alpha_1\alpha_2 - \gamma(\alpha_1 + \alpha_2)(D_+ - D_-) - 2\gamma^2 D_+ D_-, \quad i = 1,2;\ i \neq j. \end{aligned} \tag{1.13}$$

Condition $Q = 0$ defines a mean-square stability limit for system (1.10), obtained previously by Ariaratnam [6]. For second order moments $\langle V_i V_j \rangle$ a third order system of deterministic equations is obtained [2]. Numerical solution of this system for the case $D_- = 0$ (stochastic combinational resonance on the sum of natural frequencies) shows [2], that the steady-state correlation coefficient

$$\varrho_{12} = \frac{\langle V_1 V_2 \rangle - m_1 m_2}{(\langle V_1^2 \rangle - m_1^2)^{1/2}(\langle V_2^2 \rangle - m_2^2)^{1/2}} \tag{1.14}$$

always monotonically increases with D_+ from zero when $D_+ = 0$ to unity when D_+ is such that the determinant of the system vanishes, i.e. when the original system (1.10) approaches a fourth order moment stochastic stability limit. This result provides a method for detecting stochastic combinational resonance and estimating the stability margin by an analysis of measured random vibration data [2].

2. To obtain probability densities of the response for systems with random parametric excitation is much more difficult. At present only a few analytical solutions are available even for stationary Fokker-Planck-Kolmogorov (FPK) equations. For general second order linear systems such a solution, based on SHT, leads to a power-law distribution of the response, and there exists a special case for which this solution is exact [7]. Two other solutions of such type are presented in the following.

Consider the Stratonovich SDE

$$\dot{x}_i = -\beta x_i[1 + \xi(t)] + \zeta_i(t), \quad i = 1,2,\ldots,n, \tag{2.1}$$

where $\xi(t)$, $\zeta_i(t)$ are independent stationary zero-mean Gaußian white noises with

$$\langle \zeta_i(t)\, \zeta_j(t+\tau)\rangle = D_\zeta \delta_{ij}\delta(\tau), \quad \langle \xi(t)\xi(t+\tau)\rangle = D_\xi \delta(\tau) \,. \tag{2.2}$$

Here $\delta(\tau)$ is the Dirac delta function and δ_{ij} Kronecker's delta. The corresponding FPK equation for the joint probability density $p(x_1, x_2, \ldots, n_n, t)$ is

$$\frac{\partial p}{\partial t} = \beta \sum_{i=1}^{n} \frac{\partial}{\partial x_i}(x_i p) + \frac{D_\zeta}{2} \sum_{i=1}^{n} \frac{\partial^2 p}{\partial x_i^2} + \frac{D_\xi}{2} \sum_{i=1}^{n} \frac{\partial}{\partial x_i}\Big[\beta x_i \sum_{j=1}^{n} \frac{\partial}{\partial x_j}(\beta x_j p)\Big] \tag{2.3}$$

and it has the following stationary ($\partial p/\partial t = 0$) solution

$$p(x_1,\ldots,x_n) = C(\varkappa + \sum_{i=1}^{n} x_i^2)^{-\delta}, \quad C = \text{const.}, \quad \varkappa = D_\zeta/D_\xi, \quad \delta = \frac{1}{\beta D_\xi} + \frac{n}{2} \,. \tag{2.4}$$

This solution is normalizable if $\beta > 0$, i.e., if system (2.1) is almost surely stable.

Consider now a group of identical linear second order oscillators with common broadband random parametric excitation $\xi(t)$ and independent braodband random external excitations $\zeta_i(t)$ with identical spectra $\Phi_{\zeta\zeta}(\omega)$:

$$\ddot{x}_i + 2\alpha\dot{x}_i + \Omega^2 x_i[1 + \xi(t)] = \zeta_i(t), \quad i = 1,\ldots,n. \tag{2.5}$$

Using transformations (1.11) and applying SHT for the case of small α, $\Phi_{\zeta\zeta}(\omega)$, $\Phi_{\xi\xi}(\omega)$, it is possible to obtain the following Stratonovich SDE for squared amplitudes $V_i(t)$:

$$\dot{V}_i = (-2\alpha + 4D_\xi)V_i + 2D_\zeta + 2\sqrt{V_i}\,\zeta_i'(t) - 2V_i\xi'(t), \quad i = 1,\ldots,n,$$

$$D_\xi = \tfrac{1}{8}\pi\Omega^2\Phi_{\xi\xi}(2\Omega), \quad D_\zeta = (\frac{\pi}{2\Omega^2})\Phi_{\zeta\zeta}(\Omega),$$

where $\xi'(t)$, $\zeta_i'(t)$ are now equivalent white noises with intensities $2D_\xi$, $2D_\zeta$, respectively. The joint probability density $p(V_1,\ldots,V_n,t)$ satisfies the FPK equation

$$\frac{\partial p}{\partial t} = \sum_{i=1}^{n} \frac{\partial}{\partial V_i}[(2\alpha - 4D_\xi)V_i p] - 2D_\zeta \sum_{i=1}^{n} \frac{\partial p}{\partial V_i} + 4D_\zeta \sum_{i=1}^{n} [V_i^{1/2} \frac{\partial}{\partial V_i}(V_i^{1/2} p)] +$$

$$4D_\zeta \sum_{i=1}^{n} \frac{\partial}{\partial V_i}\Big[V_i \sum_{j=1}^{n} \frac{\partial}{\partial V_j}(V_j p)\Big] \,. \tag{2.6}$$

Equation (2.6) has the following stationary solution

$$p(V_1,\ldots,V_n) = C(\varkappa + \sum_{i=1}^{n} V_i)^{-\delta}, \quad C = \text{const.}, \quad \varkappa = D_\zeta/D_\xi, \quad \delta = \alpha/2D_\xi + n - 1, \tag{2.7}$$

which is normalizable (and thus indeed defines stationary probability density) if $\delta > n$, i.e., if $\alpha/2D_\xi > 1$.

The main feature of the power-law distributions (2.4), (2.7) is the limited number of bounded moments; only in the limiting case $\delta \to \infty$, which corresponds to systems with constant parameters, all moments of these distributions are finite. Power-law distributions may even have infinite variance, so that the process is nonstationary in the usual broad sense. Thus it seems appropriate to propose a new definition of the type of stationarity (in addition to the two well-known existing definitions). Namely, a random process is called stationary in the new, generalized broad sense if it has a normalizable non-negative one-dimensional stationary probability density. There exists a wide class of processes being stationary in the new sense but nonstationary in the usual broad sense (and also are not strictly stationary); namely, the responses of stochastic systems, which are stable with respect to probability but unstable in the mean square, to external stationary random excitations.

It seems that the power-law-type distribution should be typical not only for linear systems with combined external and parametric random excitation, but also for certain systems with quadratic nonlinearity, in which one generalized coordinate provides parametric excitation for another generalized coordinate (the simplest example is a system with auto-parametric resonance).

Consider finally a certain class of nonlinear systems

$$\dot{x}_i = \frac{\partial H}{\partial y_i}, \qquad \dot{y}_i = -\frac{\partial H}{\partial x_i} - g(H)\frac{\partial H}{\partial y_i} + \zeta_i(t),$$
$$H = H(x_1,\ldots,x_n,y_1,\ldots,y_n,t), \qquad i = 1,\ldots,n, \tag{2.8}$$

where $\zeta_i(t)$ are the processes with the same properties as in (2.1). The FPK equation cooresponding to (2.8) is

$$\frac{\partial p}{\partial t} = -\sum_{i=1}^{n}\frac{\partial}{\partial x_i}\left(\frac{\partial H}{\partial y_i}p\right) + \sum_{i=1}^{n}\frac{\partial}{\partial y_i}\left[\frac{\partial H}{\partial x_i}p + g(H)\frac{\partial H}{\partial y_i}p\right] + \frac{D_\zeta}{2}\sum_{i=1}^{n}\frac{\partial^2 p}{\partial y_i^2} \tag{2.9}$$

and it has the foolowing stationary solution

$$p(x_1,\ldots,x_n y_1,\ldots,y_n) = C\exp\left[-(2/D_\zeta)G(H)\right], \qquad C = \text{const.}, \tag{2.10}$$
$$G(H) = \int_0^H g(H')\,dH' .$$

For a special case, $H(x_1,\ldots,x_n,y_1,\ldots,y_n) = V(x_1,\ldots,x_n) + \frac{1}{2}\sum_{i=1}^{n} y_i^2$, the distribution (2.10) was obtained by Caughey [8].

We apply now the solution (2.10) to a certain ecological system with randomly varying parameter - the so-called predator-prey pair in random environment, governed by Lotka-Volterra equations [9]

$$\dot{u} = k\beta uv - mu, \qquad \dot{v} = \alpha v[1 + \xi(t)] - \beta uv - \gamma v^2 . \tag{2.11}$$

Here $u(t)$, $v(t)$ are the numbers of predators and preys, respectively, α, β, γ, m, k are positive constants with $\alpha - \gamma m/k\beta > 0$ and $\xi(t)$ is a stationary zero-mean Gaussian white noise, which models fluctuations of the environment. Introducing new variables $x = \ln u$, $y = \ln v$ we reduce (2.11) to the form (2.8) with $n = 1$ (so it is possible to drop subscript i) and

$$H(x,y) = k\beta\, e^y - my + \beta\, e^x - (\alpha - \gamma m/k\beta)x, \; g(H) = \gamma/k\beta, \quad D_\zeta = D_\xi \alpha^2. \tag{2.12}$$

Inserting (2.12) into (2.10) and returning to original variables, we obtain finally after normalizing the following expression for a stationary joint probability density $w(u,v)$ of $u(t)$, $v(t)$:

$$w(u,v) = w_1(u)w_2(v), \quad w_1(u) = (\delta/k)^{\delta u_0/k}\, u^{(\delta u_0/k-1)}\, e^{-\delta u/k}\, \Gamma^{-1}(\delta u_0/k),$$
$$w_2(v) = \delta^{\delta v_0}\, v^{(\delta v_0-1)}\, e^{-\delta v}\, \Gamma^{-1}(\delta v_0). \tag{2.13}$$

Here $\delta = 2\gamma/D_\xi\alpha^2$, Γ is the gamma function and

$$u_0 = \alpha/\beta - \gamma m/k\beta^2, \qquad v_0 = m/k\beta \tag{2.14}$$

are the coordinates of a stable equilibrium point of system (2.11) with $\xi(t) \equiv 0$ in the (u,v)-plane. Thus $u(t)$ and $v(t)$ are independent and identically distributed. This distribution χ^2 suffers a major qualitative change with increasing noise intensity: it has a sharp peak when D_ξ is small, while for sufficiently high D_ξ it is monotonically decreasing and has a singularity in the origin. However, this singularity is integrable, so this stationary probability density always exists. Therefore, even strong fluctuations in the environment do not lead to the death of the population as long as the mean reproduction factor of preys α remains positive. Of course, this conclusion is valid only in the framework of the present model, which does not account for "threshold effects" due to discrete character of real processes $u(t)$, $v(t)$ (these effects should become important when u and v are comparable with unity).

References

1 Evlanov, L.G.; Konstantinov, V.M., Systems with Random Parameters (in Russian), Nauka, Moscow 1976

2 Dimentberg, M.F., Method of moments for systems with randomly varying parameters (in Russian), Priklad. mat. i mech. (1982), No.2

3 Stratonovich, R.L., Conditional Markov Processes, New York 1968

4 Hasminsky, R.S., A limit theorem for the solutions of differential equations with random right-hand part (in Russian), Teor. verojatn. i prilosh. (1966), No.3

5 Frolov, K.V., Reducing the response of resonant systems by means of controlled varying of parameters (in Russian), Maschinoved. (1965),No.3

6 Ariaratnam, S.T., Stability of mechanical systems under stochastic parametric excitation, in "Stability of Stochastic Dynamical Systems", Springer-Verlag, Berlin-Heidelberg-New York 1972

7 Dimentberg, M.F., Nonlinear Stochastic Problems of Mechanical Vibrations (in Russian), Nauka, Moscow 1980

8 Caughey, T.K., Nonlinear theory of random vibrations, Advances in appl. Mech. 11(1971)

9 Svireev, Ju.M.; Logofet, D.O., Stability of Biological Communities (in Russian), Nauka, Moscow 1978.

CONDITIONAL PROBABILITY CONCEPTS IN RANDOM VIBRATION

O. Kropáč [1)]

Different forms of conditionality in the probability sense are considered. The main part of the paper deals with dynamic systems having randomly changing parameters. The response of such a system is first considered to be conditional with respect to some particular value of the random parameter. Knowing the distribution function of the random parameter the unconditioned response is expressed applying the rule for the total probability of unconditioned events. One simple example is presented and possibilities of further directions of generalization are shortly outlined.Some questions to be discussed are also suggested.

Introduction

Conditional probabilities constitute an important part of the basic methodological apparatus being used for analyzing problems in which random phenomena occur. The conditionality may be considered in several forms. The most often used relates the distribution function of a joint distribution conditional with respect to some fixed value of one of the random vector components. In applied statistics and especially in different engineering fields the conditional distribution function is very suitable for graphical presentation of joint distributions using probability charts /11/. For the study of the nonlinear structure of stochastic processes it has proved to be advantageous to consider distribution functions conditional with respect to the state variable shifted at some time lag and to introduce the so-called dispersion function as the variance of its conditional expected value /16/. Among further useful mathematical tools, martingales should be also mentioned providing for non-Gaussian processes an interlink between processes with independent and with uncorrelated increments /17/.

Distribution function with random parameters

In this contribution we shall be concerned with an other type of conditionality, viz that with respect to the parameters of distribution functions. The methodological approach to this topic has been presented in /9/, applications relating stochastic processes have been considered in /12/,/14/. Combining deterministic and random changes of parameters of a primary normal process a unified analytical model for non-stationary and/or non-Gaussian random processes could be set up /10/.

---

1) Ing. Oldřich Kropáč, CSc, Aeronautical Research and Test Institute, 199 05 Prague - Letňany, Czechoslovakia

The basic idea of this approach may be summarized as follows: Let a continuous stationary random process X(t) be described by means of probability densities $f_1(x;a,b,\ldots)$, $f_2(x,x_\tau;a,b,\ldots,r(\tau))$. Assume now that some of the parameters of f_1 is random, e.g. the parameter Λ being described by means of the distribution function $G(\lambda;\alpha,\beta,..)$ with deterministic parameters $\alpha, \beta, \ldots$. Then $f_1(x;a,b,\ldots|\lambda)$ is to be considered as conditional with respect to some particular value λ of the random parameter Λ . The density of the total probability of the phenomenon in question which includes all possible values of Λ is given by the known relation

$$f_1^*(x;a,b,\ldots,\alpha,\beta,\ldots) = \int_{-\infty}^{\infty} f_1(x;a,b,\ldots|\lambda)\, dG(\lambda;\alpha,\beta,\ldots). \tag{1}$$

For $f_1(x;a,b,\ldots|\lambda)$, arbitrary analytical forms may be chosen in principle /9/ but for the application in the field of random processes and dynamic systems, it is advantageous to use for $f_1(x|\lambda)$ the normal distribution with random variance, thus $D_X \to \Lambda$. Then, very interesting large classes of elliptically symmetric distributions may be generated /3/, /12/ for which the joint probability densities may be expressed using the relation

$$f_2^*(x,x_\tau;\alpha,\beta,\ldots,r(\tau)) = \int_0^{\infty} f_2(x,x_\tau;r(\tau)|\lambda)\, dG(\lambda;\alpha,\beta,\ldots) \tag{2}$$

where $f_2(x,x_\tau;r(\tau)|\lambda)$ is the joint probability density of a centred normal process with random variance Λ and with normalized correlation function $r(\tau)$. These distributions which include e.g. the Pearson types II and VII and distributions of the type $x^\nu K_\nu(x)$ /14/ where $K_\nu(x)$ is the modified Bessel function of the second kind of order ν, have some properties which are the same or very similar to some important properties of normal distribution.

The random nature of parameters may be also considered for parameters of the joint distribution function being in excess to the marginal distribution. For the normal process the normalized correlation function is relevant in this respect. It is necessary to express the correlation function in a suitable analytical form with some free parameters. We write e.g. $r(\tau) = \varrho(\tau;p)\cos q\tau$ with $\varrho(\tau;p)$ having the known properties required for correlation functions. Assuming some of the parameters p, q to be random and denoting it as Λ the unconditioned correlation function $r^*(\tau)$ will be expressed in a similar way as $f_1^*(x)$, viz

$$r^*(\tau;\ldots,\alpha,\beta,\ldots) = \int_{-\infty}^{\infty} r(\tau;\ldots|\lambda)\, dG(\lambda;\alpha,\beta,\ldots). \tag{3}$$

Instead of correlation function the power spectral density is more suitable for solving particular problems of theoretical and applied dynamics. The relation for unconditioned power spectral density given the conditional one and the distribution of the random parameter Λ has the form /10/

$$S^*(\omega;\ldots,\alpha,\beta,\ldots) = \int_{-\infty}^{\infty} S(\omega;\ldots|\lambda)\, dG(\lambda;\alpha,\beta,\ldots). \qquad (4)$$

One input - one output dynamic system with one random parameter

The idea of a correlation function or a power spectral density with random parameters and their reduction into unconditioned characteristics may be very advantageous when solving dynamic systems with random input processes and random changes of parameters. For the solution of systems of this type, several methods have been proposed based usually on the averaging technique or using the approach leading to Markov processes with the evaluation of moments of the response, see e.g. /1/,/2/, /4/,/5/,/6/,/18/,/20/. There are some limitations imposed both on the properties of the problem input quantities and on the depth of obtained output informations. It is the aim of this paper to show that the concept of conditional probabilities may provide more detailed results with relatively simple analytical tools /13/.

Let us consider a stable linear dynamic system with concentrated parameters, with one input and one output. Such a system may be described in the form

$$\mathcal{L}\{Y(t)\} = X(t) \qquad (5)$$

with stationary random input process $X(t)$ which is $N(C,D_X)$ with a given covariance function $K_X(\tau) = D_X\, r_X(\tau)$ or the (one-sided) power spectral density $S_X(\omega) = (2/\pi)\, D_X \int_0^{\infty} r_X(\tau)\cos\omega\tau\, d\tau$. From the operator $\mathcal{L}$ the frequency transfer function $H(i\omega)$ of the system may be derived expressing the transformation of the harmonic input signal into a harmonic response signal. With the random input $X(t)$ as specified above the expression between the power spectral densities of input and steady-state output takes the known form

$$S_Y(\omega) = |H(i\omega)|^2\, S_X(\omega) \qquad (6)$$

where, due to the linearity of the system, $Y(t)$ is $N(C,D_Y)$ with $D_Y = \int_0^{\infty} S_Y(\omega)\, d\omega$.

It has been assumed up to now that both $\mathcal{L}$ and $H(i\omega)$ contain some constant parameters. Considering these parameters to be deterministically dependent on t we obtain an analytical model of a system with variable parameters. This case will not be discussed here. Instead, we shall consider that some parameter of $H(i\omega)$, let us denote it Λ is of random nature and is described by means of the distribution function $G(\lambda;\alpha,..)$ where $\alpha,\beta,...$ are again deterministic parameters. Let the choise of G(.) be limited by the requirement that the relevant system should remain in the stable state for all possible values of Λ. Thus the questions connected with the stability behaviour could be omitted in our considerations although they are important in general (see e.g./7/,/19/). Then the frequency transfer $H(i\omega)$ is to be considered as conditional with respect to some particular value λ of the random variable Λ, thus $H(i\omega;k,m,...|\lambda)$. The unconditioned frequency transfer function $H^*(i\omega)$ taking into account the random variability of Λ over its whole definition interval may be expressed using the rule of total probability in the form

$$H^*(i\omega) = \int_{-\infty}^{\infty} H(i\omega;k,m,...|\lambda)\, dG(\lambda;\alpha,\beta,...) = H^*(i\omega;k,m,...,\alpha,\beta,...). \tag{7}$$

For the power spectral density of the unconditioned output we obtain

$$S_Y^*(\omega) = |H^*(i\omega;k,m,...,\alpha,\beta,..)|^2\, S_X(\omega) = S_Y^*(\omega;k,m,..,\alpha,\beta,..). \tag{8}$$

Alternatively, the conditional power spectral density of the output may be expressed as

$$S_Y(\omega|\lambda) = |H(i\omega;k,m,...|\lambda)|^2\, S_X(\omega) = S_Y(\omega;k,m,...|\lambda) \tag{9}$$

and the unconditioned power spectral density of the output in the form

$$S_Y^*(\omega) = \int_{-\infty}^{\infty} S_Y(\omega;k,m,..|\lambda)\, dG(\lambda;\alpha,\beta,...) = S_Y^*(\omega;k,m,...,\alpha,\beta,..) \tag{10}$$

which gives, of course, the same result as equation (8).

We shall show further that the random character of Λ affects the analytical form of the distribution of the output process $Y^*(t)$. Following the assumption that $X(t)$ is $N(0,D_X)$, $Y(t)$ is $N(0,D_Y)$. Considering the randomness of Λ we have

$$D_Y = \int_0^{\infty} S_Y(\omega;k,m,...|\lambda)\, d\omega = D_Y(k,m,...|\lambda).$$

The probability density of the unconditioned distribution of the output process $Y^*(t)$ is then to be expressed as follows:

$$f_Y^*(y) = \int_{-\infty}^{\infty} [2\pi D_Y(k,m,...|\lambda)]^{-1/2} \exp\left[-(y^2/2) D_Y^{-1}(k,m,...|\lambda)\right] \cdot dG(\lambda;\alpha,\beta,...) = f_Y^*(y;k,m,...,\alpha,\beta,...). \tag{11}$$

Likewise, an equivalent relation must hold:

$$D_Y^*(k,m,...,\alpha,\beta,...) = \int_{-\infty}^{\infty} D_Y(k,m,...|\lambda)\, dG(\lambda;\alpha,\beta,...) = \int_{-\infty}^{\infty} y^2 f_Y^*(y;k,m,...,\alpha,\beta,...)\, dy. \tag{12}$$

One may state that the parameters α,β of the random parameter Λ not only enter the variance D_Y^* of the output process but the randomness of Λ causes structural modification of the output probability density f_Y^* in such a way that $D_Y^* \geqq D_Y$ and the kurtosis $B_Y^* \geqq 0$ (for details see /9/). This feature may have great importance especially when solving various strength and reliability problems.

An illustrative example

Let be given the differential equation

$$\dot{Y}(t) + a\, Y(t) = X(t) \tag{13}$$

where X(t) is $N(0,D_X)$ with the normalized correlation function $r(\tau) = \exp(-\alpha|\tau|)$, i.e. the one-sided power spectral density of the input is $S_X(\omega) = 2\, D_X(\alpha/\pi)(\alpha^2+\omega^2)^{-1}$. The power spectral density of the response will be

$$S_Y(\omega) = S_X(\omega)|H(i\omega)|^2 = 2 D_X(\alpha/\pi)\left[(\alpha^2+\omega^2)(a^2+\omega^2)\right]^{-1}. \tag{14}$$

The corresponding covariance function of the output is

$$K_Y(\tau) = D_X\, a^{-1}(\alpha^2 - a^2)^{-1}\left[\alpha \exp(-a|\tau|) - a \exp(-\alpha|\tau|)\right] \tag{15}$$

and the variance is

$$D_Y = K_Y(0) = D_X\, a^{-1}(a+\alpha)^{-1} \tag{16}$$

whereby, due to the linearity of the system, Y(t) is $N(0,D_Y)$. Now, consider the parameter a to be random, i.e. $a \rightarrow \Lambda$ with the gamma (Pearson III) probability density

$$g(\lambda) = \beta^{-\varkappa}\Gamma(\varkappa)^{-1}\lambda^{\varkappa-1}\exp(-\lambda/\beta),\ \lambda \geqq 0,\ \beta>0,\ \varkappa>0.$$

Then the power spectral density of the output will be (see e.g./15/):

$$\begin{aligned} S_Y^*(\omega) &= 2D_X\alpha\left[\pi\beta^{\varkappa}\Gamma(\varkappa)(\alpha^2+\omega^2)\right]^{-1}\int_0^\infty \lambda^{\varkappa-1}(\lambda^2+\omega^2)^{-1}\exp(-\lambda/\beta)\,d\lambda = \\ &= 2D_X\alpha\left[\pi\beta^{\varkappa}\Gamma(\varkappa)(\alpha^2+\omega^2)\right]^{-1}\Big[\omega^{\varkappa-2}\Big\{\sin[\omega/\beta-(\pi/2)(\varkappa-1)]\,\mathrm{Ci}(\omega/\beta) - \\ &\quad - \cos[\omega/\beta-(\pi/2)(\varkappa-1)]\,\mathrm{Si}(\omega/\beta)\Big\}\Big] + \beta^{-\varkappa}\sum_{k=1}^{[(\varkappa-1)/2]}(\varkappa-1-2k)!\,(-\omega^2/\beta^2)^{k-1} \end{aligned} \tag{17}$$

Upon taking into account that the variance D_Y also changes with respect to the random parameter $a \rightarrow \Lambda$ the probability density of $Y^*(t)$ can be expressed in the form

$$f_Y^*(y) = \int_0^\infty (2\pi D_Y)^{-1/2}\exp\left[-y^2/(2D_Y)\right]\beta^{-\varkappa}\Gamma(\varkappa)^{-1}\lambda^{\varkappa-1}\exp(-\lambda/\beta)\,d\lambda \tag{18}$$

and, with $D_Y = D_X\left[\lambda(\lambda+\alpha)\right]^{-1}$

$$\begin{aligned} f_Y^*(y) &= (2\pi D_X)^{-1/2}\beta^{-\varkappa}\Gamma(\varkappa)^{-1}\int_0^\infty \lambda^{\varkappa-1/2}(\lambda+\alpha)^{1/2}\times \\ &\quad \times\exp\left[-\lambda(\lambda+\alpha)y^2/(2D_X) - \lambda/\beta\right]d\lambda. \end{aligned} \tag{19}$$

In order to evaluate the last expression in a closed form, let us consider the particular case $\alpha \to \infty$ (process X(t) approaches white noise), i.e. $r_X(\tau) \to \delta(\tau)$, $S_X(\omega) \to (2/\pi) N_X$ where $N_X = \lim_{\alpha \to \infty}(D_X/\alpha)$ is the intensity of the white noise input. Then, instead of expressions (14) to (16), we have

$$S_Y(\omega) = (2/\pi)\, N_X\, (a^2 + \omega^2)^{-1}, \tag{14a}$$

$$K_Y(\tau) = (N_X/a) \exp(-a|\tau|), \tag{15a}$$

$$D_Y = N_X/a. \tag{16a}$$

Considering $a \to \Lambda$ and putting $D_Y = N_X/\lambda$ into (18) we obtain

$$f_Y^*(y) = (2\pi N_X)^{-1/2} \beta^{-\varkappa} \Gamma(\varkappa)^{-1} \int_0^\infty \lambda^{\varkappa - 1/2} \exp\left[- y^2\lambda/(2N_X) - \lambda/\beta\right] d\lambda$$

which after evaluation yields the Pearson VII distribution for $Y^*(t)$:

$$f_Y^*(y) = (2\pi N_X)^{-1/2} \beta^{-\varkappa} \Gamma(\varkappa + 1/2) \Gamma(\varkappa)^{-1} \left[y^2/(2\,N_X) + 1/\beta\right]^{-\varkappa - 1/2}. \tag{20}$$

For $\varkappa = 1$ and $N_X = D_X/\alpha$ we obtain the expression derived earlier in /10/. Another limit case which is of particular interest is that with $\varkappa \to \infty$. Then, the distribution of the random parameter Λ turns over into the normal one and the distribution of the response $Y^*(t)$ will also be normal.

One parameter of the system is a stationary random process

So far, we have considered the parameter Λ to be a random variable. This is equivalent with the assumption that $\Lambda(t)$ is a white noise process. In concluding further generalization of this approach, let us consider $\Lambda(t)$ to be a second-order random process described by means of its joint distribution function $G_2(\lambda, \lambda_\tau; \alpha, \beta, \ldots, r_\lambda(\tau; \eta, \xi, \ldots)$ where $r_\lambda(.)$ is the function of the correlation coupling having constant parameters η, $\xi, \ldots$. The power spectral density of the conditional output $Y(t|\lambda)$ is to be expressed in the form

$$S_Y(\omega | \lambda, \lambda_\tau) = \bar{H}(i\omega; k, m, \ldots | \lambda)\, H(i\omega; k, m, \ldots | \lambda_\tau)\, S_X(\omega) \tag{21}$$

where $\bar{H}(.)$ is the function complex conjugated with H(.).
The power spectral density of the unconditioned output $Y^*(t)$ will then be expressed in the form

$$S_Y^*(\omega) = \iint_{-\infty}^{\infty} S_Y(\omega|\lambda, \lambda_\tau)\, d^2G_2(\lambda, \lambda_\tau; \alpha, \beta, \ldots, r_\lambda(\tau; \eta, \xi, \ldots)) = \tag{22}$$
$$= S_Y^*(\omega; k, m, \ldots, \alpha, \beta, \ldots, \eta, \xi, \ldots).$$

The joint distribution of the conditional output process will be normal with the conditionality in the covariance function $K_Y(\tau|\lambda, \lambda_\tau)$ which may be derived from $S_Y(\omega|\lambda, \lambda_\tau)$ applying the inverse Fourier transformation.

Thus the joint probability density of the conditional output process may be written as follows:

$$f_2(y, y_\tau | \lambda, \lambda_\tau) = \left[2\pi D_y(..|\lambda, \lambda_\tau)\right]^{-1}\left[1 - r_Y^2(\tau|\lambda, \lambda_\tau)\right]^{-1/2} \times \tag{23}$$
$$\times \exp\left\{-\left[2(1 - r_Y^2(\tau|\lambda, \lambda_\tau)) D_Y(..|\lambda, \lambda_\tau)\right]^{-1}\left[y^2 + y_\tau^2 - 2 r_Y(\tau|\lambda, \lambda_\tau) y y_\tau\right]\right\}.$$

The joint probability density of the unconditioned output $Y^*(t)$ will then be expressed from

$$f_2^*(y, y_\tau) = \iint_{-\infty}^{\infty} f_2(y, y_\tau | \lambda, \lambda_\tau)\, d^2 G_2(\lambda, \lambda_\tau; \alpha, \beta, \dots, r_\lambda(\tau; \eta, \zeta, \dots)) = \tag{24}$$
$$= f_2^*(y, y_\tau; k, m, \dots, \alpha, \beta, \dots, \eta, \zeta, \dots)$$

which is generally non-Gaussian. Note also that $r_Y^*(\tau)$ will generally differ from $r_Y(\tau)$.

Further possibilities of generalization

The simple model considered up to now may be generalized in several directions, namely:

1. The distribution function $G_2(\lambda, \lambda_\tau)$ again contains some random parameter, say Γ, having a distribution function $\mathcal{G}(\gamma)$. This is a model for serially chained random mechanisms.
2. The frequency transfer function $H(i\omega)$ contains several random parameters Λ_i, $i = 1, 2, \dots, n$, described by means of a 2n-component distribution function $G_{2n}(\lambda_1, \lambda_{1\tau}, \dots, \lambda_n, \lambda_{n\tau})$. This is a model for parallel action of several random mechanisms.
3. The dynamic system has several inputs and outputs so that the functions $H(i\omega)$, $S_X(\omega)$, $S_Y(\omega)$ and the relations between them are to be considered in the sense of matrix algebra.
4. For the system, not only the steady-state solution but also the description of transition regimes is required. In this case the initial conditions may be considered as random variables, as well.
5. The dynamic system is time-variant, i.e. its parameters are deterministic functions of time. As a result, non-stationary responses appear on the output even when processes acting on the input are stationary.
6. The input processes are non-stationary resulting in the non-stationary outputs even when the system is time-invariant. The approach using the concept of processes with evolutionary spectra may be useful in this case.
7. The dynamic system is nonlinear. This is extremely difficult to be mastered for there is no universal method of solution. For many applied engineering problems including random vibration the approach based on the properties of Markov processes has been found to be useful.

8. Several types of generalization apply simultaneously.
It is evident that it is necessary to be concerned in problems connected with the analytical solvability of particular applied problems. In this respect the following physically motivated principle may be of some value: The conditional and unconditioned states exist simultaneously and the transition from the conditional state to the unconditioned one may be realized in an arbitrary stage of the solution process particularly in that moment in which this is most advantageous from the point of view of analytical solvability. For linear systems the validity of this procedure follows from the possibility for linear limited operators to be interchanged.

Some further subjects to be discussed

In concluding this contribution we shall outline some further suggestions which arise in connection with the possibilities for the described approach to be practically used and which deserve some discussion. For the expression of unconditioned variables or functions with various application meanings and commonly denoted here as $Q^{*}(.)$ from conditional ones $Q(.|\lambda)$ the general expression for the total probability has been used in the form

$$Q^{*}(.) = \int_{-\infty}^{\infty} Q(.|\lambda)\, dG(\lambda). \tag{25}$$

This expression is formally equivalent with the expression for the mean value of the random quantity $Q(.)$, thus $Q^{*}(.) = E[Q(.)]$. A question then arises if and under what conditions it would be useful to consider some other characteristics of location or to consider higher-order moments of $Q(.)$, particularly from the point of view of evaluation possibilities for resulting analytical expressions in question.

For the characterization of location, median and mode are most often used when not considering the mean value. When the function $Q(.|\lambda)$ is monotonic with respect to λ, then the known relation $Me[Q(.|\lambda)] = Q(Me[.|\lambda])$ holds. Its evaluation is generally simpler then that of the mean value following equation (25). The information in terms of median may thus be useful in cases when the mean value does not happen to be expressed by means of some currently used analytical functions. Note that the analytical form of $Q^{*}(.)$ is the same as that of $Q(.|\lambda)$ in this case and the consequences which follow for particular engineering problems must be carefully examined.
As related mode, it holds $Mo[\Lambda] = \max_{\lambda \in R} g(\lambda)$ and $Mo[Q(.|\lambda)] = \max_{Q \in R} g[\lambda(Q)] \times |d\lambda/dQ|$ provided $\lambda(Q)$ is monotonic. Its evaluation is generally more difficult as compared with the median estimation, moreover its use is

limited to distributions which are not too non-symmetric.
As related higher-order moments the dispersion

$$D[Q(.)] = \int_{-\infty}^{\infty} \{Q(.|\lambda) - E[Q(.|\lambda)]\}^2 \, dG(\lambda) \qquad (26)$$

should be considered first. It is evident, however, that its evaluation would not be easy in general. Alternatively, the quartile deviation may be taken into account, i.e. half of the difference of the upper and the lower quartiles $(\lambda_{0,75} - \lambda_{0,25})/2$ and $(Q(.)_{0,75} - Q(.)_{0,25})/2$, respectively. These quartiles may be evaluated under the same conditions as medians but with the distinction whether the function $Q(.|\lambda)$ is increasing or decreasing.
Considerations which have been presented for the case with one random variable Λ may be adequately generalized for a vector random variable.

Conclusions

In this contribution we have shortly outlined the possibilities of how to exploit the concept of conditional and unconditional (total) probabilities for the solution of random vibration problems in dynamic systems with random parameters. The particular realization of the basic idea has been shown for linear systems but it is believed that the approach presented may be easily adapted for nonlinear systems, as well. In this connection, it would be highly desirable to discuss the relation of the given approach to the solution exploiting properties of Markov processes. It seems that the methodology presented here enables, as compared with the standard approach, the solution of more generally formulated problems to be obtained in a larger and deeper extent and with less requirements laid on necessary mathematical tools. On the other hand, the present method requires for the vector of external inputs and the vector of random parameters to be statistically independent. Evidently, all indicated application possibilities and generalizations of the simple primary model are to be elaborated in details necessary for particular applied problems.

References

/1/ Ariaratnam S.T.-Graefe P.W.U.: Linear systems with stochastic coefficients. Int.J.Control, 1, 1965, pp. 239 - 248, ibid. 2, 1965, pp. 161 - 169, 205 - 211.

/2/ Bunke H.: Gewöhnliche Differentialgleichungen mit zufälligen Parametern. Berlin, Akademie-Verlag 1972.

/3/ Chmielewski M.A.: Elliptically symmetric distributions: A review and bibliography. Internat.Statist.Review, 49, 1981, pp. 67 - 74.

/4/ Clarkson B.L.(ed.): Stochastic problems in dynamics. London, Pitman 1977.

/5/ Dimentberg M.F.: Nonlinear stochastic problems of mechanical vibration (in Russian). Moscow, Nauka 1980.

/6/ Drexler J.-Kropáč O.: One class of nonlinear stochastic differential equations characterized by random excitation. In: Hetényi M.-Vincenti W.G.(eds.):Applied Mechanics. Proceed.12th IUTAM Congress, Stanford University, Aug.26-31, 1968. Berlin, Springer 1969, pp. 179 - 191.

/7/ Hasminskii R.Z.: Stability of differential equation systems under random excitation of their parameters (in Russian). Moscow, Nauka, 1969.

/8/ Hennig K.-Heinrich W.: Zufallsschwingungen mechanischer Systeme. Berlin, Akademie-Verlag 1977.

/9/ Kropáč O.: Probability distributions with random parameters and their engineering applications (in Czech). Strojnícky Časopis (Bratislava), 31, 1980, pp. 597 - 622.

/10/ Kropáč O.: A unified model for non-stationary and/or non-Gaussian random processes. J.Sound Vibrat., 79, 1981, pp. 11 - 21.

/11/ Kropáč O.: Solution of stochastic dynamics problems using conditional characteristics. In: Proceed. 13th International Conference on Machine Dynamics INTERDYNAMICS 81, Warszawa, April 6 - 10, 1981, pp. 251 - 258.

/12/ Kropáč O.: Some general properties of elliptically symmetric and some related random processes. Kybernetika (Prague), 17, 1981, pp. 401 - 412.

/13/ Kropáč O.: Response of systems with random external and parametric excitation: A conditional probability approach. In: Proceed. IX th International Conference on Nonlinear Oscillations, Kiev, Aug.30-Sept.6, 1981.

/14/ Kropáč O.: Some properties and applications of probability distributions based on MacDonald function. Aplikace Matematiky (Prague), 27, 1982, pp. 285 - 302.

/15/ Prudnikov A.P.-Brychkov Yu.A.-Marichev O.I.: Integrals and series. Elementary functions (in Russian). Moscow, Nauka 1981.

/16/ Raibman N.S.(ed.): Dispersional identification (in Russian). Moscow, Nauka, 1982.

/17/ Segall A.: Stochastic processes in estimation theory. IEEE Trans. on Information Theory, IT-22, 1976, pp. 275 - 286.

/18/ Schmidt G.: Vibrating mechanical systems with random parameter excitation. In: Koiter W.T.(ed.): Theoretical and applied mechanics. Proceed. 14th IUTAM Congress, Delft, Aug.30 - Sept.4, 1976. Amsterdam, North-Holland 1977.

/19/ Wedig W.: Stabilitätsbedingungen für Schwingungssysteme mit zufälliger Parametererregung. Dissertation Universität Karlsruhe 1969.

/20/ Yevlanov L.G.-Konstantinov V.M.: Systems with random parameters (in Russian). Moscow, Nauka 1976.

RANDOM VIBRATION OF MULTI-STORY BUILDING ON COMPLIANT SOIL UNDER EARTHQUAKE AND WIND EXCITATIONS

Y. K. Lin[#]

SUMMARY

An analytical solution is given for the statistics of structural response of an N-story building to either earthquake or wind excitations, taking into account the soil compliancy under the footing. The earthquake input is a horizontal free-field ground acceleration, modeled as an evolutionary random process. The wind load inputs are lumped at the floor levels, and are assumed to be stationary random processes characterized by their spectral densities and cross-spectral densities.

INTRODUCTION

Recognizing that soil compliance can change significantly the response of a building under dynamic excitations, there has been considerable interest in developing analytical methods to take this effect into account. Such an analysis is difficult for a multistory building involving many degrees of freedom since stiffness and damping attributed to the soil mass under the footing are frequency dependent[2] and the combined building-soil system does not possess a set of classical normal modes. In the present paper the problem is simplified by assuming that the super-structure is composed of N identically constructed stories, making it possible to obtain some closed form solutions for an arbitrary N. The advantage of closed form solutions includes (a) the ease with which the effect of changing any single parameter can be examined, (b) much smaller computation effort required to obtain numerical results.

HORIZONTAL EARTHQUAKE EXCITATION

As shown in Fig. 1, the structural model consists of an N-story super-structure founded on a rigid footing which, in turn, is supported by a flexible soil mass. The combined structure-soil system is linear. It is assumed further that (a) the mass of the super-structure is concentrated at the floor levels; (b) the deformation of the dynamic system can be characterized by the horizontal translations of various floors and the footing, and the rotation of the footing; (c) linear elasticity is provided by massless columns or shear wall between neighboring floors; (d) damping due to the motion of the super-structure is linear and viscous , and (e) the interaction between the footing and the supporting soil can be described by a set of frequency-domain impedance functions.

The motion of the jth story-unit in the super-structure is governed by

$$m\ddot{u}_j^t + \beta\dot{u}_j^t = U_j^+ - U_j^-, \tag{1}$$

[#]Professor of Aeronautical and Astronautical Engineering and Civil Engineering University of Illinois at Urbana-Champaign, Urbana, Illinois 61801, U.S.A.

$$U_j^- = U_{j-1}^+ = k(u_j - u_{j-1}), \tag{2}$$

$$u_j^t = u_j + u_g + u_a + jh\phi \tag{3}$$

where j = number of stories above the footing (when used as a subscript, it indicates the jth floor), m = floor mass, u^t = total floor translation, β = viscous damping coefficient, U^+ = shear force from above, U^- = shear force from below, k = stiffness of the shear-wall, u = translation relative to the footing (which is treated also as the 0th floor), u_g = free field horizontal ground translation during an earthquake, h = single story height, u_a = footing translation in addition to the free field ground motion, φ = footing rotation. Substituting (3) into (1), taking the Fourier transforms of (1) and (2), and rearranging terms, we obtain a frequency-domain matrix equation

$$\begin{Bmatrix} \overline{u}_j \\ \overline{U}_j^+ \end{Bmatrix} = [T] \begin{Bmatrix} \overline{u}_{j-1} \\ \overline{U}_{j-1}^+ \end{Bmatrix} + (-m\omega^2 + i\beta\omega) \begin{Bmatrix} 0 \\ \overline{u}_g + \overline{u}_a + jh\ \overline{\phi} \end{Bmatrix} \tag{4}$$

where an over-bar denotes the Fourier transform, ω is the frequency, and

$$[T] = \begin{bmatrix} 1 & k^{-1} \\ (-m\omega^2 + i\beta\omega) & k^{-1}(-m\omega^2 + i\beta\omega) + 1 \end{bmatrix}. \tag{5}$$

Repeated application of Eq. (4) results in

$$\begin{Bmatrix} \overline{u}_j \\ \overline{U}_j^+ \end{Bmatrix} = [T]^j \begin{Bmatrix} 0 \\ \overline{U}_0^+ \end{Bmatrix} + (-m\omega^2 + i\beta\omega) \sum_{r=1}^{j} [T]^{j-r} \begin{Bmatrix} 0 \\ \overline{u}_g + \overline{u}_a \end{Bmatrix}$$

$$+ (-m\omega^2 + i\beta\omega) \sum_{r=1}^{j} r[T]^{j-r} \begin{Bmatrix} 0 \\ h\ \overline{\phi} \end{Bmatrix} \quad ; \quad j = 1, 2, \ldots, N, \tag{6}$$

where the boundary condition $\overline{u}_0 = 0$ has been imposed. Letting j=N and imposing another boundary condition $\overline{U}_N^+ = 0$ we obtain from the second row of (6),

$$\overline{U}_0^+ = \frac{m\omega^2 - i\beta\omega}{\tau_{22}(N)} [(\overline{u}_g + \overline{u}_a) \sum_{r=1}^{N} \tau_{22}(N-r)$$

$$+ h\overline{\phi} \sum_{r=1}^{N} r\tau_{22}(N-r)] \tag{7}$$

where $\tau_{pq}(n)$ represents the (p,q) element of matrix $[T]^n$.

Matrix [T] is a transfer matrix which has the properties that its determinant is equal to one and its eigenvalues are inverse of each other. Denoting the eigenvalues of [T] by exp(iψ), it can be shown that[5]

$$\cos\psi = 1 + \frac{-m\omega^2 + i\beta\omega}{2k}, \tag{8}$$

$$\tau_{12}(j) = \frac{\sin j\psi}{k\sin\psi}, \tag{9}$$

$$\tau_{22}(j) = \frac{\cos(j+\frac{1}{2})\psi}{\cos(\frac{\psi}{2})}. \tag{10}$$

Using (10), the summations in Eq. (7) can be obtained in closed form, resulting in[5],

$$\frac{\overline{U}_0^+}{k} = \frac{2}{\cos(N+\frac{1}{2})\psi}\left[(\overline{u}_g + \overline{u}_a)\sin N\psi \sin(\frac{\psi}{2}) + h\overline{\phi}\sin(\frac{N+1}{2}\psi)\sin(\frac{N\psi}{2})\right]. \tag{11}$$

The summations in (6) can also be carried out, yielding after substituting (11),

$$\overline{u}_j = \left[\frac{\cos(N-j+\frac{1}{2})\psi}{\cos(N+\frac{1}{2})\psi} - 1\right](\overline{u}_g + \overline{u}_a) + \left[\frac{\sin j\psi}{2\sin\frac{\psi}{2}\cos(N+\frac{1}{2})\psi} - j\right]h\overline{\phi} \tag{12}$$

and $\overline{U}_j^+ = k(\overline{u}_{j+1} - \overline{u}_j)$.

The translation of the footing is governed by

$$m_0(\ddot{u}_g + \ddot{u}_a) = U_0^+ - U_0^- \tag{13}$$

where m_0 = the mass of the footing and U_0^- = the reactive horizontal shear force from the soil. The rotation of the super-structure and footing as a whole is governed by

$$\sum_{j=1}^{N} mjh(\ddot{u}_j + \ddot{u}_g + \ddot{u}_a + jh\ddot{\phi}) + I_t\ddot{\phi} + M_0 = 0 \tag{14}$$

in which $I_T = \sum_{j=0}^{N} I_j$ and I_j is the rotational inertia of the jth floor about its own centroidal axis, M_0 = reactive base moment from the soil. The frequency-domain versions of (13) and (14) are

$$-m_0\omega^2(\overline{u}_g + \overline{u}_a) = \overline{U}_0^+ - U_0^-, \tag{15}$$

$$-m\omega^2\sum_{j=1}^{N} j\overline{u}_j - \frac{N(N+1)}{2}m\omega^2 h(\overline{u}_g + \overline{u}_a)$$

$$-\left[I_t + \frac{N(N+1)(2N+1)}{6}mh^2\right]\omega^2\overline{\phi} + \overline{M}_0 = 0. \tag{16}$$

The summation in (16) is found to be

$$\sum_{j=1}^{N} j\bar{u}_j = \left[\frac{\sin(\frac{N\psi}{2})\sin(\frac{N+1}{2}\psi)}{2\sin^2(\frac{\psi}{2})\cos(N+\frac{1}{2})\psi} - \frac{N(N+1)}{2}\right](\bar{u}_g + \bar{u}_a)$$

$$+ \left[\frac{(N+1)\sin N\psi - N\sin(N+1)\psi}{8\sin^3(\frac{\psi}{2})\cos(N+\frac{1}{2})\psi} - \frac{N(N+1)(2N+1)}{6}\right] h\bar{\phi}. \tag{17}$$

The force $\bar{U}_0^-$ and moment M_0 exerted upon the footing by the soil are linearly related to the displacements u_a and ϕ of the footing in the frequency domain as follows

$$\begin{Bmatrix} \bar{U}_0^- \\ \bar{M}_0 \end{Bmatrix} = \begin{bmatrix} K_{HH} & K_{HM} \\ K_{MH} & K_{MM} \end{bmatrix} \begin{Bmatrix} \bar{u}_a \\ \bar{\phi} \end{Bmatrix} \tag{18}$$

where K is an impedance matrix. This matrix or its inverse is available in the literature for a number of footing shapes and soil properties[1,3,7,9,10,11].

In an earthquake excitation problem, $\bar{u}_g$ is given, and $\bar{u}_a$, $\bar{\phi}$, $\bar{u}_j$, $\bar{U}_j^+$, $\bar{U}_0^-$, $\bar{M}_0$ are unknown variables. However, by combining Eqs. (15)-(18), we obtain two equations involving only $\bar{u}_g + \bar{u}_a$ and $h\bar{\phi}$ as the remaining two unknowns,

$$[A-B] \begin{Bmatrix} \bar{u}_g + \bar{u}_a \\ h\bar{\phi} \end{Bmatrix} = - \begin{Bmatrix} B_{11} \\ B_{21} \end{Bmatrix} \bar{u}_g \tag{19}$$

where

$$[A] = \begin{bmatrix} \frac{2\sin N\psi \sin(\frac{\psi}{2})}{\cos(N+\frac{1}{2})\psi} + \frac{m_0\omega^2}{k} & \frac{2\sin(\frac{N+1}{2}\psi)\sin\frac{N\psi}{2}}{\cos(N+\frac{1}{2})\psi} \\ \frac{\sin(\frac{N+1}{2}\psi)\sin(\frac{N\psi}{2})}{2\sin^2(\frac{\psi}{2})\cos(N+\frac{1}{2})\psi} & \frac{(N+1)\sin N\psi - N\sin(N+1)\psi}{8\sin^3(\frac{\psi}{2})\cos(N+\frac{1}{2})\psi} + \frac{I_t}{mh^2} \end{bmatrix}, \tag{20}$$

$$[B] = \begin{bmatrix} K_{HH}/k & K_{HM}/(kh) \\ K_{MH}/(mh\omega^2) & K_{MM}/(mh^2\omega^2) \end{bmatrix}. \tag{21}$$

Equation (19) is solved readily, and the other unknowns are obtained from Eqs. (11), (12), etc.

For earthquake response calculations, the free field ground acceleration is treated as the input to the dynamic system. Then the frequency response functions corresponding to various response variables are precisely $\bar{u}_a$, $\bar{\phi}$, $\bar{U}_0$, $\bar{M}_0$, $\bar{u}_j$, $\bar{U}_j^+$ obtained for the case $\bar{u}_g = -\omega^{-2}$. For example, the frequency response function for $u_j(t)$ is given by

$$H_{u_j}(\omega) = -\frac{1}{\omega^2\Delta}\{[\frac{\cos(N-j+\frac{1}{2})\psi}{\cos(N+\frac{1}{2})\psi} - 1]\ [|B| + B_{11}(\frac{N}{4\sin^2(\frac{\psi}{2})} - \frac{I_t}{mh^2})]$$

$$- [\frac{\sin(\frac{N}{2})\psi\sin(\frac{j}{2})\psi\sin(\frac{N-j}{2})\psi}{2\sin^3(\frac{\psi}{2})\cos(N+\frac{1}{2})\psi} + j\frac{\sin(\frac{N}{2})\psi\sin(\frac{N+1}{2})\psi}{2\sin^2(\frac{\psi}{2})\cos(N+\frac{1}{2})\psi}]\ B_{11}$$

$$+ [1 + \frac{jm_0\omega^2}{k} + j\frac{2\sin N\psi\sin(\frac{\psi}{2})}{\cos(N+\frac{1}{2})\psi} - \frac{m_0\omega^2\sin j\psi}{2k\sin(\frac{\psi}{2})\cos(N+\frac{1}{2})\psi}$$

$$- \frac{2\sin(\frac{j}{2})\psi\sin(\frac{j-1}{2})\psi + \cos(N-j+\frac{1}{2})\psi}{\cos(N+\frac{1}{2})\psi}]B_{21}\} \qquad (22)$$

where $|B|$ = determinant of [B] and Δ = determinant of [A-B].

It has been suggested that earthquake acceleration may be modeled as an evolutionary process, which is a nonstationary random process having a Stieltjes integral representation[8]

$$\ddot{u}_g(t) = \int_{-\infty}^{\infty} \psi(t,\omega)e^{i\omega t}\, d\tilde{P}(\omega) \qquad (23)$$

where $\tilde{P}(\omega)$ is another random process with uncorrelated increments[12]. Eq. (23) is a generalization of a stationary random process representation for which function $\psi(\omega,t)$ reduces to a constant. The uncorrelated-increment random process $\tilde{P}(\omega)$ has the property

$$E[d\tilde{P}(\omega_1)dP^*(\omega_2)] = \begin{cases} 0 & \omega_1 \neq \omega_2 \\ \Phi(\omega)d\omega\ , & \omega_1 = \omega_2 = \omega \end{cases} \qquad (24)$$

where E[ ] denotes an ensemble average, an asterisk represents the complex conjugate, and $\Phi(\omega)$ is the spectral density of some stationary random process. It is simple to show that the autocorrelation function of a general response variable $\xi(t)$ can be computed as follows:

$$E[\xi(t_1)\xi(t_2)] = \int_{-\infty}^{\infty} M(t_1,\omega)M^*(t_2,\omega)e^{i\omega(t_1-t_2)}\Phi(\omega)d\omega \qquad (25)$$

in which

$$M(t,\omega) = \int_{-\infty}^{\infty} e^{-i(\omega-\omega')t}\ Q(t,\omega,\omega')H_\xi(\omega')d\omega' \qquad (26)$$

and

$$Q(t,\omega,\omega') = \frac{1}{2\pi} \int_0^t \psi(\tau,\omega) e^{i(\omega-\omega')} d\tau . \tag{27}$$

WIND EXCITATIONS

Restrict our consideration to the along-wind motion, the governing equations for the jth story-unit are given by

$$m \ddot{u}_j^t + \beta \dot{u}_j^t = U_j^+ - U_j^- + F_j, \tag{1'}$$

$$U_j^- = U_{j-1}^+ = k(u_j - u_{j-1}), \tag{2'}$$

$$u_j^t = u_j + u_a + jh\phi . \tag{3'}$$

These are obtained by adding the wind load term F_j to Eq. (1) and deleting the earthquake excitation u_g from Eq. (3). The wind load problem is more complicated than the earthquake problem since the excitations occur at more than one location. On the other hand, the wind loads are usually modeled as stationary random processes whereas earthquakes are clearly nonstationary. Therefore, we may assume that

$$F_r(t) = \int_{-\infty}^{\infty} e^{i\omega t} d\tilde{F}_r(\omega) \tag{28}$$

where $\tilde{F}_r(\omega)$, $r = 1,2,\ldots,N$, are random processes with uncorrelated increments, having the property

$$E[d\tilde{F}_r(\omega_1)d\tilde{F}_\ell(\omega_2)] = \begin{cases} 0, & \omega_1 \neq \omega_2 \\ \Phi_{F_rF_\ell}(\omega)d\omega , & \omega_1 = \omega_2 = \omega \end{cases} , \tag{29}$$

$\Phi_{F_rF_\ell}$ is the cross-spectral density of $F_r(t)$ and $F_\ell(t)$. We shall seek the stationary solutions of the response variables, of the type

$$u_j(t) = \int_{-\infty}^{\infty} e^{i\omega t} d\tilde{u}_j(\omega) . \tag{30}$$

Following a procedure similar to the one leading to Eq. (6), we obtain

$$\begin{Bmatrix} d\tilde{u}_j \\ d\tilde{U}_j^+ \end{Bmatrix} = [T]^j \begin{Bmatrix} 0 \\ d\tilde{U}_0^+ \end{Bmatrix} + (-\omega^2 m + i\omega\beta) \sum_{r=1}^{j} [T]^{j-r} \begin{Bmatrix} 0 \\ d\tilde{u}_a \end{Bmatrix}$$

$$+ (-\omega^2 m + i\omega\beta) \sum_{r=1}^{j} r[T]^{j-r} \begin{Bmatrix} 0 \\ hd\tilde{\phi} \end{Bmatrix} - \sum_{r=1}^{j} [T]^{j-r} \begin{Bmatrix} 0 \\ d\tilde{F}_r \end{Bmatrix}. \tag{31}$$

Letting j=N and imposing $d\tilde{U}_N^+ = 0$, an expression analogous to Eq. (11) can be obtained, but with an additional series involving the wind loads $d\tilde{F}_r$. Using equations similar to (15)-(18) the unknowns $d\tilde{u}_a$ and $d\tilde{\phi}$ can be eliminated from (31), arriving at expressions for $d\tilde{u}_j$ and $d\tilde{U}_j^+$ in terms of $d\tilde{F}_r$. Finally, these can be used to obtain the spectral densities and the cross-spectral densities of $u_j(t)$ and $U_j^+(t)$ in terms of wind load spectra. For example, the spectral density of $u_j(t)$ is given by

$$\Phi_{u_j u_j}(\omega) = |C_1(j)|^2 \sum_{r=1}^{N} \sum_{r'=1}^{N} \tau_{22}(N-r)\tau_{22}^*(N-r')\Phi_{F_r F_{r'}}(\omega)$$

$$+ |C_2(j)|^2 \sum_{s=1}^{N} \sum_{s'=1}^{N} ss' \sum_{\ell=1}^{s} \sum_{\ell'=1}^{s'} \tau_{12}(s-\ell)\tau_{12}^*(s'-\ell')\Phi_{F_\ell F_{\ell'}}(\omega)$$

$$+ \frac{|C_2(j)|^2}{m^2\omega^4} \sum_{p=1}^{N} \sum_{p'=1}^{N} pp' \Phi_{F_p F_{p'}}(\omega) + \sum_{q=1}^{j} \sum_{q'=1}^{j} \tau_{12}(j-q)\tau_{12}^*(j-q')\Phi_{F_q F_{q'}}(\omega)$$

$$+ 2\,\mathrm{Re}\{C_1(j)C_2^*(j) \sum_{r=1}^{N} \sum_{s=1}^{N} s \sum_{\ell=1}^{s} \tau_{22}(N-r)\tau_{12}^*(s-\ell)\Phi_{F_r F_\ell}(\omega)$$

$$- \frac{|C_2(j)|^2}{m\omega^2} \sum_{s=1}^{N} s \sum_{\ell=1}^{s} \sum_{p=1}^{N} p\, \tau_{12}(s-\ell)\Phi_{F_\ell F_p}(\omega) - \frac{C_2(j)C_1^*(j)}{m\omega^2} \sum_{p=1}^{N} \sum_{r=1}^{N} p\tau_{22}^*(N-r)\Phi_{F_p F_r}(\omega)$$

$$- C_1(j) \sum_{r=1}^{N} \sum_{q=1}^{j} \tau_{22}(N-r)\tau_{12}^*(j-q)\Phi_{F_r F_q}(\omega)$$

$$- C_2(j) \sum_{s=1}^{N} s \sum_{\ell=1}^{s} \sum_{q=1}^{j} \tau_{12}(s-\ell)\tau_{12}^*(j-q)\Phi_{F_\ell F_q}(\omega)$$

$$+ \frac{C_2(j)}{m\omega^2} \sum_{p=1}^{N} \sum_{q=1}^{j} p\, \tau_{12}^*(j-q)\Phi_{F_p F_q}(\omega)\} \tag{32}$$

where

$$C_1(j) = \frac{1}{k\Delta} \{ [\frac{\cos(N-j+\frac{1}{2})\psi}{\cos(N+\frac{1}{2})\psi} - 1] [(A_{12} - B_{12})(A_{22} - \frac{I_t}{mh^2}) - \frac{A_{22} - B_{22}}{\tau_{22}(N)}]$$

$$+ [\frac{\sin j\psi}{2\sin\frac{\psi}{2}\cos(N+\frac{1}{2})\psi} - j] [\frac{A_{21} - B_{21}}{\tau_{22}(N)} - (A_{11} - B_{11})(A_{22} - \frac{I_t}{mh^2})]\}$$

$$+\frac{\cos\frac{\psi}{2}\sin j\psi}{k\sin\psi\cos(N+\frac{1}{2})\psi}, \qquad (33)$$

$$C_2(j) = \frac{1}{\Delta}\{[\frac{\sin j\psi}{2\sin\frac{\psi}{2}\cos(N+\frac{1}{2})\psi} - j](A_{11} - B_{11}) - [\frac{\cos(N-j+\frac{1}{2})\psi}{\cos(N+\frac{1}{2})\psi} - 1](A_{12}-B_{12})\}. \qquad (34)$$

NUMERICAL EXAMPLE

For the case of earthquake excitation, calculations have been carried out for an 8-story building with the following physical properties: $h = 3.6$ m, $m = 3.456 \times 10^6$ kg, $k = 3.404 \times 10^9$ N/m, $\beta = 1.00 \times 10^6$ N/m/s , $m_0 = 0.5$ m. The soil under the footing was assumed to be a linear elastic half space characterized by shear velocity = 300 m/s, mass density = 1800 kg/m^3, Poisson's ratio = 1/3. The graphical results[9] for circular footings was used to compute the impedance matrix K, assuming an equivalent footing radius of 7.5 m. The free field ground acceleration was modeled as an evolutionary process, Eq. (23).

The computed root-mean-square value for the top floor displacement u_8 is shown in Fig. 2. The results corresponding to a rigid or non-compliant soil are also included for comparison. As seen from this figure, the soil compliancy reduces the level of response. The reader is referred to Ref. 5 for more details and the mean-square values for other response variables.

For the case of wind excitation, a 40-story super-structure with the following properties is chosen as an example: $h = 4$ m, $m = 1.290 \times 10^6$ kg, $m_0 = (1/2)m$, $k = 10^9$ N/m, $\beta = 21{,}550$ N/m/s . The soil impedance matrix has been computed using the same soil properties as those of the earthquake case, but with an equivalent footing radius of 15 m. It is assumed that the mean wind velocity obeys a power law and that the random fluctuation in the wind has a Davenport spectrum. The solid line in Fig. 3 shows the computed frequency spectrum for the variance of the displacement at the top floor, which is the same as the mean-square spectrum but without a delta function at $\omega = 0$. Only the positive ω-domain is shown in this figure; the negative ω-domain can be obtained as its mirror image. The result obtained for a rigid soil is also shown for comparison. It is seen that the soil compliancy does not change the first peak of the spectrum, which occurs at $\omega = 0.04$ rad/s , coinciding with the dominant peak in the Davenport spectrum. As expected, the location of the second peak, which is near the fundamental natural frequency of the super-structure, is shifted lower due to soil compliancy. Somewhat unexpected, however, is the fact that soil compliancy does not reduce the height of this second peak; it is increased slightly instead. More details and additional numerical results may be found in Ref. 6.

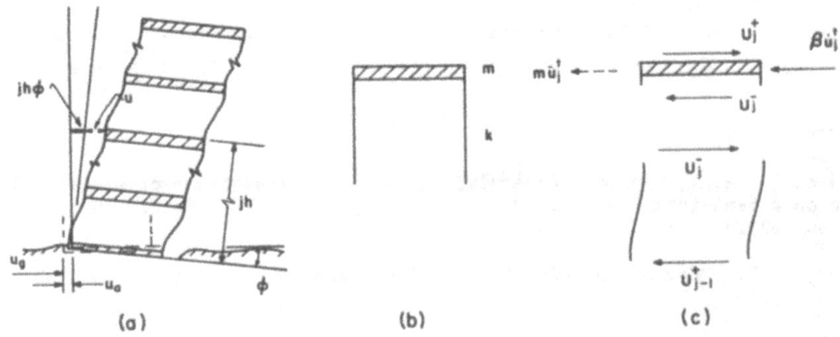

Fig. 1. Structural Model: (a) N-Story Building; (b) jth Story-Unit; (c) Forces on jth Unit.

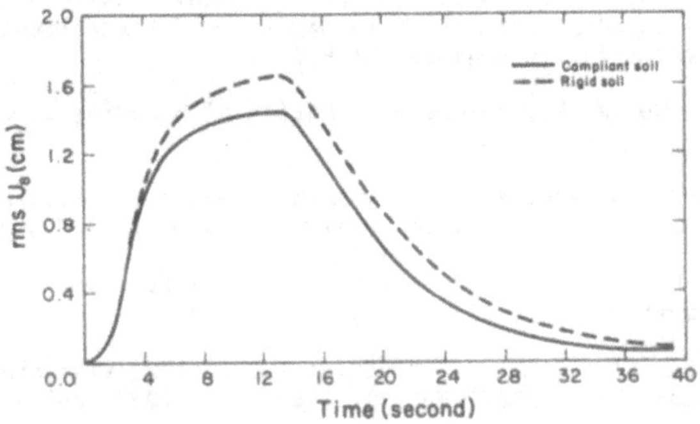

Fig. 2. Root-Mean-Square Displacement of the 8th Floor Relative to Footing due to Model Earthquake Excitation.

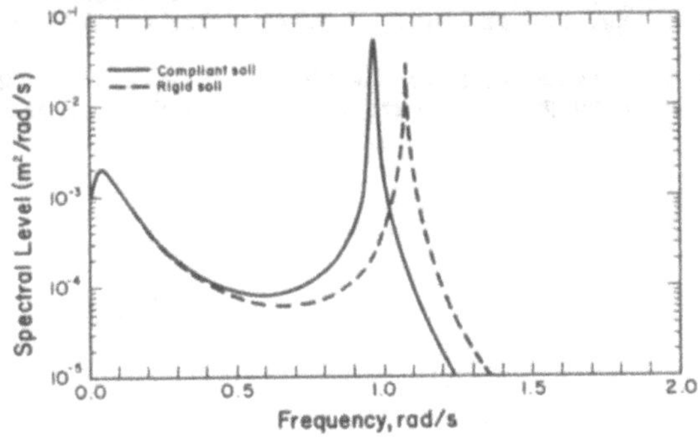

Fig. 3. Frequency Spectrum for the Variance of the Top Floor Displacement of a 40-Story Building under Wind Load Excitation.

ACKNOWLEDGMENT

The research is supported by the US National Science Foundation under Grants NSF CEE-8024882 and CEE-8120438. Opinions, findings, and conclusions or recommendations expressed are those of the writers and do not necessarily reflect the view of NSF.

REFERENCES

1. Gladwell, G.M.L., "Forced Tangential and Rotatory Vibrations of a Rigid Circular Disc on a Semi-infinite Solid," Int. J. Engng. Sci., Vol. 6, No. 10, October 1968, pp. 591-607.

2. Hsieh, T. K., "Foundation Vibrations," Proc. Inst. Civil Engrs., Vol. 22, 1962, pp. 211-225.

3. Iguchi, M. and Luco, J. E., "Dynamic Response of Flexible Rectangular Foundations on an Elastic Half-space," Earthqu, Eng. Struct. Dyn., Vol. 9, 1981, pp. 239-249.

4. Lin, Y. K. and McDaniel, T. J., "Dynamics of Beam-Type Periodic Structures," J. Engng. for Industry, Vol. 91, 1969, pp. 1133-1141.

5. Lin, Y. K. and Wu, W. F., "A Closed Form Earthquake Response Analysis of Multistory Building on Compliant Soil," to appear in International Journal of Probability and Statistics in Engineering R & D.

6. Lin, Y. K. and Wu, W. F., "Along-Wind Motion of Building on Compliant Soil," to appear.

7. Luco, I. E. and Westmann, R. A., "Dynamic Response of Circular Footings," J. Engng. Mech. Div., ASCE, Vol. 97, No. EM5, Proc. Paper 8416, 1971, pp. 1381-1395.

8. Priestley, M. B., "Evolutionary Spectra and Nonstationary Processes," Journal of Royal Statistical Society, B27, 1965, pp. 204-228.

9. Veletsos, A. S. and Wei, Y. T., "Lateral and Rocking Vibration of Footings," J. Soil Mech. Found. Div., ASCE, Vol. 97, 1971, pp. 1227-1248.

10. Veletsos, A. S. and Verbic, B., "Vibration of Viscoelastic Foundations," Earthquake Engineering and Structural Dynamics, Vol. 2, 1973, pp. 87-102.

11. Wong, H. L. and Luco, J. E., "Dynamic Response of Rigid Foundations of Arbitrary Shape," Earthqu. Eng. Struct. Dyn., Vol. 4, 1976, pp. 579-587.

12. Yaglom, A. M., An Introduction to the Theory of Stationary Random Functions, English translation by R. A. Silverman, Prentice-Hall, Englewood Cliffs, NJ, 1962.

STOCHASTIC STABILITY OF FLUID CONVEYING TUBES

S. NARAYANAN [1)]

Summary

A theoretical study of the stochastic stability of fluid conveying tubes with random velocity fluctuations is presented. The stability is based on the finiteness of the second moment of the pipe wall response. The stochastic averaging procedure of Stratonovich and Khasminskii is made use of to establish the mean square stability conditions. For a broad-band random parametric excitation expressions for the drift and diffusion coefficients in the corresponding Ito's equations are derived. Results for simply-supported, cantilevered and fixed-fixed tubes and white noise parametric excitation are presented.

Introduction

It is well known that if the velocity of fluid flowing through a pipe has harmonic fluctuations over a mean velocity, then the pipe can experience parametric instabilities depending on the amplitude and frequency of the harmonic velocity fluctuation. The regions of parametric instabilities for a single span pipe with different end conditions have been determined by Paidoussis and Issid [3].

This paper presents a theoretical analysis of the stochastic stability of a single span pipe when the fluid velocity fluctuation is a stationary wide band random process. The Stratonovich-Khasminskii [4, 1] stochastic averaging procedure is made use of in the analysis to derive a set of Ito's differential equations interms of the normal coordinates of the tube wall response. The Ito's equations are made use of to establish the moment stability conditions by a set of ordinary differential equations containing the statistics of the random parametric excitation.

The results of the general analysis show that the second moment stability of the tube depends on the power spectral density of the random velocity fluctuation at twice the natural frequencies and combinations of the sums and differences of the natural frequencies of the tube, a result somewhat analogous to the deterministic stability analysis of the pipe. Results for simply-supported, cantilevered and fixed-fixed tubes are presented.

Equation of Motion

The equation of motion of a single span tube in nondimensional form is given by [3]

$$\frac{\partial^4 \bar{y}}{\partial \xi^4} + \{u^2 + \gamma + \beta \frac{\partial u}{\partial \tau}(1-\xi)\} \frac{\partial^2 \bar{y}}{\partial \xi^2} + 2\beta u \frac{\partial^2 \bar{y}}{\partial \xi \partial \tau} + \delta \frac{\partial \bar{y}}{\partial \tau} + \frac{\partial^2 \bar{y}}{\partial \tau^2} = 0 . \tag{1}$$

In equation (1) $u = (m_f/EI)^{\frac{1}{2}} VL$, $\gamma = p A_t L^2/EI$,

$\beta = [m_f/(m_f+m_t)]^{\frac{1}{2}}$, $\delta = CL^2/[EI(m_f + m_t)]^{\frac{1}{2}}$, $\bar{y} = y/L$,

1) Department of Applied Mechanics, Indian Institute of Technology, Madras 600 036, INDIA.

$\tau = t[EI/(m_f + m_t)]^{\frac{1}{2}}/L^2$, $\xi = x/L$ are nondimensional parameters, where L is the length and EI the flexural rigidity of the tube, m_f and m_t, respectively, are the mass per unit length of the fluid and the tube wall, V is the fluid velocity, p is the fluid pressure, A_t is the cross-sectional area of the tube wall, C is the viscous damping coefficient per unit length, x the coordinate along the length of the tube and y is the tube wall response. In deriving equation (1) it has been assumed that conditions of plug flow exist and the tube wall is purely elastic and the effects of axial tension and gravity are negligible.

Consider the nondimensional flow velocity to be given by

$u = u_o[1 + \eta(\tau)]$, where $\eta(\tau)$ is a stationary wide band process. For small $\eta(\tau)$ $u^2 \simeq u_o^2 [1 + 2\eta(\tau)]$ and hence equation (1) can be approximated by the expression

$$\frac{\partial^4\bar{y}}{\partial\xi^4} + \{u_o^2[1 + 2\eta(\tau)] + \gamma + \beta u_o \dot{\eta}(\tau)(1-\xi)\} \frac{\partial^2\bar{y}}{\partial\xi^2} + 2\beta u_o[1+\eta(\tau)] \frac{\partial^2\bar{y}}{\partial\xi\partial\tau} \quad \delta \frac{\partial \bar{y}}{\partial\tau} + \frac{\partial^2\bar{y}}{\partial\tau^2} = 0 \tag{2}$$

where the number of dots represents the order of differentiation with respect to τ. Assuming a normal mode expansion for $\bar{y}(\xi,\tau)$ in the form $\bar{y}(\xi,\tau) = \sum_{n=1}^{\infty} q_n(\tau)\, \alpha_n(\xi)$ (3)

and substituting in equation (3) one obtains

$$\ddot{q}_i + \Omega_i^2 q_i + \sum_j \delta_{ij} \dot{q}_j + \sum_j N_{ij} q_j + \eta(\tau) \sum_j (P_{ij} \dot{q}_j + R_{ij} q_j) + \dot{\eta}(\tau) \sum_j D_{ij} q_j = 0 \tag{4}$$

where

$$\Omega_i^2 = \int_0^1 \alpha_i \alpha_i'''' d\xi + (u_o^2 + \gamma) \int_0^1 \alpha_i \alpha_i'' d\xi, \qquad \delta_{ij} = \delta \int_0^1 \alpha_i \alpha_j d\xi + 2\beta u_o \int_0^1 \alpha_i \alpha_j' d\xi,$$

$$N_{ij} = (u_o^2 + \gamma) \int_0^1 \alpha_i \alpha_j'' d\xi \,, \quad P_{ij} = 2\beta u_o \int_0^1 \alpha_i \alpha_j' d\xi,$$

$$R_{ij} = 2u_o^2 \int_0^1 \alpha_i \alpha_j'' d\xi, \quad D_{ij} = \beta u_o \int_0^1 (1-\xi)\alpha_i \alpha_j'' d\xi$$

and $\alpha_i(\xi)$ is the ith normal mode of the tube and the primes indicate differentiation with respect to ξ. The expressions for the modal constants can be obtained from [3].

Ito's equations

Assuming that the viscous damping parameter δ and the parametric excitation $\eta(\tau)$ are of appropriate orders that would satisfy the conditions of Stratonovich-Khasminskii limit theorem, the normal coordinates can be expressed as narrow band random process given by

$$q_i(\tau) = A_i \cos (\Omega_i\tau + \psi_i), \; i = 1,2... \tag{5}$$

Substituting equation (5) into equation (4), with the condition that $q_i(\tau) = -A_i\Omega_i \sin(\Omega_i\tau + \psi_i)$, equation (4) can be expressed by the set of first order equations

$$\dot{A}_i = Z_i \sin(\Omega_i\tau + \psi_i), \quad i = 1, 2 \ldots; \quad \dot{\psi}_i = Z_i \cos(\Omega_i\tau + \psi_i)/A_i, \; i = 1, 2 \ldots; \tag{6}$$

where

$$Z_i = -\frac{1}{\Omega_i}\{\sum_j [\delta_{ij} A_j \Omega_j - \eta(\tau) P_{ij} A_j\Omega_j] \sin(\Omega_j\tau + \psi_j) - \sum_j[\eta(\tau)R_{ij} A_j] \cos(\Omega_j\tau + \psi_j)$$

$$- \sum_{j\neq i} N_{ij} A_j \cos(\Omega_j\tau + \psi_j) + \sum_j \dot{\eta}(\tau) D_{ij} A_j \cos(\Omega_j\tau + \psi_j)\}. \tag{7}$$

Applying the Stratonovich-Khasminskii limit theorem for the terms containing the stochastic process $\eta(\tau)$ and $\dot{\eta}(\tau)$ and the conventional averaging procedure of Krylov-Bogoliubov for the other terms and dropping oscillatory terms, A_i and ψ_i tend to a diffusive Markov vector process governed by the set of Ito's equations

$$d A_i = m_i \, d\tau + \sum_j \sigma_{ij} \, d B_j,$$

$$d \psi_i = n_i \, d\tau + \sum_j \mu_{ij} \, d B_j', \quad i = 1, 2 \ldots, \tag{8}$$

where B_j and B_j' are independent Wiener processes with unit variance, m_i and n_i are the corresponding drift coefficients and σ_{ij} and μ_{ij} are the diffusion coefficients given by

$$m_i = -\delta_{ii} A_i/2 + \lim_{\Delta\to\infty} \frac{1}{\Delta} \int_0^\Delta d\tau \, [\int_{-\infty}^0 E\{\sum_j (\frac{\partial G_{si}}{\partial A_j})_\tau (G_{sj})_{\tau+\zeta}\} d\zeta$$

$$+ \int_{-\infty}^0 E\{\sum_j (\frac{\partial G_{si}}{\partial \psi_j})_\tau (G_{cj})_{\tau+\zeta}\} d\zeta], \tag{9}$$

$$n_i = \lim_{\Delta\to\infty} \frac{1}{\Delta} \int_0^\Delta d\tau \, [\int_{-\infty}^0 E\{\sum_j (\frac{\partial G_{ci}}{\partial A_j})_\tau (G_{sj})_{\tau+\zeta}\} d\zeta$$

$$+ \int_{-\infty}^0 E\{\sum_j (\frac{\partial G_{ci}}{\partial \psi_j})_\tau (G_{cj})_{\tau+\zeta}\} d\zeta], \tag{10}$$

$$(\underline{\sigma}\, \underline{\sigma}^T)_{ij} = \lim_{\Delta\to\infty} \frac{1}{\Delta} \int_0^\Delta d\tau \int_{-\infty}^\infty E[(G_{si})_\tau (G_{sj})_{\tau+\zeta}] d\zeta, \tag{11}$$

$$(\underline{u}\, \underline{u}^T)_{ij} = \lim_{\Delta\to\infty} \frac{1}{\Delta} \int_0^\Delta d\tau \int_{-\infty}^\infty E[(G_{ci})_\tau (G_{cj})_{\tau+\zeta}] d\zeta \tag{12}$$

where

$G_{si} = Z_i' \sin(\Omega_i\tau + \psi_i)$, $G_{ci} = Z_i' \cos(\Omega_i\tau + \psi_i)/A_i$ and

$$Z_i' = Z_i + \Sigma\delta_{ij}A_j\Omega_j \sin(\Omega_j\tau + \psi_j)/\Omega_i .$$

The indicated integrations in the foregoing equations involve tedious calculations and are given in the work of Narayanan and Chandrasekaran [2].

Expressions for m_i and $(\underline{\sigma}\,\underline{\sigma}^T)_{ii}$ which are used further in the analysis are given by,

$$\begin{aligned} m_i = & -\delta_{ii} A_i/2 + \pi P_{ii}^2 A_i S_{\eta\eta}(0)/4 \\ & + 3\pi A_i [(P_{ii} - 2D_{ii})^2 + R_{ii}^2/\Omega_i^2] S_{\eta\eta}(2\Omega_i)/8 \\ & + \pi A_i \sum_{j\neq i} S_{ji}^+/4 \, [P_{ij}P_{ji} + \{(R_{ij}R_{ji}/\Omega_{ji}^+ + D_{ij}D_{ji}\Omega_{ji}^+)/\Omega_i\Omega_j \\ & - P_{ij}D_{ji}/\Omega_i - D_{ij}P_{ji}/\Omega_j\}\, \Omega_{ji}^+ + A_j^2 \{(P_{ij}\Omega_j - D_{ij}\Omega_{ji}^+)^2 + R_{ij}^2\}/2 A_i^2\Omega_i^2] \\ & + \pi A_i \sum_{j\neq i} S_{ji}^-/4 [P_{ij}P_{ji} - \{R_{ij}R_{ji}/\Omega_{ji}^- + D_{ij}D_{ji}\Omega_{ji}^-)/\Omega_i\Omega_j \\ & + D_{ij}P_{ji}/\Omega_j - P_{ij}D_{ji}/\Omega_i\}\, \Omega_{ji}^- + A_j^2 \; (P_{ij}\Omega_j - D_{ij}\Omega_{ji}^-)^2 + R_{ij}^2\} \; 2A_i^2\Omega_i^2] + \\ & + \pi A_i \sum_{j\neq i} Q_{ji}^+/4 [P_{ij}R_{ji}/\Omega_i - R_{ij}P_{ji}/\Omega_j + \Omega_{ji}^+ (R_{ij}D_{ji} - D_{ij}R_{ji})/\Omega_i\Omega_j] \\ & - \pi A_i \sum_{j\neq i} Q_{ji}^- /4 [P_{ij}R_{ji}/\Omega_i + R_{ij}P_{ji}/\Omega_j + \Omega_{ji}^- (R_{ij}D_{ji} - D_{ij}R_{ji})/\Omega_i\Omega_j], \end{aligned} \tag{13}$$

$$\begin{aligned} (\underline{\sigma}\,\underline{\sigma}^T)_{ii} = & \pi A_i^2 P_{ii}^2 S_{\eta\eta}(0)/2 + \pi A_i^2 S_{\eta\eta}(2\Omega_i)/4 [(P_{ii} - 2D_{ii})^2 + R_{ii}^2/\Omega_i^2] \\ & + \pi S_{ji}^+ /4\Omega_i^2 \sum_{j\neq i} A_j^2 [(P_{ij}\Omega_j - D_{ij}\Omega_{ji}^+)^2 \\ & + R_{ij}^2] + \pi S_{ji}^-/4\Omega_i^2 \sum_{j\neq i} A_j^2 [(P_{ij}\Omega_j - D_{ij}\Omega_{ji}^-)^2 + R_{ij}^2] \end{aligned} \tag{14}$$

where $\Omega_{ji}^+ = \Omega_j + \Omega_i$, $\Omega_{ji}^- = \Omega_j - \Omega_i$, $S_{ji}^+ = S_{\eta\eta}(\Omega_j+\Omega_i)$ $S_{ji}^- = S_{\eta\eta}(\Omega_j-\Omega_i)$, $Q_{ji}^+ = Q_{\eta\eta}(\Omega_j+\Omega_i)$ and

$Q_{ji}^- = Q_{\eta\eta}(\Omega_j-\Omega_i)$,

$$S_{\eta\eta}(\Omega) = \frac{1}{\pi} \int_{-\infty}^{0} R_{\eta\eta}(\zeta) \cos\Omega\zeta \, d\zeta \text{ and } Q_{\eta\eta}(\Omega) = \frac{1}{\pi} \int_{-\infty}^{0} R_{\eta\eta}(\zeta) \sin\Omega\zeta \, d\zeta$$

and $R_{\eta\eta}(\tau)$ is the autocorrelation function of $\eta(\tau)$.

It can be shown that the equations for the amplitude are uncoupled from those of the phases and hence $A(\tau)$ itself tends to a Markov vector.

Stability Conditions

For conditions of stability in the second moment the norm of the response vector $(q, \dot{q})$ is defined by

$$||q|| = [\sum_i \{q_i^2 + \dot{q}_i^2/\Omega_i^2\}]^{\frac{1}{2}} = [\sum_i A_i^2]^{\frac{1}{2}}. \tag{15}$$

From equation (15)

$E[||q||^2] = \sum_i E[A_i^2]$, hence $E[||q||^2] \to 0$ if and only if $E[A_i^2] \to 0$ $i = 1,2 \ldots$ Thus the necessary and sufficient conditions for second moment stability are that all the eigenvalues of the second moment coefficient matrix with amplitudes A_i's have negative real parts.

From the Ito equations (8) and using the differential generator

$$L(.) = \frac{\partial}{\partial\tau}(.) + \sum_i m_i \frac{\partial}{\partial A_i}(.) + \frac{1}{2}\sum_i\sum_j (\sigma\sigma^T)_{ij}\frac{\partial^2}{\partial A_i\,\partial A_j}(.) \tag{16}$$

it can be shown that the second moments of the amplitudes are governed by the following set of first order ordinary differential equations

$$\frac{d}{d\tau}E[A_i^2(\tau)] = \sum_i E[2A_i m_i] + E[(\underline{\sigma}\,\underline{\sigma}^T)_{ii}] \quad i = 1,2 \ldots \tag{17}$$

which can be expressed in the matrix form

$$\frac{d}{d\tau}E[A^2(\tau)] = [B]\;E[A^2(\tau)]. \tag{18}$$

The elements of the coefficient matrix B depend on the power spectral density function of the parametric excitation $S_{\eta\eta}$ and the function $Q_{\eta\eta}$ at twice the natural frequencies of the pipe and combination of these frequencies of the form $\Omega_i \pm \Omega_j$. Since the necessary and sufficient conditions for stability in the second norm of the tube wall response in its different normal modes require that all the eigen values of the coefficient matrix [B] have negative real parts, the stochastic stability depends on $S_{\eta\eta}$ and $Q_{\eta\eta}$ at Ω_i & $\Omega_i \pm \Omega_j$. This result is somewhat analogous to the parametric instability of tubes with harmonic velocity fluctuations which occurs at the frequencies of the flow velocity fluctuations being in the vicinity of these frequencies.

Examples, Results and Discussion

The cases of simply-supported, cantilevered and fixed-fixed tubes are considered as examples. The random parametric fluctuation of fluid velocity $\eta(\tau)$ is considered as a white noise for simplicity. For white noise excitation $Q_{\eta\eta} = 0$ and $S_{\eta\eta}(\Omega) = S_o$, a constant.

For simply supported end conditions and for white noise excitation the elements of the coefficient matrix [B] are given by and

$$B_{ij} = \pi S_o[(P_{ij}-D_{ij})^2 \quad \Omega_j^2/\Omega_i^2 + D_{ij}^2\ ]. \tag{19}$$

Considering only the first mode of the pipe, the mean square stability criterion reduces to

$$\pi S_o < \delta\Omega_1^2\ /[4\pi^4 u_o^4(1 + \beta^2\Omega_1^2/4u_o^2)]. \tag{20}$$

It is seen from the above expression that for constant values of β, u_o, γ, the power spectral density at which the pipe becomes unstable with respect to the second norm in the first mode increases linearly with δ. It is also seen from the inequality (20) that corresponding to the values of u_o, γ which make Ω_1 vanish, equation (20) can never be satisfied indicating that the pipe is inherently unstable. Even for steady flow velocities such a condition represents the divergence type of instability (3). With increase in β, the stochastic instability occurs at progressively lower spectral density values.

For establishing the stability conditions taking into consideration higher order modes, the eigen values of [B] are computed. For a given δ the limiting value of the power spectral density at which the real part of at least one of the eigen values of the matrix [B]becomes zero before becoming positive is determined by a regula-falsi bracketing procedure.

Results upto including four modes in the analysis have been plotted in the S - δ space for $u_o = 2$ and $\gamma = 2$ and $\beta = 0.4$ in Figure 1 for both simply supported and fixed end conditions. In both the cases it is observed that the power spectral density values at which the tube becomes unstable in the second norm progressively decrease with increasing number of modes considered in the analysis. Also the variation of the limiting value of S_o with δ is linear in all cases. For the fixed end conditions the tube becomes unstable at a higher value of the power spectral density than for the simply supported case for the same value of the damping parameter δ.

For the cantilevered pipe the variation of the limiting value of S_o with δ is shown in Table 1 for $u_o = 6$, $\beta = 0.4$ and $\gamma = 2$. It is seen that the power spectral density at which the tube wall becomes unstable in the second moment does not significantly vary with the damping parameter. The power spectral density values for stability even for a higher u_o in the case of the cantilever tubes are considerably higher than those for the simply supported or fixed tubes. This behaviour can be attributed to the fact that in the case of the cantilevered tube, the steady flow upto a certain critical velocity induces damping even in the absence of dissipation (3). Moreover, damping in the case of the cantilevered pipe is intimately connected with the dynamics of the system as it arises by the action of Coriolis forces. Consequently its effect on the stochastic stability is not uniform.

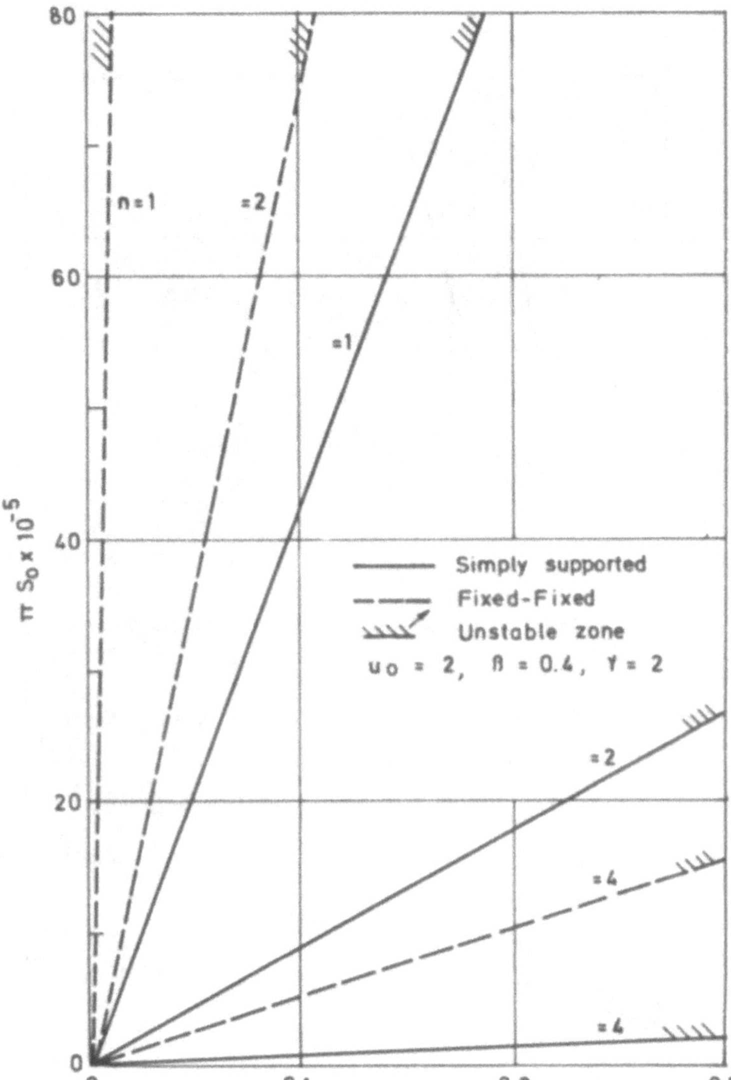

FIG. 1 STOCHASTIC STABILITY BOUNDARIES FOR DIFFERENT NUMBER OF MODES n

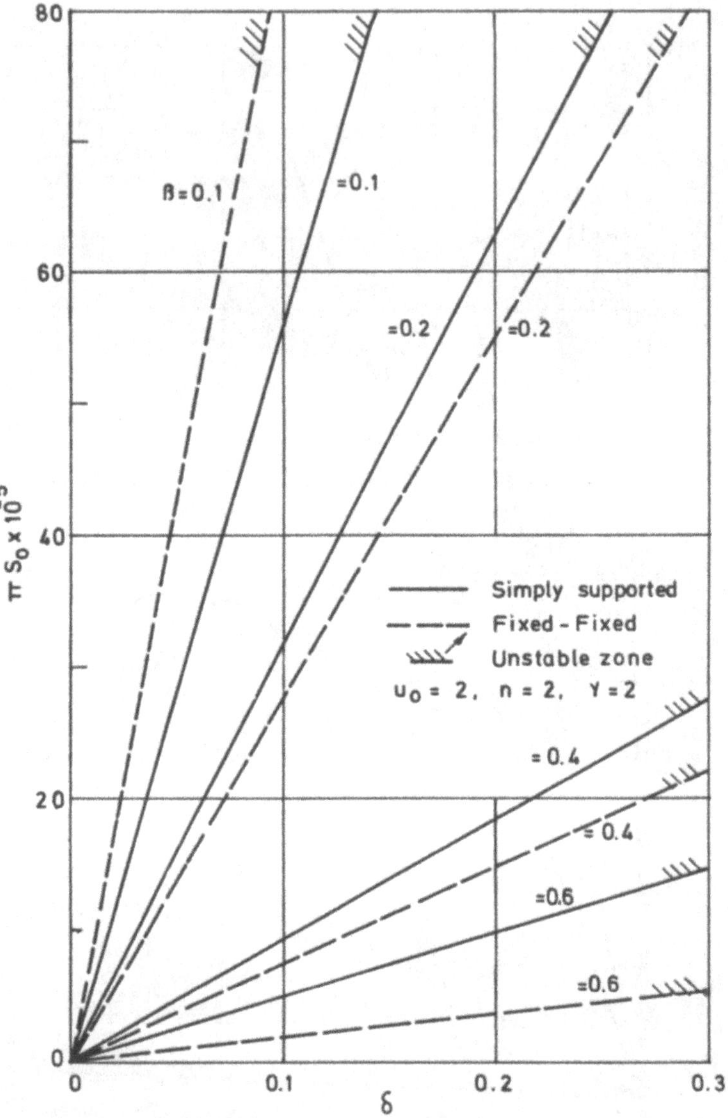

FIG. 2 STOCHASTIC STABILITY BOUNDARIES FOR DIFFERENT β

Table 1 Limiting values of S_o for stochastic stability of cantilever tube (u_o = 6, β = 0.4, γ = 2)

No. of modes	δ = 0.1	δ = 0.2	δ = 0.3
1	0.1088	0.1099	0.1111
2	0.1960	0.1980	0.2000
3	0.5614×10^{-33}	0.5671×10^{-3}	0.5729×10^{-3}
4	0.1572×10^{-3}	0.1588×10^{-3}	0.1604×10^{-3}

Figure 2 shows the variation of stochastic stability boundaries with β for a two mode analysis and u_o = 2 and γ = 0.4 for both simply supported and fixed end conditions. In both the cases the critical spectral density values decrease with increase in β.

Figure 3 shows the stability boundaries for different nondimensional steady flow velocities u_o, β = 0.4, γ = 2 for the simply supported and fixed end conditions and when two modes are taken into consideration in the analysis. The stability zones reduce with increasing u_o for both the cases. While in the case of simply supported end conditions a divergence type of instability occurs for a critical steady flow velocity of u_o = 2.806, for the fixed-fixed tube finite stability zones in the S - δ space exist even for higher steady flow velocities u_o, as divergence in this case occurs at a comparatively higher u_o.

For the cantilever pipes the variation of the limiting power spectral density values for stochastic instability with u_o and β are shown in Tables 2 and 3 respectively. The values are considerably higher as compared to the corresponding values for the simply supported or fixed end conditions. The variation with δ of the spectral density values is also not linear as in the previous cases. Another point to be noted is that the values of S_o for stochastic stability increase with increase in β for the cantilever tube which is quite opposite to what is observed in the other two cases.

Table 2 Limiting values of S_o for stochastic stability of cantilever tube (β = 0.4. γ = 2.0, No. of modes 2)

u_o	δ = 0.1	δ = 0.2	δ = 0.3
6	0.1960	0.1980	0.2000
8	0.1149	0.1158	0.1167
10	0.0817	0.0822	0.0827
12	0.0638	0.0641	0.0644

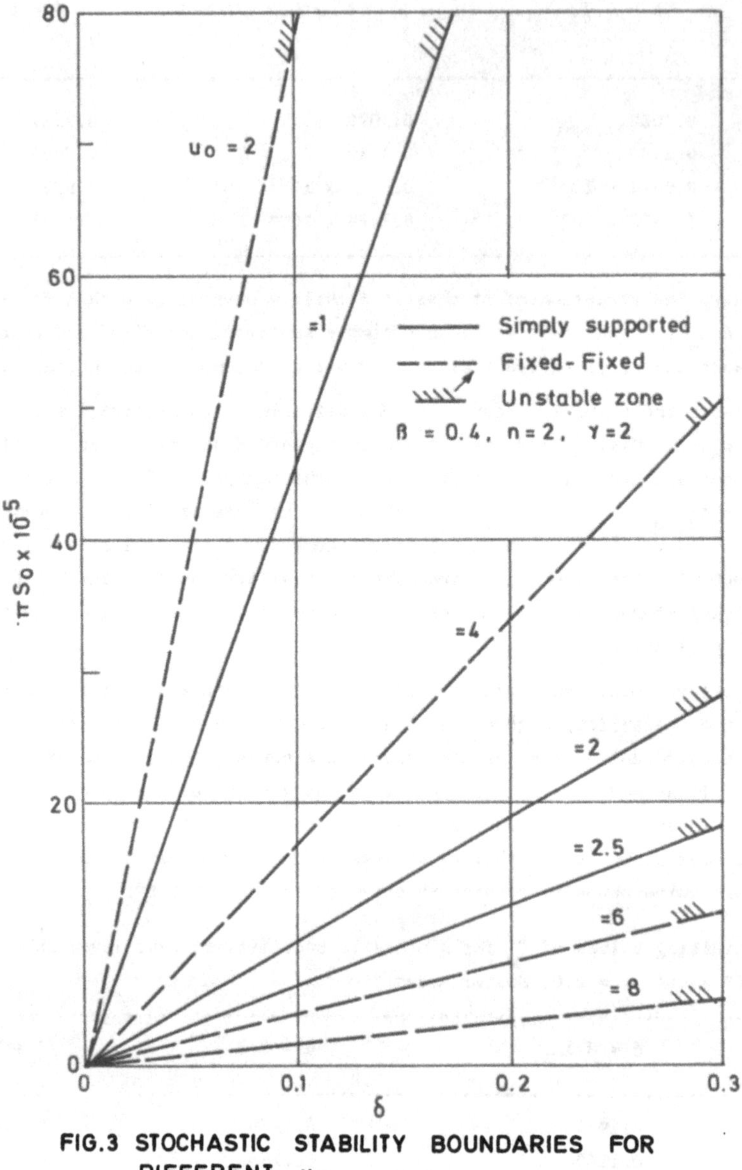

FIG.3 STOCHASTIC STABILITY BOUNDARIES FOR DIFFERENT u_0

Table 3 Limiting values of S_o for stochastic stability of cantilever tube (u_o = 6, γ = 2.0, No. of modes 2)

β	δ = 0.1	δ = 0.2	δ = 0.3
0.2	0.0644	0.0657	0.0671
0.3	0.1123	0.1138	0.1154
0.4	0.1960	0.1980	0.2000
0.5	0.4092	0.4125	0.4160

Conclusion

The stochastic stability of a single span tube with random fluctuations in the flow velocity is analysed by means of the stochastic averaging procedure. General conditions for the stochastic stability in the second moment of the norm of the tube response are derived. Results show the dependence of stochastic stability on the power spectral densities of the velocity fluctuation at twice the natural frequencies and the combinations of the sums and differences of these frequencies. Simply supported, fixed-fixed and cantilevered tubes are considered. While for the first two conditions the stability behaviour is similar, for the cantilever tube the behaviour is significantly different.

Acknowledgement

The author wishes to gratefully acknowledge Mr.M. Chandrasekaran for helping him in carrying out some of the computations.

References

1 Khasminskii, R.Z., 'A limit theorem for the solution of differential equations with random right hand sides', Theory of Probability and its Applications, 11, 1966, 390-406.

2 Narayanan, S. and Chandrasekaran,M., 'Stochastic Stability of Pipes Conveying Fluid' Submitted for publication in The Journal of Applied Mechanics, Trans ASME, 1982.

3 Paidoussis, M.P. and Issid, N.T., 'Dynamic Stability of Pipes Conveying Fluid' Journal of Sound and Vibration, 33(3), 1974, 267-294.

4 Stratonovich, R.L., 'Topics in the Theory of Random Noise', Vol.1, Gordon and Breach, New York, 1963.

ENERGY METHODS FOR NON-LINEAR SYSTEMS WITH NON-WHITE EXCITATION

J.B. Roberts*

SUMMARY

It is shown that, for a lightly damped non-linear oscillator, driven by random excitation, an approximate Fokker-Planck equation can be derived for the energy envelope process. The stationary solution of this equation is compared with simulation estimates for the case of a Duffing oscillator with linear damping.

1 INTRODUCTION

It has been shown by Stratonovitch [9] and the author [4,5] that the energy envelope, V(t) of a lightly damped non-linear oscillator can be approximated as a one-dimensional Markov process, provided that the input process can be modelled as a white noise. This enables an expression for the joint distribution of the displacement and velocity to be found, from which a wide variety of response statistics can be obtained [6,7]. In the special case of an oscillator with a linear restoring force, but non-linear damping, the Markov property of V(t) can be verified by the method of stochastic averaging, which has a rigorous foundation [2,3].

In principle V(t) should converge to a Markov process as the damping approaches zero, irrespective of the shape of the input spectrum (although, of course, the nature of the input will affect the rate of convergence). This suggests that, for the case of light damping, it should be possible to derive a Fokker-Planck-Kolmogorov (FPK) equation for V(t) which is valid for non-white excitation - i.e. takes into account the shape of the input spectrum. In this paper an intuitive approach is outlined which leads, after certain approximations, to the derivation of such an equation. The result agrees with that found recently by the author, through a rather different approach [8] and, moreover, reduces to Stratonovitch's result [9] in the case of white noise excitation.

The validity of the theoretical result is tested through a comparison with corresponding simulation estimates, for the case of a Duffing oscillator with linear damping.

2 THE EQUATIONS OF MOTION

We consider a stochastic equation of motion of the form

$$\ddot{x} + \varepsilon^2 f(x,x) + g(x) = \varepsilon\, y(t) \tag{1}$$

where x is the displacement response of the oscillator, f(x,x) is an arbitrary damping function and g(x) is an arbitrary restoring force. ε is a scaling parameter which, it will be assumed at the outset, is <u>small</u>. The scaling of the excitation, as in equation (1), ensures that the standard deviation of the response is of order ε^0. y(t) is a stationary random process, with zero mean, and a spectrum, $S(\omega)$, defined by the relation

$$S(\omega) = \frac{1}{2\pi}\int_{-\infty}^{\infty} w(\tau)\cos\omega\tau\, d\tau \tag{2}$$

* School of Engineering and Applied Sciences, University of Sussex, Falmer, Brighton, Sussex, BN1 9QT, U.K.

where

$$w(\tau) = E\{y(t)\ y(t+\tau)\}. \tag{3}$$

The energy envelope, associated with x(t), is defined by

$$V(t) = \frac{\dot{x}^2}{2} + U(x) \tag{4}$$

where

$$U(x) = \int_0^x g(\xi)\, d\xi . \tag{5}$$

It is also convenient to introduce a phase process, $\phi(t)$, defined such that

$$[U(x)]^{\frac{1}{2}} = V^{\frac{1}{2}} \cos\phi; \quad \dot{x} = -(2V)^{\frac{1}{2}} \sin\phi . \tag{6}$$

One can then picture the response in a modified phase plane, with $\dot{x}/\sqrt{2}$ and $[U(x)]^{\frac{1}{2}}$ as the coordinates. In the case of free, undamped motion ($\varepsilon = 0$) $V(t) = V_o$, a constant, and the motion appears as a circular orbit in the modified phase plane.

Equation (1) can be recast as a pair of coupled, first-order equations, in terms of V and ϕ. Thus,

$$\dot{V} = \varepsilon^2 F(V,\phi)(2V)^{\frac{1}{2}} \sin\phi - \varepsilon(2V)^{\frac{1}{2}} \sin\phi\ y(t), \tag{7}$$

$$\dot{\phi} = \frac{1}{(2V)^{\frac{1}{2}}} \left[\varepsilon^2 F(V,\phi) \cos\phi - \varepsilon \cos\phi\ y(t) + \frac{G(V,\phi)}{\cos\phi}\right] \tag{8}$$

where

$$F(V,\phi) = f(x,\dot{x}) \ ; \quad G(V,\phi) = g(x). \tag{9}$$

For free, undamped motion ($\varepsilon = 0$) equation (8) reduces to

$$\dot{\phi} = \frac{1}{(2V)^{\frac{1}{2}}} \cdot \frac{G(V,\phi)}{\cos\phi} . \tag{10}$$

The solution to this equation, denoted $\phi_o(t)$, is a function of the energy level V_o.

Equation (6) can be interpreted as a power equation, i.e.

$$\dot{V} = P_{in}(t) - P_{dis}(t) \tag{11}$$

where

$$P_{dis}(t) = -\varepsilon^2 F(V,\phi)(2V)^{\frac{1}{2}} \sin\phi; \quad P_{in}(t) = -\varepsilon(2V)^{\frac{1}{2}} \sin\phi\ y(t). \tag{12}$$

$P_{dis}(t)$ is the rate of energy loss due to damping and $P_{in}(t)$ is the rate of energy gain due to the input process, y(t).

3. AVERAGING THE ENERGY EQUATION

When ε is small V(t) is a slowly varying process (in a macroscopic sense) and equation (7) can be approximated by averaging over time, by generalising the well-known method of Krylov and Bogoliubov [1].

In the free undamped case ($\varepsilon = 0$) the period of oscillation is

$$T(V) = 2\sqrt{2} \int_0^b \frac{dx}{\sqrt{V - U(x)}} \tag{13}$$

where b is such that U(b) = V. The expression for $P_{dis}(t)$ can be approximated, for small ε, by averaging over a period of one oscillation, assuming that V(t) is constant over this period and that $\phi(t) = \phi_o(t)$ - i.e.

$$P_{dis}(t) = - \frac{\varepsilon^2 (2V)^{\frac{1}{2}}}{T(V)} \int_0^{T(V)} F(V,\phi_o) \sin\phi_o dt = \varepsilon^2 B(V) \tag{14}$$

and this is, in fact, correct to order ε^2. Equation (7) can now be approximated as

$$\dot{V} = - \varepsilon^2 B(V) + P_{in}(t). \tag{15}$$

The next stage involves approximating P_{in} (t). Suppose that y(t) has a correlation time scale τ_{cor}; this can be compared with the correlation time scale of V(t), which is of order $1/\varepsilon\omega_o$. If $\tau_{cor} << 1/\varepsilon\omega_o$ then $P_{dis}(t)$ can be approximated as an uncorrelated process. However the mean (which is non-zero) and the mean square of this uncorrelated process will depend on the energy level, V.

3.1 Mean of $P_{in}(t)$

If ε is small enough one can define an interval of time Δt such that $\Delta t >> \tau_{cor}$, but small enough for V(t) to change by only a small amount, ΔV. During this interval

$$\phi(t) = \phi_o(t) + \lambda(t) \tag{16}$$

where $\phi_o(t)$ corresponds to the energy level at the beginning of the interval and λ(t) is due to the damping and to the excitation process. If Δλ is the change in λ(t) over the interval Δt, we can write

$$P_{in}(t) = \varepsilon\beta_{t-\Delta t}\, y(t) + \varepsilon \left[\left(\frac{\partial\beta}{\partial V}\right)_{t-\Delta t} V + \left(\frac{\partial\beta}{\partial\lambda}\right)_{t-\Delta t} \Delta\lambda \right] y(t) \tag{17}$$

+ higher order terms in ΔV and Δλ

where

$$\beta_t = -(2V)^{\frac{1}{2}} \sin (\phi_o + \lambda). \tag{18}$$

Since $\Delta t >> \tau_{cor}$, $\beta_{t-\Delta t}$ and y(t) are statistically independent and the mean of the first term in equation (17) is thus zero. Further, since ΔV and Δλ are of order ε, then we have, correct to order ε^2

$$E\{P_{in}(t)\} = \varepsilon\, E\{[\left(\frac{\partial\beta}{\partial V}\right)_{t-\Delta t} \Delta V + \left(\frac{\partial\beta}{\partial\lambda}\right)_{t-\Delta t} \Delta\lambda]\}y(t) \tag{19}$$

for the mean of $P_{in}(t)$.

Expressions for ΔV and Δλ, correct to order ε, can be found by integrating equations (7) and (8). Thus for ΔV we have

$$\Delta V = -\varepsilon\, (2V)^{\frac{1}{2}} \int_{t-\Delta t}^{t} \sin \phi_o(\tau)\;\; y(\tau)\, d\tau. \tag{20}$$

The corresponding expression for Δλ is more difficult to obtain, due to the presence of the last term in equation (8). Here we make the simplifying approximation that

$$\int_{t-\Delta t}^{t} \frac{G(V,\phi)}{\cos\phi}\, d\tau \sim \int_{t-\Delta t}^{t} \dot{\phi}_o\, d\tau = \Delta\phi_o$$

where $\Delta\phi_o$ is the change in $\phi_o(t)$ in the interval Δt. There is an error, of order ε, in this approximation and thus the validity of this step must be tested (here, by a comparison of the final theoretical results with simulation results). There are, however, some theoretical grounds for believing that the error incurred through this approximation is likely to be relatively small, in most applications [8]. We can now write (ignoring terms of order higher than ε)

$$\Delta\lambda = -\frac{\varepsilon}{(2V)^{\frac{1}{2}}} \int_{t-\Delta t}^{t} \cos\phi_o(\tau)\, y(\tau)\, d\tau\,. \tag{21}$$

On combining equations (18) to (21) we obtain,

$$E\{P_{in}(t)\} = \varepsilon^2 \int_{-\infty}^{0} [\sin\phi_o(t)\sin\phi_o(t+\tau) + \cos\phi_o(t)\cos\phi_o(t+\tau)]\, w(\tau)\, d\tau\,. \tag{22}$$

Here the quantity in the square bracket is a fluctuating term and can be averaged over one cycle, of period T (dependent on V). Expanding $\sin\phi_o(t)$ and $\cos\phi_o(t)$ as Fourier series, we have

$$\sin\phi_o(t) = \sum_{n=1}^{\infty} s_n \sin\frac{2\pi nt}{T}\,; \qquad \cos\phi_o(t) = \sum_{n=1}^{\infty} c_n \cos\frac{2\pi nt}{T} \tag{23}$$

where the coefficients s_n and c_n are functions of the energy level V. On substituting these expressions into equation (22), and averaging, we obtain

$$E\{P_{in}(t)\} = \varepsilon^2 \frac{\pi}{2} \sum_{n=1}^{\infty} (s_n^2 + c_n^2)\, S[n\omega(V)] \tag{24}$$

where

$$\omega(V) = \frac{2\pi}{T(V)}\,. \tag{25}$$

3.2 Correlation function of $P_{in}(t)$

The correlation function of $P_{in}(t)$ is defined by

$$w_p(\tau) = E\{P_{in}(t)\, P_{in}(t+\tau)\}\,. \tag{26}$$

If terms of order higher than ε^2 are ignored again one finds that (see equations (17) and (18))

$$w_p(\tau) = \varepsilon^2 .\ 2V \sin\phi_o(t)\sin\phi_o(t+\tau)\, w(\tau)\,. \tag{27}$$

Here the harmonic terms can be time averaged over one cycle of oscillation, of period T (dependent on V), in a manner consistent with the treatment of the damping term and the mean of $P_{in}(t)$. Using the Fourier expansion for $\sin\phi_o(t)$ again (see equation (23) we have, on average,

$$w_p(\tau) = \varepsilon^2\, V \sum_{n=1}^{\infty} s_n^2 \cos[n\omega(V)\tau]\, w(\tau)\,. \tag{28}$$

Now if τ_{cor} is small relative to the correlation time scale of $V(t)$, the excitation term $P_{in}(t)$ in the energy equation behaves approximately as a white noise, as already stated. Thus one can approximate $w_p(\tau)$ as

$$w_p(\tau) = I(V)\ \delta(\tau) \tag{29}$$

where

$$I(V) = \int_{-\infty}^{\infty} w_p(\tau)\ d\tau\,. \tag{30}$$

Hence one finds that

$$I(V) = 2\pi\varepsilon^2 V \sum_{n=1}^{\infty} s_n^{\ 2}\ S[n\omega(V)]\ . \tag{31}$$

3.3 FPK equation for V(t)

From the above analysis it follows that $V(t)$ is governed, approximately, by the Itô equation

$$dV = \varepsilon^2 m(V)\ dt + \varepsilon\ D^{\frac{1}{2}}(V)\ dW \tag{32}$$

with drift coefficient

$$m(V) = -B(V) + \frac{\pi}{2} \sum_{n=1}^{\infty} (s_n^{\ 2} + c_n^{\ 2})\ S[n\omega(V)] \tag{33}$$

and diffusion coefficient

$$D(V) = 2\pi V \sum_{n=1}^{\infty} s_n^{\ 2}\ S[n\omega(V)]. \tag{34}$$

W is a unit Wiener (or Brownian) process. The FPK equation for $V(t)$ is thus

$$\frac{\partial p}{\partial t} = -\ \varepsilon^2 \frac{\partial}{\partial V}\ [m(V)p]\ \ + \frac{\varepsilon^2}{2}\ \frac{\partial^2}{\partial V^2}\ \ [D(V)p] \tag{35}$$

where $p(V|V_o;t)$ is the transition density function for $V(t)$.

The well-known stationary solution to equation (35) is

$$p(V) = \frac{C}{D(V)}\ \ \exp\{\ 2 \int_o^V \frac{m(\xi)}{D(\xi)}\ \ d\xi\} \tag{36}$$

where C is a normalisation constant.

4 APPLICATION TO THE DUFFING OSCILLATOR

To illustrate the method we consider a Duffing oscillator with linear damping - i.e. the equation of motion is

$$\ddot{x} + 2\zeta\omega_o x + \omega_o^{\ 2} x(1 + \kappa x^2) = Y(t) \tag{37}$$

where ζ is the usual non-dimensional damping coefficient and ω_o is the undamped, linear natural frequency ($\kappa = 0$). It is convenient to introduce the non-dimensional time

$$\tau = \omega_o t \tag{38}$$

in place of t. Equation (37) then becomes

$$\ddot{x} + 2\zeta\dot{x} + x(1 + \kappa x^2) = Y(\tau)/\omega_o^2 \tag{39}$$

where now (and henceforth) differentiation is with respect to τ. Further, we can write

$$Y(\tau)/\omega_o^2 = A.\ Z(\tau) \tag{40}$$

where A has the same dimensions as x and $Z(\tau)$ is non-dimensional. If $X = x/A$, equation (39) may be written in non-dimensional form as

$$\ddot{X} + 2\zeta\dot{X} + X(1 + \mu X^2) = Z(\tau) \tag{41}$$

where $\mu = \kappa A^2$.

The excitation process which will be considered here can be obtained by first filtering white noise, $n(\tau)$, through a first order linear filter. The output, $\eta(\tau)$, is obtained from

$$\dot{\eta} + \beta\eta = \beta n(\tau) \tag{42}$$

and has the spectrum

$$S_\eta(\Omega) = \frac{\beta^2}{2\pi(\beta^2 + \Omega^2)} \tag{43}$$

where Ω is the non-dimensional frequency $\Omega = \omega/\omega_o$.

If two such filtering operations are applied to independent white noises one obtains two first-order processes, $\eta_1(\tau)$ and $\eta_2(\tau)$. These two can be combined to form the input process

$$Z(\tau) = h[\eta_1(\tau) \cos \Omega_p\tau + \eta_2(\tau) \sin\Omega_p\tau] \tag{44}$$

where h is a variable amplitude scaling parameter. The spectrum of $Z(\tau)$, so defined, is given by

$$S(\Omega) = \frac{\sigma^2\beta}{2\pi}\left[\frac{1}{\beta^2 + (\Omega - \Omega_p)^2} + \frac{1}{\beta^2 + (\Omega + \Omega_p)^2}\right] \tag{45}$$

where σ is the standard deviation of the process.

Fig. 1 shows this spectrum for various values of β($\sigma = \Omega_p = 1$).

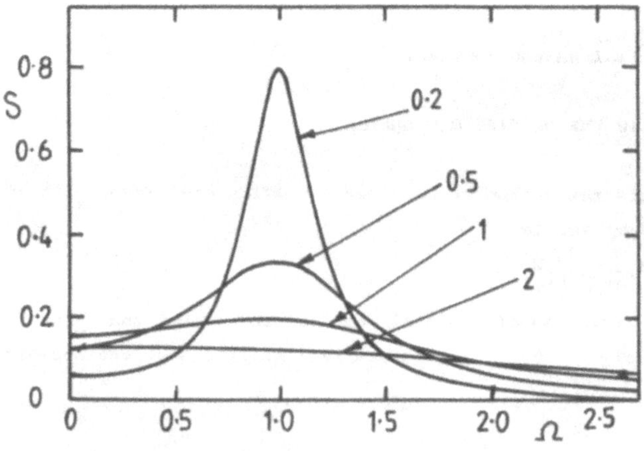

Fig.1 Spectrum of the excitation process.

For small β the spectrum has a single peak in the neighbourhood of frequency Ω_p. There are thus two useful disposable parameters for this spectrum - (a) bandwidth of the peak (measured by β) and (b) position of the peak relative to ω_o (measured by Ω_p).

Here we will compare the cumulative probability function for V(t),

$$P(V) = \int_0^V p(\xi)\, d\xi \tag{46}$$

obtained from the analytical solution (see equation (36)) with corresponding digital simulation estimates. The parameters which may be varied are as follows: ζ, μ, σ, β and Ω_p.

4.1 Analytical evaluation of P(V)

As an initial step in the application of the general theory we need to solve the equation for free, undamped oscillation - i.e.

$$\ddot{X} + X(1 + \mu X^2) = 0. \tag{47}$$

The solution to this equation can be expressed as

$$X = X_o \,\mathrm{cn}(\omega'\tau | m) \tag{48}$$

where cn is the Jacobian elliptic function, X_o is the amplitude of oscillation and

$$\omega' = [1 + \mu X_o^2]^{\frac{1}{2}}, \qquad m = \frac{\mu X_o^2}{2[1 + \mu^2 X_o^2]}. \tag{49),(50}$$

The period of oscillation, T(V), measured in units of τ, is found to be

$$T(V) = \frac{4K(m)}{[1 + \mu X_o^2]^{\frac{1}{2}}} \tag{51}$$

where K(m) is the complete elliptic integral.

To evaluate the terms m(V) and D(V) in the FPK equation one also requires an expression for $\sin \phi_o(\tau)$. This may be found by differentiating equation (39) and combining the result with equation (6). Thus

$$\sin \phi_o(\tau) = \frac{1}{(1 - m)^{\frac{1}{2}}} \,\mathrm{sn}(\omega'\tau | m)\, \mathrm{dn}(\omega'\tau | m) \tag{52}$$

where sn and dn are further Jacobian functions; hence one can obtain $\cos \phi_o(\tau)$. As $m \to 0$ $\sin \phi_o(\tau) \to \sin \tau$ and $\cos \phi_o(\tau) \to \cos \tau$, which is the linear result ($\mu = 0$). With increasing m, deviations from harmonic motion become more pronounced, with maximum distortion at m = 0.5 (the highest possible value for m (see equation (50)).

From equation (52) one can calculate the coefficient s_n and c_n which relate, respectively, to the Fourier expansions of $\sin \phi_o(\tau)$ and $\cos \phi_o(\tau)$ (see equation (23)). An expression for s_n can be found analytically - the result is

$$s_n = \frac{n\pi^2\, q^{n/2}}{K^2(m) m^{\frac{1}{2}}\, (1 - m)^{\frac{1}{2}} (1 + q^n)} \quad (n \text{ odd}), \qquad s_n = 0 \quad (n \text{ even}) \tag{53}$$

where

$$q = \exp\left\{\frac{-\pi\, K(1 - m)}{K(m)}\right\}. \tag{54}$$

Unfortunately, it does not seem possible to obtain c_n analytically - however, these coefficients are easy to compute numerically, using standard sub-routines for the evaluation of the Jacobian elliptic functions. Again only the odd coefficients are non-zero. For small m one finds, from equation (53), that

$$s_1 = 1 + \frac{3}{16} m, \tag{55}$$

with an error of order m^2, and, for small m, a good approximation to c_1 is

$$c_1 = 1 - \frac{3}{16} m. \tag{56}$$

The term B(V) is also required in the calculation of P(V) (see equation (14)). Here $\varepsilon^2 = 2\zeta$ and $F(V,\phi_o) = -(2V)^{\frac{1}{2}} \sin\phi_o$; hence

$$B(V) = \frac{2V}{T(V)} \int_0^{T(V)} \sin^2\phi_o(\tau)\, d\tau . \tag{57}$$

This can be evaluated numerically by using equation (52) for $\sin \phi_o$. In the linear case ($\mu = 0$) we have B(V) = V. For small m the following asymptote can be found:

$$B(V) \to V\{1 + \frac{3}{8} m\}. \tag{58}$$

This has an error of order m.

To compute m(V) and D(V), according to equations (33) and (34), it is necessary to truncate the summations. In the cases studied here (and in most other cases) the first term (n = 1) is completely dominant and the remaining terms in the summations can be omitted, without significant error.

4.2 Simulation estimates of P(V)

As a first step in simulating realisations of the response process $X(\tau)$ the two first order processes $\eta_1(\tau)$ and $\eta_2(\tau)$ must be generated, at equi-spaced intervals of time, $\Delta\tau$ say. For this purpose the following exact relationship is useful.

$$\eta_1(\tau + \Delta\tau) = \eta_1(\tau) \exp(-\beta\Delta\tau) + \{\frac{\beta}{2} [1 - \exp(-2\beta\Delta\tau)]\}^{\frac{1}{2}} \tag{59}$$

(and similarly for $\eta_2(\tau)$). The excitation process $Z(\tau)$ can be generated fairly rapidly, by combining $\eta_1(\tau)$ and $\eta_2(\tau)$ according to equation (44), again at equi-spaced time intervals, $\Delta\tau$. Hence the response process, $X(\tau)$, can be generated by numerically integrating the equations of motion, by means of a Runge-Kutta routine. From the samples of $X(\tau)$, and $\dot{X}(\tau)$, so obtained one can compute sample values of $V(\tau)$, where here

$$V(\tau) = \frac{\dot{X}^2}{2} + \frac{X^2}{2} + \frac{\mu X^4}{4} . \tag{60}$$

The sample values can be used to form a histogram of V(t), from which an estimate of the cumulative distribution, P(V), is obtained.

In the present work a value of $\Delta\tau = 2\pi/50$ (i.e. 50 samples per "cycle") was used throughout, and a total of 10^6 consecutive sample values were generated for each chosen set of parameters.

4.3 Comparison of theoretical and simulation results

We now present a few typical comparisons between theoretical and simulation estimates of P(V).

Fig. 2 shows the variation of P(V) with V, plotted on normal probability graph paper for the set of parameter values shown. The full circles are the simulation estimates and the full line shows the theoretical prediction, obtained by the method given earlier. Evidently there is very good agreement in this case.

In this example the input process has a fairly flat spectrum (see Fig.1) and one can expect a white noise approximation to be quite accurate. For ideal white noise $S(\omega) = S_o$ (say) where S_o is a constant; equations (33) and (34) then simplify to

$$m(V) = -B(V) + \pi S_o; \quad D(V) = 2\pi S_o B(V). \tag{61}$$

The present theory then gives a result in agreement with the exact result, which can be obtained from the well-known exact expression for the stationary joint distribution of $X(\tau)$ and $\dot{X}(\tau)$. The broken line in Fig.2 shows the result of using a white noise approximation for $S(\omega)$, with $S_o = S(\omega_o)$. Evidently in this particular case the effect of the input spectrum is small, but is correctly accounted for in the present theory.

Figs. 3, 4 and 5 show further similar comparisons between simulation results, the present theory and the white noise approximation. It is noted that there is a tendency for the agreement between theory and simulation to deteriorate as the input bandwidth decreases but in each case the present theory is a substantial improvement on the white noise approximation. The trends shown in these results can be predicted from simple physical reasoning. When $\Omega_p = 2.0$ the resonant frequent ($\mu = 0$) lies to the left of the peak in the input spectrum. In this case, increasing amplitude in the response is

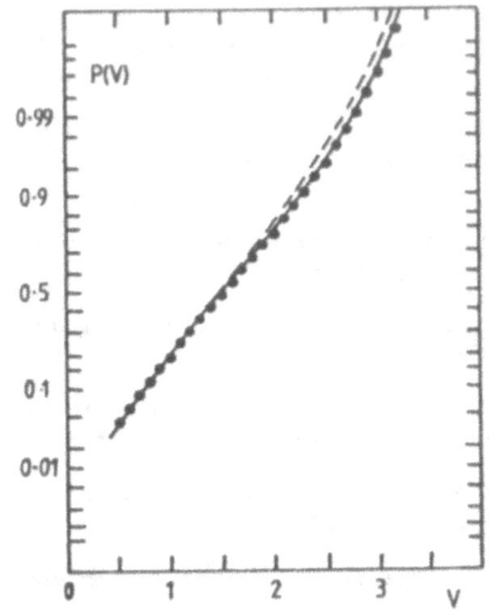

Fig.2. $\zeta = 0.005$, $\Omega_p = 2.0, \mu = 1.0, \sigma = 0.4$

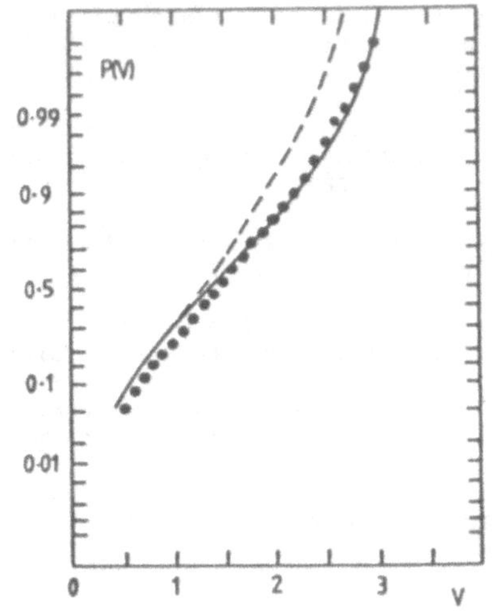

Fig.3. $\zeta = 0.005$, $\Omega_p = 2.0, \mu = 1.0, \sigma = 0.3$.

associated with increasing frequency of oscillation, which in turn is associated with higher levels in the input spectrum. One can then expect that the response is more likely to reach high amplitude levels than the white noise approximation predicts (see Figs. 2, 3 and 4). Conversely, when Ω_p = 0.5 the resonant frequency lies to the right of the input spectrum peak and the likelihood of reaching high amplitudes is less than the white noise approximation predicts.

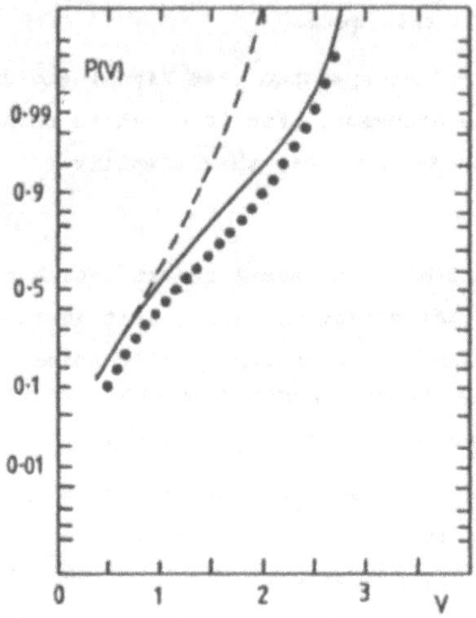

Fig. 4. ζ = 0.005, Ω_p = 2.0, μ = 1.0, σ = 0.2

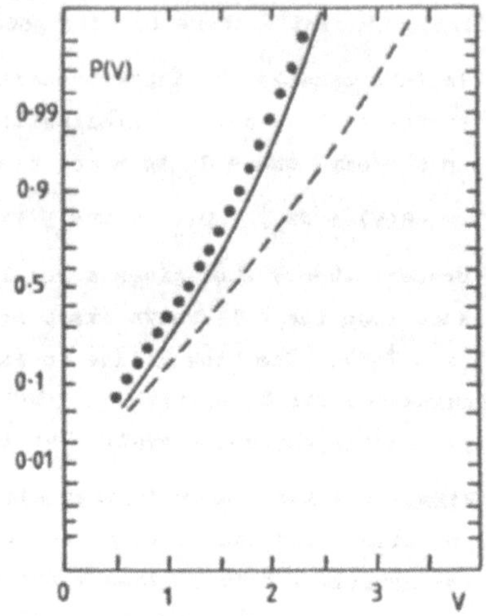

Fig.5. ζ = 0.005, Ω_p = 0.5, μ = 1.0, σ = 0.3

REFERENCES

1. **Bogoliubov, N.N. and Mitropolsky, Y.A., "Asymptotic Methods in the Theory of Non-linear Oscillations", Gordon and Breach, New York, 1961.**

2. **Khasminskii, R.Z., "A Limit Theorem for Solutions of Differential Equations with a Random Right Hand Side", Theory of Probability and its Applications, Vol.11, 1966, pp.390-406.**

3. **Papanicolaou, G.C. and Kohler, W., "Asymptotic Theory of Mixing Stochastic Ordinary Differential Equations", Communications in Pure and Applied Mathematics, Vol.27, 1974, pp.641-668.**

4. **Roberts, J.B.,"First Passage Probability for Non-linear Oscillators", Journal of the Engineering Mechanics Division, ASCE, Vol.102, 1976, pp.851-866.**

5. **Roberts, J.B., "The Energy Envelope of a Randomly Excited Non-linear Oscillator", Journal of Sound and Vibration, Vol.60, No.2, 1978, pp.177-185.**

6. **Roberts, J.B., "Stationary Response of Oscillators with Non-linear Damping to Random Excitation", Journal of Sound and Vibration, Vol.50, No.1, 1977, pp.145-156.**

7. **Roberts, J.B., "Transient Response of Non-linear Systems to Random Excitation", Journal of Sound and Vibration, Vol.74, No.1, 1981, pp.11-29.**

8. **Roberts, J.B., "A Stochastic Theory for Non-linear Ship Rolling in Irregular Seas", to be published in the Journal of Ship Research, SNAME, 1982.**

9. **Stratonovitch, R.L., "Topics in the Theory of Random Noise, Volume 1", Gordon and Breach, New York, 1963, pp.177-185.**

NONSTATIONARY RANDOM VIBRATIONS

Werner O. Schiehlen[1)]

Summary: Earthquake and vehicle engineering are increasingly important applications of nonstationary random vibrations. The covariance analysis of nonstationary random vibrations results in a differential Ljapunov matrix equation, usually solved by numerical integration. The paper presents a completely algebraical method for the computation of the time-varying covariance matrix with respect to excitation processes modulated by polynomials.

1. Introduction

In structural and vehicle dynamics random vibrations occur due to stochastic excitation processes. Typical examples are structures excited by earthquakes and vehicles running over rough ground. The excitation processes may be white or colored noise with stationary or nonstationary nature. The discrete mechanical models of structures and vehicles are given by linear or nonlinear differential equations. The rating of the stochastic responses requires statistical characteristics, at least the variances.

Random vibrations of linear systems may be computed by the spectral density analysis or by the covariance analysis, see e.g. /13/. The spectral density analysis has been originally restricted to stationary excitations but Priestley /9/ has then extended this method to nonstationary excitation processes. However, in the nonstationary case, one double matrix integral or two single matrix integrals with infinite boundaries, respectively, have to be determined in the frequency or time domain to obtain the required variances. Thus, lengthy computations occur. Nevertheless, many problems have been solved by the spectral density analysis in the last years. Earthquake engineering applications are due, e.g., to Grossmayer /1/, Vanmarcke /16/, Parkus /8/, Höllinger and Ziegler /3, 4/, vehicle responses are investigated by Yadav and Nigam /17/. On the other hand, the covariance analysis includes by definition nonstationary excitations. Then, in the nonstationary case, the algebraical Ljapunov matrix equation is replaced by an ordinary differential matrix equation, or one single matrix integral, respectively, to be integrated by any numerical procedure. The covariance analysis is also applied in structural

1) Institut B für Mechanik, Universität Stuttgart
Pfaffenwaldring 9, D - 7000 Stuttgart 80, F.R.G.

dynamics, see Spanos /15/, and successfully used in vehicle dynamics, see Hammond and Harrison /2/ and Rill /10, 11/.

In vehicle dynamics two problems are known:

i) homogeneous ground profil and variable vehicle velocity,

ii) inhomogeneous ground profil and constant speed.

The first problem results in a time-varying system with nonstationary excitation and has been discussed in detail by Sobczyk and Macvean /14/, Hammond and Harrison /2/ and Rill /11/. In general, the difference between the stationary and nonstationary analysis was found to be not significant. The second problem represents a time-invariant system with nonstationary excitation as it occurs in structural dynamics.

In this paper the computation of the covariance matrix of linear time-invariant systems with nonstationary white noise excitation is treated. In particular, the remaining numerical integration of the differential Ljapunov matrix equation will be replaced by a completely algebraical procedure for excitation processes modulated by polynomials.

2. Linear Systems with Nonstationary Excitations

The state equation of a linear time-invariant asymptotically stable system driven by white noise[1)] reads as

$$\dot{x}(t) = A\,x(t) + B(t)\,w(t), \quad x(t_0) \sim (0, P_0), \tag{1}$$

where $x(t)$ is the nx1-state vector, A the nxn-system matrix, $B(t)$ the mxl-excitation matrix and $w(t) \sim (0,Q)$ a 1x1-white noise vector process with the lxl-intensity matrix Q. The state equation (1) is discussed in detail in /7/.

The response $x(t)$ of system (1) is a nonstationary Gauss-Markov-process described uniquely by the nx1-mean vector

$$m(t) = 0 \tag{2}$$

and the nxn-correlation matrix

$$N(t,\tau) = P(t)\,\phi^T(\tau,t), \quad \tau > t \quad , \tag{3}$$

---

1) Colored noise processes are also admitted. Then, the corresponding shape filter has to be included in the system.

where $\phi(\tau,t) = e^{A(\tau-t)}$ is the nxn-state transition matrix and $P(t)$ represents the symmetrical nxn-covariance matrix given by the differential Ljapunov matrix equation

$$\dot{P}(t) = A\,P(t) + P(t)\,A^T + B(t)\,Q\,B^T(t)\ , \quad P(t_o) = P_o\ . \tag{4}$$

This means the covariance matrix $P(t)$ summarizes the total information of the nonstationary response process under consideration. The general solution of (4) reads as, see e.g. /7/,

$$P(t) = \phi(t,t_o)\,P_o\,\phi^T(t,t_o) + \\ + \int_o^t \phi(t,\tau)\,B(t)\,Q\,B^T(t)\,\phi^T(t,\tau)\,d\tau\ . \tag{5}$$

The state transition matrix $\phi(t,t_o)$ can be computed analytically via the eigenvalue problem or the theorem of Cayley-Hamilton. The matrix integral in (5) requires, however, for arbitrary excitation matrices $B(t)$ a numerical integration.

3. Covariance Analysis under Polynomial Excitations

In structural dynamics and in the second problem of vehicle dynamics a nonstationary excitation matrix $B(t)$ exists only during a finite time interval $t_A \leq t \leq t_E$. Within this interval the given excitation intensity $[B(t)\,Q\,B^T(t)]_{given}$ will be approximated by a polynomial of degree k,

$$[B(t)\,Q\,B^T(t))]_{approx} = \sum_{i=o}^{k} Q_i\,\{t-t_A,t_E\}^i\ , \tag{6}$$

while the excitation intensity outside of the interval vanishes[1]. The approximation results in nxn-coefficient matrices Q_i which are found by comparison of $[B(t)\,Q\,B^T(t)]_{given}$ and $[B(t)\,Q\,B^T(t)]_{approx}$ at a limited number of characteristic points or by balancing of the errors at an arbitrary number of points,

$$\left|[B(t)\,Q\,B^T(t)]_{given} - [B(t)\,Q\,B^T(t)]_{approx}\right|^{\perp} \text{Min}, \tag{7}$$

---

[1] The excitation intensity outside of the intervall may also be constant. Then, a constant matrix has to be added to (6) .

respectively. For more details see e.g. Zurmühl /18/. The accuracy of the approximation is increasing with the degree of the polynomial. However, only polynomials of low degree are of practical importance.

In earthquake engineering, Kozin /6/ has proposed to identify the non-stationary excitation matrix $B(t)$ by spline fitting of the measured records. However, the approximation of a record isn't identical with the approximation of the excitation intensity.

Further, the window function used in (6) is defined as follows:

$$\{t-t_A,t_E\}^i = \begin{cases} 0 & \text{if} \quad t < t_A \;, \\ (t-t_A)^i & \text{if} \quad t_A < t < t_E \;. \\ 0 & \text{if} \quad t_E < t \;. \end{cases} \tag{8}$$

The window function is an extension of the Heaviside function written as

$$\{t-t_A\}^i = \begin{cases} 0 & \text{if} \quad t < t_A \;, \\ (t-t_A)^i & \text{if} \quad t_A < t \;. \end{cases} \tag{9}$$

Then, the differential Ljapunov matrix equation (4) reads as

$$\dot{P}(t) = A\,P(t) + P(t)\,A^T + \sum_{i=o}^{k} Q_i\,\{t-t_A,t_E\}^i \;,\; P(t_o) = P_o . \tag{10}$$

Now, there exists an explicit analytical solution of the integral in (5) . It yields

$$\int_o^t \phi(t,\tau) \sum_{i=o}^{k} Q_i\,\{t-t_A,t_E\}^i\,\phi^T(t,\tau)\,d\tau$$

$$= \phi(t,t_A)\,R_o\,\phi^T(t,t_A)\,\{t-t_A\}^o - \sum_{i=o}^{k} R_i\,\{t-t_A,t_E\}^i$$

$$- \phi(t,t_E) \sum_{i=o}^{k} R_i\,(t_E-t_A)^i\,\phi^T(t,t_E)\,\{t-t_E\}^o \tag{11}$$

where the time-invariant auxiliary nxn-matrix R_i follows from the formula:

$$A\,R_i + R_i\,A^T = Q_i + (i+1)\,P_{i+1}\ , \quad i = k(-1)0\ , \quad P_{k+1} = 0\ . \qquad (12)$$

Thus, in addition to the well-known computation of the state transition matrix $\phi(t,t_o)$ the algebraical Ljapunov matrix equation (12) has to be solved repeatedly. The numerical solution of the algebraical Ljapunov matrix equation can be performed by economic procedures, see Koupan and Müller /5/.

The general solution of (10) is obtained according to (5) by adding the homogeneous solution.

$$\begin{aligned} P(t) = {} & \phi(t,t_o)\,P_o\,\phi^T(t,t_o) + \\ & + \phi(t,t_A)\,R_o\,\phi^T(t,t_A)\,\{t-t_A\}^o - \sum_{i=o}^{k} R_i\,\{t-t_A,t_E\}^i - \\ & - \phi(t,t_E)\sum_{i=o}^{k} R_i\,(t_E-t_A)^i\,\phi^T(t,t_E)\,\{t,t_E\}^o\ . \end{aligned} \qquad (13)$$

The nonstationary covariance matrix (13) shows three characteristic time intervals. In the first interval, $0 < t < t_A$, there is a free motion due to the initial matrix P_o. The second interval, $t_A < t < t_E$, represents a free motion with a changed initial matrix and a forced motion with a polynomial response. In the third interval there remains a free motion with an again changed initial matrix.

The stationary response of the covariance matrix

$$P(t \to \infty) = 0 \qquad (14)$$

is vanishing due to the system's asymptotic stability.

4. Mechanical Systems with Nonstationary Excitation

Mechanical systems occur, e.g., in structural dynamics and they are characterized by a second order vector differential equation of motion,

$$M\,\ddot{y}(t) + D\,\dot{y}(t) + K\,y(t) = w(t)\ , \qquad (15)$$

where $y(t)$ is the fx1-position vector describing the f degrees of freedom, M the symmetrical positive definite fxf-inertia matrix, D the fxf-damping matrix and K the fxf-stiffness matrix. Further, $w(t) \sim (0,Q(t))$ means a fx1-white noise excitation process with the time-varying fxf-intensity matrix $Q(t)$. The equation (15) can be also written in the form of the state equation (1) resulting in special matrices:

$$x(t) = \begin{bmatrix} y(t) \\ \dot{y}(t) \end{bmatrix}, \quad A = \begin{bmatrix} 0 & E \\ -M^{-1}K & -M^{-1}D \end{bmatrix}, \quad B = \begin{bmatrix} 0 \\ M^{-1} \end{bmatrix} . \tag{16}$$

For the dimension it yields $n = 2f$. Further, the covariance matrix may be partitioned as

$$P(t) = \begin{bmatrix} P_I(t) & P_{II}(t) \\ P_{II}^T(t) & P_{III}(t) \end{bmatrix} , \tag{17}$$

where $P_I(t)$ $P_{II}(t)$, $P_{III}(t)$ are fxf-submatrices. Then, the differential Ljapunov matrix equation may be written, see also /18/, as

$$\dot{P}_I(t) = P_{II}(t) + P_{II}^T(t) , \tag{18}$$

$$\dot{P}_{II}(t)\, M = -\, K\, P_I(t) - D\, P_{II}(t) + M\, P_{III}(t) , \tag{19}$$

$$\begin{aligned} M\, \dot{P}_{III}(t)\, M = & -\, K\, P_{II}(t)\, M - D\, P_{III}(t)\, M \\ & -\, M\, P_{II}^T(t)\, K - M\, P_{III}(t)\, D + Q(t) . \end{aligned} \tag{20}$$

In particular, it follows from (18) for the diagonal elements of the submatrices

$$\dot{P}_{Ijj}(t) = 2\, P_{IIjj}(t) , \quad j = 1(1)f . \tag{21}$$

Regarding (16) the scalar variances of the position coordinates y_j are given by

$$\sigma^2_{yjyj}(t) = P_{Ijj}(t) , \quad j = 1(1)f . \tag{22}$$

This means that the extrema of variances $\sigma^2_{yjyj}(t)$ appear immediately at the zeros of $P_{IIjj}(t)$.

In contrary to mechanical systems with stationary excitations, the special structure (16) of the matrices does not reduce the number of essential equations in the nonstationary case. There isn't any difference between the $n(n+1)/2$ scalar differential equations obtained from (4) and from (18) to (20).

5. Example

As an example a single degree of freedom system, $f = 1$, will be treated. Such a system represents either a single story structure, Fig. 1, or a simple vehicle model, Fig. 2. The white noise excitation is due to the acceleration by the ground, $\ddot{y}_e(t) = w(t) \sim (0, q(t))$.

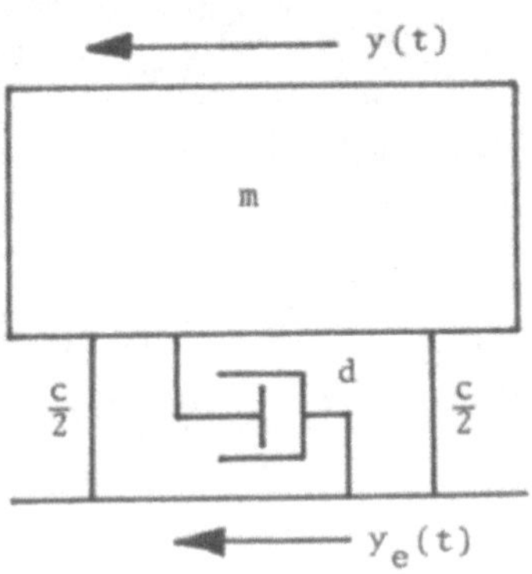

Fig. 1 Single story structure

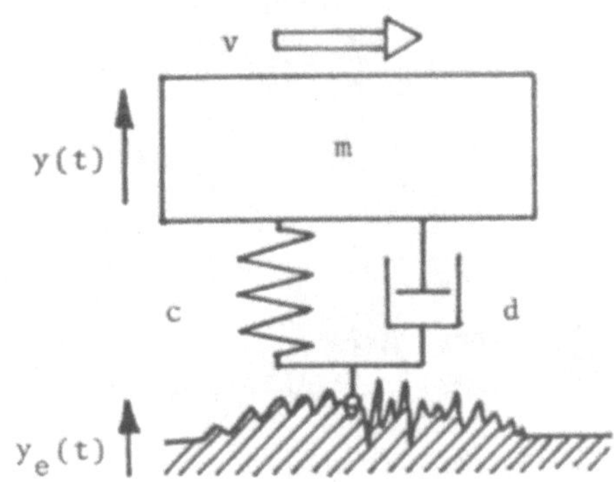

Fig. 2 Simple vehicle model

Then, the equation of motion in normalized form reads as

$$\ddot{y}(t) + 2\,D\,\dot{y}(t) + y(t) = w(t) \quad , \quad y(0) = 0 \quad , \quad \dot{y}(0) = 0 \quad . \tag{23}$$

The time-varying scalar excitation intensity may be

$$q(t) = \sin t \,\{t-0,\pi\}^{o} \quad . \tag{24}$$

The intensity (24) is approximated according to (6) by

$$q_{approx}(t) = \sum_{i=o}^{k} q_i \,\{t-0,\pi\}^{i} \quad . \tag{25}$$

Two cases are considered. For a polynomial of the second degree, $k = 2$, the coefficients are $q_o = 0$, $q_1 = 4/\pi$, $q_2 = -4/\pi^2$ and for a polynomial of the fourth degree, $k = 4$, the following coefficients are used, $q_o = 0$, $q_1 = 1$, $q_2 = (16-5\pi)/\pi^2$, $q_3 = -8(4-\pi)/\pi^3$, $q_4 = 4(4-\pi)/\pi^4$. For comparison the differential equations according to (18) to (20) are also integrated numerically. The results are shown in Fig. 3.

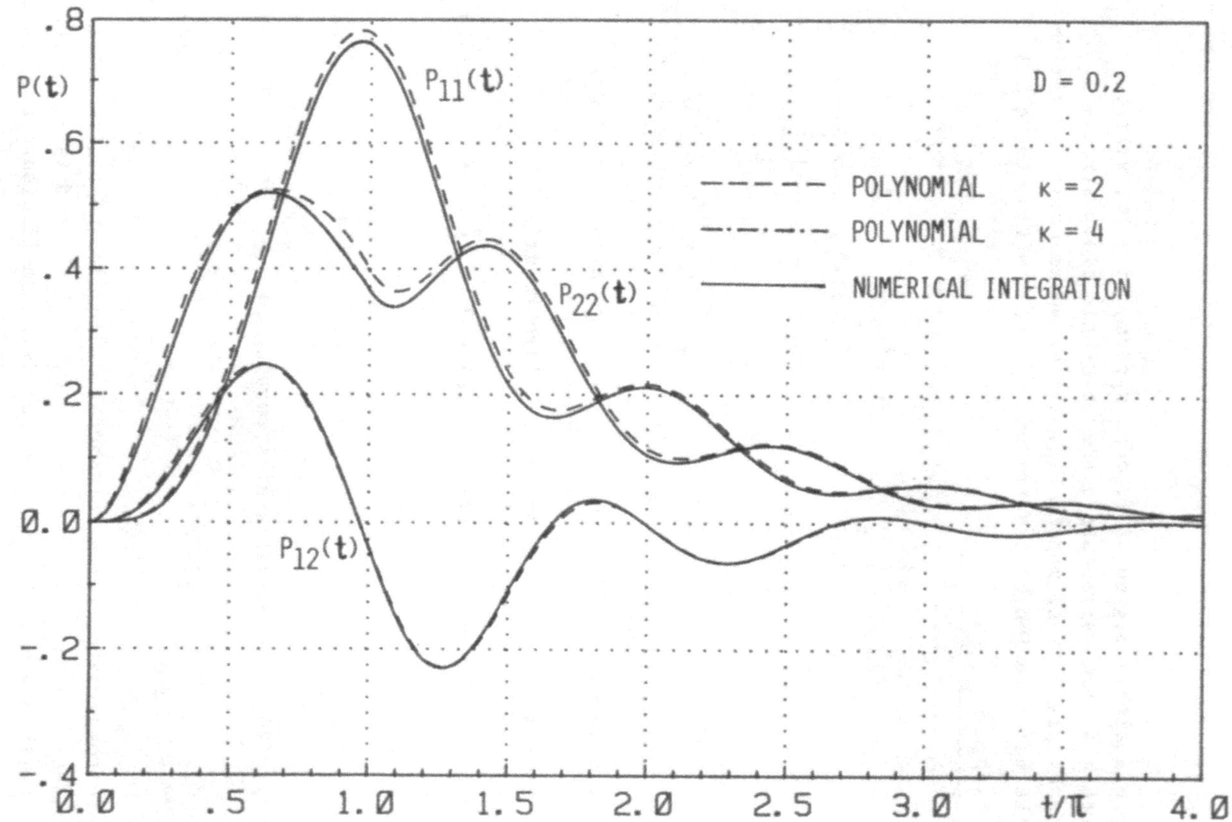

Fig.3 Comparison of the results

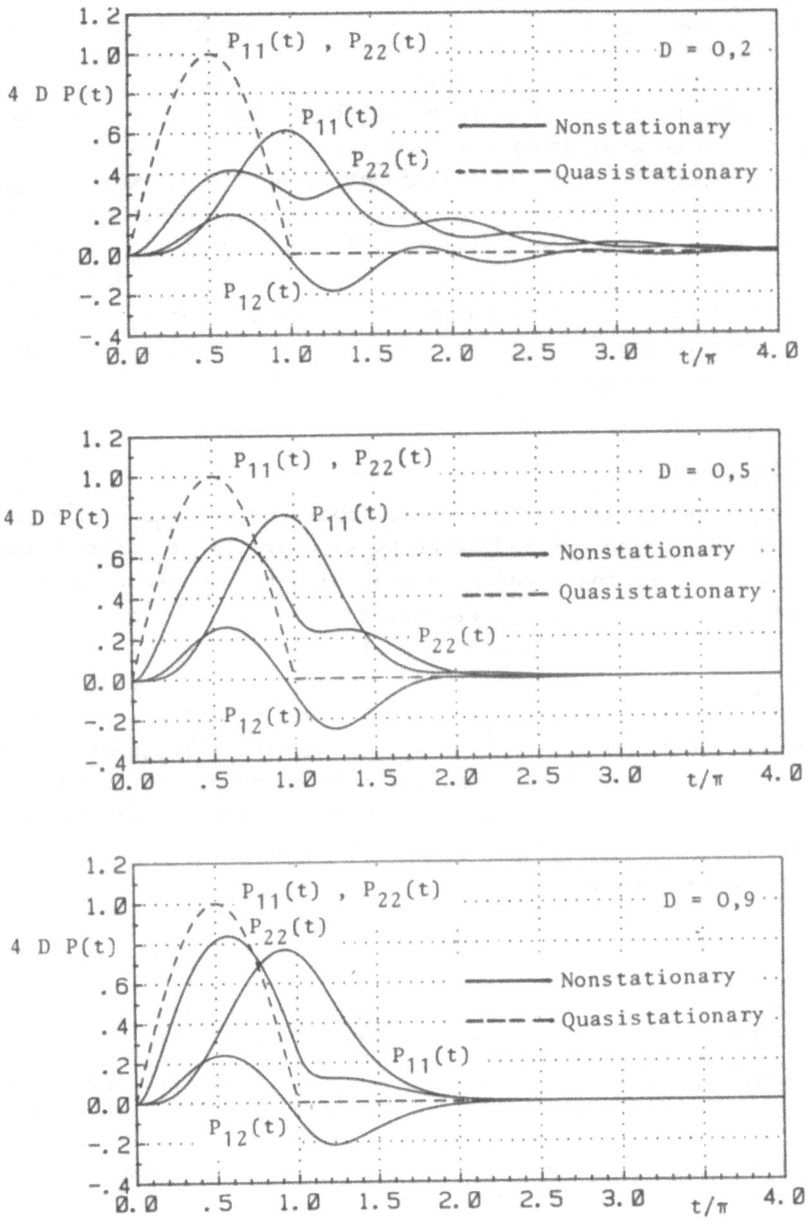

Fig.4 Quasistationary and nonstationary response

It turns out that the numerical integration and the algebraic solution by the polynomial of the fourth degree coincide completely while the solution by the polynomial of the second degree shows only small deviations.

This example is also used to consider the question if a quasistationary analysis or a nonstationary analysis results in lower variances. Fig. 4 shows algebraically computed solutions, $k = 4$, for various damping coefficients. The nonstationary analysis results independently of the damping in lower variances. The absolute maximum of the variance $\sigma^2_{yy}(t) = P_{11}(t)$ of the position coordinate appears at the first zero of $\sigma^2_{y\dot{y}}(t) = P_{12}(t)$ according to (21), (22) . For large damping coefficients the variance $\sigma^2_{\dot{y}\dot{y}}(t) = P_{22}(t)$ of the velocity coordinate tends to the stationary response.

6. Conclusion

The covariance matrix of linear time-invariant systems with nonstationary excitation processes modulated by polynomials can be computed completely by solving the eigenvalue problem once and the algebraical Ljapunov matrix equation repeatedly. The proposed method is purely algebraical and avoids the numerical integration at all. In the case of nonstationary excitations mechanical systems don't offer a reduced number of essential equations as it is known from the stationary case. However, the extrema of the time-varying position variances are easily obtained for mechanical systems. A simple example shows a sufficient accuracy of the proposed method even with polynomials of low degree. Compared with the nonstationary analysis a quasi-stationary investigation leads to worse results.

References

/1/ Grossmayer, R.: Stochastische Zuverlässigkeitsanalyse erdbebenerregter Strukturen, Dissertation, TU Wien 1975.

/2/ Hammond, J.K.; Harrison, R.F.: Nonstationary Response of Vehicles on Rough Ground - A State Space Approach, ASME J. Dyn. Syst. Meas. Control 103 (1981) 245-250.

/3/ Höllinger, F.: Harmonical and Nonstationary Random Vibrations of Dam-Reservoir-Systems, Acta Mechanica 46 (1983) to appear.

/4/ Höllinger, F.; Ziegler, F.: Instationäre Zufallsschwingungen einer elastischen Gewichtsmauer bei allgemeiner Beckenform, Z. angew. Math. Mech. 63 (1983) 49 54

/5/ Koupan, A.; Müller, P.C.: Zur numerischen Lösung der Ljapunovschen Matrizengleichung $A^T P + P A = - Q$, Regelungstechnik <u>24</u> (1976) 167-169.

/6/ Kozin, F.: An Approach to Characterizing, Modeling and Analyzing Earthquake Excitation Records. In: Random Excitation of Structures by Earthquakes and Atmospheric Turbulence, Ed. H. Parkus, CISM Courses and Lectures No. 225, Wien-New York, Springer-Verlag 1977.

/7/ Müller, P.C.; Schiehlen, W.: Lineare Schwingungen, Wiesbaden, Akad. Verlagsges. 1976.

/8/ Parkus, H.: Überschwappwahrscheinlichkeit für einen Flüssigkeitsbehälter unter Erdbebeneinwirkung, Ing. Arch. <u>49</u> (1980) 179-185.

/9/ Priestley, M.B.: Power Spectral Analysis of Non-Stationary Random Processes, J. Sound Vibr. <u>6</u>-1 (1967) 86-97.

/10/ Rill, G.; Macvean, D.B.: Diskussion zum Beitrag von D.B. Macvean: Durch Fahrbahnunebenheiten erzeugte Schwingungen eines beschleunigten Fahrzeugs, Ing.-Arch. <u>52</u> (1982) 91-94.

/11/ Rill, G.: Instationäre Fahrzeugschwingungen bei stochastischer Erregung, Dissertation, Universität Stuttgart 1983.

/12/ Schiehlen, W.: Zur Untersuchung von Zufallsschwingungen, Z. angew. Math. Mech. <u>54</u> (1974) T64-T65.

/13/ Schiehlen, W.: Mechanische Zufallsschwingungen, Z. angew. Math. Mech. <u>63</u> (1983) T14-T20.

/14/ Sobczyk, K.; Macvean, D.B.: Non-Stationary Random Vibrations of Systems Travelling with Variable Velocity. In: Stochastic Problems in Dynamics, Ed. B.L. Clarkson, London, Pitman 1977.

/15/ Spanos, P.D.: Stochastic Linearization in Structural Dynamics, Appl. Mech. Rev. <u>34</u> (1981) 1-8.

/16/ Vanmarcke, E.H.: Structural Response to Earthquakes. In: Seismic Risk and Eng. Decisions, Ed. C. Lomnitz, E. Rosenblueth, Amsterdam-Oxford-New York, Elsevier 1976.

/17/ Yadav, D.; Nigam, N.C.: Ground Induced Non-Stationary Response of Vehicles, J. Sound Vibr. <u>61</u>-1 (1978) 117-126.

/18/ Zurmühl, R.: Praktische Mathematik, 4. Aufl., Berlin-Heidelberg-Göttingen, Springer 1963.

NONLINEAR RANDOM VIBRATIONS OF SYSTEMS WITH SEVERAL DEGREES OF FREEDOM

G. Schmidt[1)], R. Schulz[1)]

As an example for coupling vibrations in a two-degree-of-freedom system we consider systems with autoparametric coupling. In such systems a forced excitation of one mode acts, because of a nonlinear connection between different modes, as a parametric excitation of another mode. An example is the autoparametric vibration absorber system investigated by R. S. Haxton and A. D. S. Barr [1] and R. A. Ibrahim and J. W. Roberts [2,3]. It consists of a main mass and a weightless beam with a concentrated end mass mounted on the main mass (Fig.1). Forced vertical vibrations of the main mass impose forced axial vibrations and, in case of a parametric response, additional lateral vibrations on the beam which may react on the main mass through inertial nonlinearities to give a vibration absorber effect.

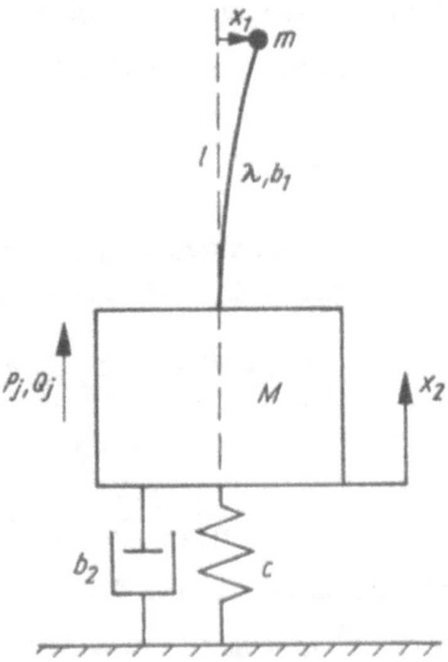

The equations of motion of typical systems with autoparametric coupling following from the Lagrangian formalism are

$$\ddot{x}_1 + \Omega_1^2 x_1 = -B_1\dot{x}_1 - C_1 x_1^2 \dot{x}_1 - D_1 x_1^3 - E x_1(\dot{x}_1^2 + x_1\ddot{x}_1) + J x_1 x_2 + K_1 x_1 \ddot{x}_2 + L_1(\dot{x}_2^2 + x_2\ddot{x}_2) - \sum_{j=1}^{\infty}(G_j \cos j\omega t + H_j \sin j\omega t) x_1,$$

$$\ddot{x}_2 + \Omega_2^2 x_2 = \sum_{j=1}^{\infty}(P_j \cos j\omega t + Q_j \sin j\omega t) - B_2\dot{x}_2 - C_2 x_2^2 \dot{x}_2 - D_2 x_2^3 + F x_2^2 + K_2 \ddot{x}_1 x_2 + L_2(\dot{x}_1^2 + x_1\ddot{x}_1)$$

where some terms have only been included for other examples of autoparametric coupling which can be considered in the same way; for the autoparametric vibration absorber we can put

---

1) Institut für Mechanik der Akademie der Wissenschaften der DDR, Rudower Chaussee 5, DDR-1199 Berlin-Adlershof

$$F = G_j = H_j = J = K_2 = L_1 = 0,$$

and it is especially

$$\Omega_1^2 = \frac{\lambda}{m}, \quad \Omega_2^2 = \frac{c}{M+m}, \quad B_1 = \frac{b_1}{m}, \quad B_2 = \frac{b_2}{M+m},$$

where (only here) m is the absorber mass, M the main mass, l the length of the beam, c the vertical and $\lambda = EI/l^3$ the lateral spring stiffness with the flexural rigidity EI, and b_i the viscous damping acting on the absorber mass and the main mass respectively.

We introduce a dimensionless time $\tau = \omega t$ and write $x_i(t) = y_i(\tau)$ and

$$\lambda_1 = \frac{\Omega_1^2}{\omega_1^2}, \quad \lambda_2 = \frac{\Omega_2^2}{\omega_2^2}.$$

In the non-resonance case, neither λ_1 nor λ_2 is the square of an integer for $\omega_1 \approx \omega_2 \approx \omega$, and the first approximative solution yields to be

$$y_{110} = 0, \quad y_{210} = \frac{1}{\omega^2}\sum_{j=1}^{\infty} \frac{P_j \cos j\tau + Q_j \sin j\tau}{\vartheta_{\lambda_2}^{j^2}\lambda_2 - j^2}.$$

Now we assume a two-fold resonance

$$\lambda_1 = n^2, \quad \lambda_2 = n_2^2$$

with integers n, n_2. The case of a single resonance is included.We write

$$\omega = \omega_1(1+\alpha) = \omega_2(1+\alpha_2), \quad \alpha_2 = \alpha - \varepsilon$$

with the frequency variation α and the detuning ε.

The first approximation is

$$y_{11} = r\cos n\tau + s\sin n\tau, \quad y_{21} = r_2\cos n_2\tau + s_2\sin n_2\tau + y_{210}.$$

The second approximation of y_1, y_2 and by introducing the latter the third approximation of y_1 have been evaluated but can not be given here. It is not easy to interpret these approximations which include many informations on the time dependence of the solutions. The periodicity conditions stemming from them are, in vector form,

$$a_1\binom{r}{s} - b_1\binom{s}{-r} + p_1\binom{r}{-s} + q_1\binom{s}{r} - k\left[r_2\binom{r}{-s} + s_2\binom{s}{r}\right] + b\left[r_2\binom{s}{r} + s_2\binom{-r}{s}\right]$$
$$+ c\left[s_2(s_2^2-3r_2^2)\binom{r}{-s} + r_2(r_2^2-3s_2^2)\binom{s}{r}\right] + d\left[r_2(3s_2^2-r_2^2)\binom{r}{-s} + s_2(s_2^2-3r_2^2)\binom{s}{r}\right]$$
$$= 0 \qquad (1)$$

and

$$a_2\binom{r_2}{s_2} - b_2\binom{s_2}{-r_2} - l\binom{r^2-s^2}{2rs} + \binom{p_2}{q_2} = 0 \qquad (2)$$

with the abbreviations

$$a_1 = n^2\alpha - \left(\frac{3D_1}{4\omega^2} - \frac{n^2E}{2} - \frac{2n^4K_1L_2}{4n^2-\vartheta_{n_2}^{2n}n_2^2}\right)A^2, \quad a_2 = n_2^2\alpha_2 - \frac{3D_2}{4\omega^2}A_2^2,$$

$$b_1 = \frac{n}{\omega}(B_1+\frac{C_1}{4}A^2), \quad b_2 = \frac{n_2}{\omega}(B_2+\frac{C_2}{4}A_2^2), \quad b = \delta_{n_2}^{2n}\frac{b_2K_1}{2},$$

$$p_1 = \frac{2n^2K_1}{(4n^2-\delta_{n_2}^{2n}n_2^2)\omega^2}P_{2n}, \quad q_1 = \frac{2n^2K_1}{(4n^2-\delta_{n_2}^{2n}n_2^2)\omega^2}Q_{2n},$$

$$p_2 = \frac{1}{\omega^2}P_{n_2}, \quad q_2 = \frac{1}{\omega^2}Q_{n_2}, \quad k = \delta_{n_2}^{2n}(2n^2 - \frac{a_2}{2})K_1,$$

$$l = \delta_{n_2}^{2n}n^2L_2, \quad c = \delta_{3n_2}^{2n}\frac{3nC_2K_1}{32\omega}, \quad d = \delta_{3n_2}^{2n}\frac{9D_2K_1}{64\omega^2},$$

where $A^2 = r^2+s^2$, $A_2^2 = r_2^2+s_2^2$ are the squares of the partial amplitudes (more exactly: of the resonance parts of the first approximation of the partial amplitudes) in directions y_1 and y_2 respectively.

The equations (2) are linear in r_2, s_2, their solution is

$$\begin{aligned} Br_2 &= -a_2p_2 - b_2q_2 + a_2l(r^2-s^2)\ 2b_2lrs, \\ Bs_2 &= b_2p_2 - a_2q_2 + 2a_2lrs - b_2l(r^2-s^2), \end{aligned} \tag{3}$$

when we abbreviate $B = a_2^2 + b_2^2$.

Squaring the equations (3), adding and dividing by B yields

$$BA_2^2 = p_2^2 + q_2^2 + l^2A^4 - 2lp_2(r^2-s^2) - 4lq_2rs.$$

The equations (1) and (2), complicated as they are, already show the coupling between the vertical vibrations of the main mass, represented by r_2, s_2, A_2, and the lateral absorber vibrations, represented by r, s, A. If $l = 0$, i.e. if the coupling coefficient L_2 vanishes or the resonance condition $n_2 = 2n$ does not hold, the vertical vibration is (in the approximation at hand) not influenced by the lateral vibration but always influences the lateral vibration. The latter influence is given already by L_2, n_2 and the forced excitation terms P_{2n}, Q_{2n} in a_1, p_1, q_1, but is much more complicated by the linear r_2, s_2 terms in (1), if k, b do not disappear, i.e. if $n_2 = 2n$ and $K_1 \neq 0$, and by the nonlinear r_2, s_2 terms, if c, d do not disappear, i.e. if $3n_2 = 2n$ and K_1, C_2, D_2, respectively, are not zero.

We abbreviate $a_1 = a-eA^2$, $b_1 = f+gA^2$. For $n_2 \neq 2n$ (no internal resonance) we can find the amplitude formula

$$(e^2+g^2)A^2 = ae - fg \pm \sqrt{(e^2+g^2)(p_1^2+q_1^2) - (ef+ag)^2}$$

which shows again that the motion of the main mass influences the absorber amplitudes only by L_2, P_{2n}, Q_{2n}, n_2 in e and p_1, q_1. The formula for the main mass amplitude now reads

$$BA_2^2 = p_2^2 + q_2^2 ,$$

it is independent of the absorber motion.

For internal resonance, $n_2 = 2n$, the absorber amplitude equation yields

$$(k^2l^2+2b_2klg+Bg^2)A^2 = (aa_2-fb_2)kl - Bfg \pm$$
$$\pm \sqrt{(k^2l^2+2b_2klg+Bg^2)k^2(p_2^2+q_2^2) - [(ab_2+fa_2)kl+aBg]^2}$$

whereas the formula for the main mass amplitudes is, for linear damping

$$k^2A_2^2 = a^2 + f^2 \; .$$

It shows that, in contrast to the formula above, the main mass amplitudes only depend (besides of ω) on the absorber motion parameters n, K_1, α_1, B_1.

In what follows we will assume that not a periodic excitation, but a stationary narrow-band random excitation with mean value zero

$$P(t) = \pi(t) \cos[2n\omega t + \delta(t)]$$

acts on the main mass. The excitation amplitude $\pi(t)$ and the excitation phase $\delta(t)$ are slowly varying functions. The excitation amplitude is assumed as Rayleigh distributed with the probability density

$$w(\pi) = \frac{\pi}{\sigma^2} \exp(-\frac{\pi^2}{2\sigma^2}) \; ,$$

so that the excitation process P(t) is Gaussian distributed. The response amplitudes corresponding with a strip $w(\pi)d\pi$ of excitation amplitudes can be found by the transformation

$$w(\pi)d\pi = dW(\pi) = dV(A) = v(A)dA \; . \tag{4}$$

The amplitude formula can now be written in the form

$$K_1^2\pi^2 = n^2\omega^2(\alpha B_2+2\alpha_2B_1)^2 + (n^2\omega^2K_1L_2A^2-2n^2\omega^2\alpha\alpha_2+B_1B_2)^2$$

so that (4) yields

$$V(A) = e^{-\varphi} \; ,$$
$$\varphi = \frac{1}{2K_1^2\sigma^2}\left[n^2\omega^2(\alpha B_2+2\alpha_2B_1)^2 + (n^2\omega^2K_1L_1A^2-2n^2\omega^2\alpha\alpha_2+B_1B_2)^2\right] .$$

By differentiation we get the probability density of the vibration amplitude

$$v(A) = \frac{2n^2\omega^2L_2A}{K_1\sigma^2}(n^2\omega^2K_1L_2A^2 - 2n^2\omega^2\alpha\alpha_2 + B_1B_2)\, e^{-\varphi} \; . \tag{5}$$

The factor in parentheses vanishes on the skeleton line $A = A_s$ following from the amplitude equation for vanishing radicand. Because amplitudes belonging to parts of the response curve below the skeleton line do not appear, we have to assume $v(A) = 0$ for negative values of the above factor in parentheses, so that always $v(A) \geq 0$ holds.

The whole factor preceding $e^{-\varphi}$ in (5) ensures that the probability density disappears if the amplitude approach $A_s > 0$ or zero whereas $e^{-\varphi}$ causes the probability density to disappear for great amplitudes.

It is of interest to evaluate the probability $V_1 = V(A_1 < A < \infty)$ for exceeding a given amplitude level A_1. From (4) there follows

$$V_1 = \int_{A_1}^{\infty} v(A)dA \leq 1.$$

Because of $\alpha_2 = \alpha - \varepsilon$ we get for $n^2\omega^2K_1L_2A_1^2 \geq 2n^2\omega^2\alpha(\alpha-\varepsilon) - B_1B_2$

$$V_1 = \exp - \frac{1}{2K_1^2\sigma^2}\left\{n^2\omega^2\left[(2B_1+B_2)\alpha - 2B_1\varepsilon\right]^2 + \left[n^2\omega^2K_1L_2A_1^2 - 2n^2\omega^2\alpha(\alpha-\varepsilon)+B_1B_2\right]^2\right\} \tag{6}$$

and for $n^2\omega^2K_1L_2A_1^2 < 2n^2\omega^2\alpha(\alpha-\varepsilon)-B_1B_2$, i.e. for $2n^2\omega^2\alpha(\alpha-\varepsilon) > B_1B_2$ and sufficiently small A_1, independently of A_1,

$$V_{[1]} = \exp - \frac{n^2\omega^2}{2K_1^2\sigma^2}\left[(2B_1+B_2)\alpha - 2B_1\epsilon\right]^2 . \tag{7}$$

For a discussion of V_1 and $V_{[1]}$ we first state that the maximum value $V_{[1]} = 1$ would appear for $(2B_1+B_2)\alpha = 2B_1\varepsilon$ (but then (7) does not hold, because $A_1^2 < 0$ would follow) and $V_{[1]}$ momotonously diminishes if α removes from this value. In the existence interval of V_1 the inequality $V_1 \leq V_{[1]}$ is valid. In every case, $V_1 < 1$ and $V_{[1]} < 1$ hold, so that the probability for vanishing amplitude A is not zero.

The condition $dV_1/d\alpha = 0$ for extreme values of V_1 leads to a cubic equation in α. In order to enable an analytical discussion we confine ourselves to vanishing detuning, $\varepsilon = 0$. Then the three extreme values for $\alpha = \alpha_0 = 0$ are

$$V_{1o} = \exp - \frac{1}{2K_1^2\sigma^2}(n^2\omega^2K_1L_2A_1^2 + B_1B_2)^2 \tag{8}$$

and for $2\alpha^2 = 2\alpha_m^2 = K_1L_2A_1^2 - \frac{1}{n^2\omega^2}(B_1^2+\frac{1}{4}B_2^2)$ (only real and different from $\alpha_0 = 0$, if

$$n^2\omega^2K_1L_2A_1^2 > B_1^2 + \tfrac{1}{4}B_2^2 \tag{9}$$

holds)

$$V_{1m} = \exp - \frac{(B_1+\frac{1}{2}B_2)^2}{2K_1^2\sigma^2}\left[2n^2\omega^2K_1L_2A_1^2 - (B_1 - \tfrac{1}{2}B_2)^2\right]. \tag{10}$$

It can be shown that $V_{1o} < V_{1m}$ because of (9), so that V_1 has a minimum V_{1o} for $\alpha = 0$ and maxima V_{1m} for $\alpha = \pm\alpha_m$. These maxima really appear because the condition for the validity of (6) holds for $\alpha = \pm\alpha_m$.

If the condition (9) for maxima in $\pm\alpha_m \neq 0$ does not hold, V_{1o} is the maximum value of V_1, as the factor

$$B_1^2 + \tfrac{1}{4}B_2^2 - n^2\omega^2K_1L_2A_1^2$$

of α^2 in (6) shows.

Vice versa, the amplitude level A_1 amplitudes above which have only the (may be small) probability V_1 follows from (10)

$$n^2\omega^2K_1L_2A_1^2 = -\frac{K_1^2\sigma^2 \ln V_1}{(B_1+\frac{1}{2}B_2)^2} + \frac{1}{2}(B_1-\frac{1}{2}B_2)^2 \text{ if } n^2\omega^2K_1L_2A_1^2 > B_1^2+\frac{1}{4}B_2^2 \tag{11}$$

and from (8)

$$n^2\omega^2K_1L_2A_1^2 = \sqrt{-2K_1^2\sigma^2 \ln V_1} - B_1B_2 \text{ if } n^2\omega^2K_1L_2A_1^2 \le B_1^2+\frac{1}{4}B_2^2 \, . \tag{12}$$

For the equality sign in the last condition both formulae coincide. They show the dependence of A_1 on the different parameters. Whereas (12) causes a diminuation of A_1 for any increase of the damping coefficients B_1 or B_2, (11) leads to a minimum A_1 for a ceratin damping ratio which follows from

$$\left(\frac{B_2}{2B_1} - 1\right)\left(\frac{B_2}{2B_1} + 1\right)^3 = -\frac{2K_1^2\sigma^2 \ln V_1}{B_1^4}$$

for (say) B_1 given.

In the last part we investigate a white noise forced excitation $\gamma\dot{\xi}(t)$ of the main mass and a white noise parametric excitation $\varepsilon\dot{\xi}(t)x_1$ of the absorber. Transforming the equations of motion by

$$x_i = A_i\cos(\Omega_i t+\Theta_i), \quad A_1 = A,$$

into standard form and understanding them as physical or Stratonovic ones, the corresponding Itô equations are written in the form

$$dA_i = P_i dt + Q_i d\xi, \quad d\Theta_i = R_i dt + S_i d\xi$$

where the Stratonovic iterative averaging method is used to successively determine the coefficients P_i, Q_i, R_i, S_i. The Fokker-Planck equation for the stationary probability density $v(A_1,A_2)$ of the amplitudes leads to the partial differential equation

$$Tv + T_1\frac{\partial v}{\partial A_2} + T_2\frac{\partial v}{\partial A_2} = \frac{Q_1^2}{2}\frac{\partial^2 v}{\partial A_1^2} + \frac{Q_2^2}{2}\frac{\partial^2 v}{\partial A_2^2} + Q_1Q_2\frac{\partial^2 v}{\partial A_1 \partial A_2} \tag{13}$$

where now, if there is no internal resonance,

$$\left.\begin{aligned}
T &= -\frac{B_1+B_2}{2} - \frac{3C_1A_1^2}{8} + \frac{(\varepsilon-K_1\gamma)^2}{16\Omega_1^2} - \frac{\gamma^2}{4\Omega_2^2A_2^2} + \frac{3EB_1A_1^2}{8} + \frac{5C_1D_1A_1^4}{32\Omega_1^2},\\
T_1 &= -\frac{B_1A_1}{2} - \frac{C_1A_1^3}{8} - \frac{(\varepsilon-K_1\gamma)^2A_1}{16\Omega_1^2} + \frac{EB_1A_1^3}{8} + \frac{C_1D_1A_1^5}{32\Omega_1^2},\\
T_2 &= -\frac{B_2A_2}{2} - \frac{\gamma^2}{4\Omega_2^2A_2}, \quad Q_1^2 = \frac{(\varepsilon-K_1\gamma)^2A_1^2}{8\Omega_1^2}, \; Q_2^2 = \frac{\gamma^2}{2\Omega_2^2}, \quad Q_1Q_2 = 0.
\end{aligned}\right\} \tag{14}$$

Generalizing the one-degree-of-freedom solution of the absorber motion, we write the solution of (13) in the form

$$v = cA_1^i A_2^j \exp\Big(\sum_{k,l=0,1,\ldots} c_{kl} A_1^k A_2^l\Big). \tag{15}$$

Insertion into (13) with (14) and comparison of the coefficients yields coupled nonlinear equations for i, j, c_{kl} which have been solved analytically up to exponents $k+l \geq 8$. It is especially

$$i = 1 - \frac{8\Omega_1^2 B_1}{(\epsilon - K_1\gamma)^2}, \qquad j = 1 .$$

In general the normalization constant c has to be determined numerically. Under certain conditions we can confine ourselves on c_{20}, c_{02} and c_{42} as the only non-vanishing coefficients c_{kl}. Then we get

$$c = - \frac{4 c_{02}^{\frac{3-i}{4}} c_{42}^{\frac{1+i}{4}} \sin\left(\frac{1+i}{2}\pi\right)}{\pi L_{\frac{1+i}{2}}\left(-c_{20}\sqrt{\frac{c_{02}}{c_{42}}}, 0\right)}$$

where L_n are the Lommel functions of two variables.

The two-dimensional probability density (15) yields the probability for exceeding given amplitude levels, for instance the probability

$$V_1(\alpha_1) = \int_{\alpha_1}^{\infty}\int_{0}^{\infty} v\,dA_1\,dA_2$$

for A_1 to exceed a level α_1 or the probability

$$V_2(\alpha_2) = \int_{0}^{\infty}\int_{\alpha_2}^{\infty} v\,dA_1\,dA_2$$

for A_2 to exceed a level α_2.

If $A_1^4 \gg c_{02}/c_{42}$ we get

$$V_1(\alpha_1) = - \frac{c}{4c_{42}}(-c_{20})^{\frac{3-i}{2}} \Gamma\left(\frac{i-3}{2}, -c_{20}\alpha_1^2\right)$$

with the incomplete gamma function Γ.

Using asymptotic developments of L_n and Γ for $|c_{20}|\alpha_1^2 \gg 1$, we can find

$$V_1(\alpha_1) = \frac{c_{02}\Gamma\left(\frac{1-i}{2}\right)\sin\left(\frac{1+i}{2}\pi\right)}{\pi c_{42}(-c_{20})^{\frac{1-i}{2}} \alpha_1^{5-i} e^{-c_{20}\alpha_1^2}} \tag{16}$$

Neglecting c_{42} from the very beginning yields $V_2(\alpha_2) = e^{c_{20}\alpha_2^2}$ and for $|c_{20}|\alpha_1^2 \gg 1$

$$V_1(\alpha_1) \approx \frac{1}{\Gamma\left(\frac{1+i}{2}\right)(-c_{20})^{\frac{1-i}{2}} \alpha_1^{1-i} e^{-c_{20}\alpha_1^2}} . \tag{17}$$

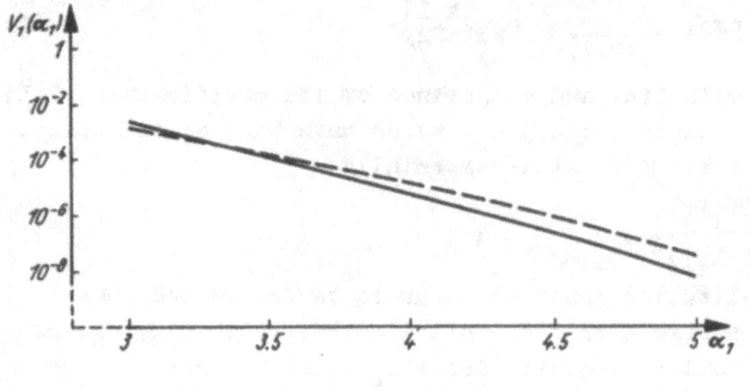

Figure 2

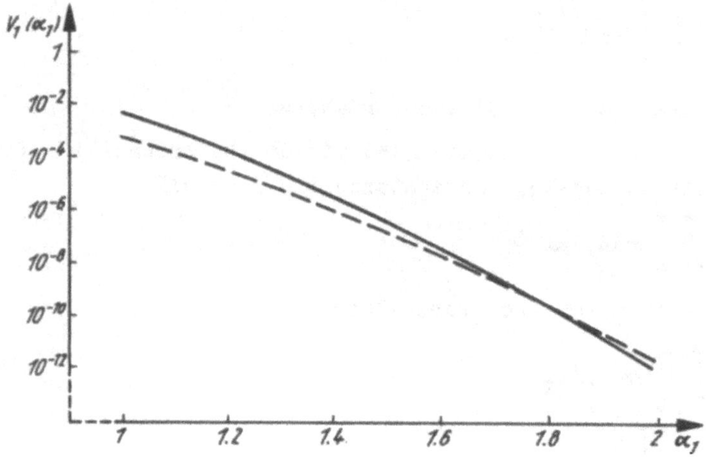

Figure 3

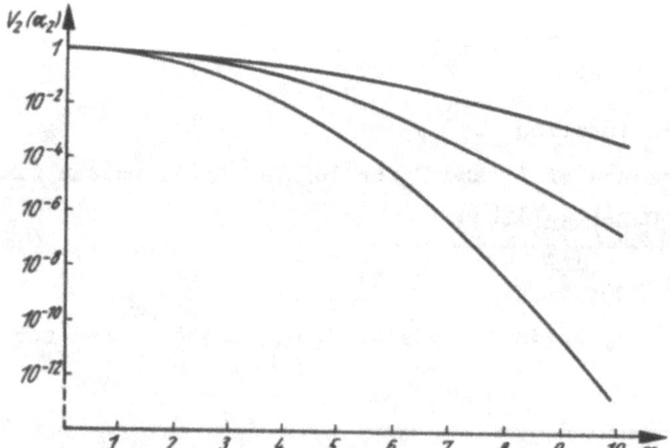

Figure 4

For an example corresponding with experomental examples in R.S. Haxton and A.D.S. Barr [1], the probability (16) resp., by dashed line, (17) is given in Fig. 2 for nonlinear damping $C_1 = 10^{-6}$, in Fig. 3 for $C_1 = 10^{-5}$.

For smaller values α_1 than the ones drawn, the asymptotic formulae (16), (17) do not hold. It shows that V_1 rapidly diminishes with the level value α_1 and with increasing nonlinear damping. It shows further that the neglection of c_{42} yields probabilities of about the same order of magnitude. The probability $V_2(\alpha_2)$ is given in Fig. 4 for different values of linear damping B_2. Because c_{02} is proportional to B_2/γ^2, doubling B_2 corresponds with dividing the forced excitation intensity by $\sqrt{2}$, so that Fig. 4 shows how the probability V_2 diminishes with B_2 respectively augments with γ, depending on the level α_2.

References

1 Haxton, R.S.; Barr, A.D.S.: The autoparametric vibration absorber, J. Engg. Industr., Trans. ASME, Ser.B 94(1972),119-125

2 Ibrahim, R.A.; Roberts, J.W.: Broad band random excitation of a two-degree-of-freedom system with autoparametric coupling, J. Sound Vibrations 44(1976),335-348

3 Ibrahim, R.A.; Roberts, J.W.: Stochastic stability of the stationary response of a system with autoparametric coupling, ZAMM 57(1977), 643-649.

ON THE NORMAL APPROXIMATION IN STOCHASTIC DYNAMICS

K. Sobczyk[1)]

Summary: The paper is devoted to the problems of departure of the responses of linear systems from Gaussian behaviour. After some remarks about various measures of departure from normality, the applicability of asymptotic results (based on the central limit theorem) to real systems is discussed. A quantitative information is given about the effect of system parameters and spectrum of non-Gaussian excitation to the distance from normality. Linear systems which deteriorate normality are characterized and then the departure from normality caused by parametric noises is discussed.

1. Introduction. A significant part of the results in stochastic dynamics has been obtained under the assumption that the processes under consideration are normal (Gaussian). In the majority of cases this assumption is a very convenient one, since it simplifies the analysis. However, in many real situations it turns out to be very difficult to justify correctly the normality of the process in question. Often normality is justified (or verified) only by results. For instance, in the engineering theory of random signals it is usually believed that, if a stationary stochastic process is the input of a linear system, then the response is "approximately normally distributed" as the bandwidth of the system tends to zero. Obviously, such a statement can not be true in general and in the cases when it could be justified one should be able to sketch out a domain within which such a result would in fact be valid. In spite of this we need not only the asymptotic result but also the estimation of the departure of the responses from normality in any practical situation. It is, therefore, important to establish the appropriate criteria of Gaussian assumption in stochastic analysis of dynamical systems and to give a measure of departure of the real processes from Gaussian behaviour. Even when general answers are impossible, it seems to be very useful to analyse the problem of normality and non-normality taking some particular problems of practical importance.

In the literature there exists now a number of papers constituting a valuable contribution to better understanding of the problems mentioned above (cf. [1,2,3,4,5,6]). The purpose of this study is to give further (and more detailed) information concerning the normality of the response of dynamical linear systems.

1) Institute of Fundamental Technological Research, Polish Academy of Sciences, ul. Świętokrzyska 21, 00-049 Warsaw, Poland

2. Measures of departure from normality. The definition and properties of Gaussian stochastic processes imply that measures of departure from normality should be expressed in terms of the joint probability distributions or in terms of moments of order greater than two. In general, departure from normality can be characterized in the following way.

Let $X(t)$ be a non-Gaussian second order stochastic process and $X_G(t)$ a Gaussian process associated with $X(t)$. Let I be any n-dimensional interval in R_n. Let us denote by $\underline{X}$ the n-dimensional random vector $[X(t_1),\ldots,X(t_n)]$, obtained from the process $X(t)$ for an arbitrary set $\{t_1,\ldots,t_n\}$ of the values of parameter t. Let $\underline{X}_G$ be the n-dimensional Gaussian vector associated with $\underline{X}$ (having the same first and second moments). The quantity

$$\Delta = P(\underline{X} \in I) - P(\underline{X}_G \in I) \tag{2.1}$$

where P denotes the probability of event (.), characterizes fully the departure of the process $X_G(t)$ from normality. Unfortunately the quantity (2.1) - including all joint probability distributions - is too complicated to serve as a tool in measuring non-normality of real processes.

Historically, first measures of departure from normality have been introduced in connection with central limit theorems and some problems of mathematical statistics (cf. [5]). Simpler measures are mainly based on one-dimensional distributions. As far as random processes are concerned, the statements obtained by use of one-dimensional distributions provide sufficient information when the departure from normality is significant. Small departure from normality indicated by a one-dimensional distribution does not mean, of course, that the process as a whole is nearly Gaussian. The most popular measures of departure from normality can be defined as follows:

$$\varepsilon_x = |F(x) - \Phi(x)| \quad \text{or} \quad \varepsilon = \sup_x |F(x) - \Phi(x)| \tag{2.2}$$

where $F(x)$ is the distribution function of a non-Gaussian variable and $\Phi(x)$ is the Gaussian distribution with the same two first moments as $F(x)$.

If the stochastic process under consideration has finite moments of order $p \geq 1$, then a convenient measure of departure from normality can be defined as a distance in the metric space L_p, i.e.

$$\varrho_p(F,\Phi) = \left\{ \int_{-\infty}^{+\infty} |F(x) - \Phi(x)|^p \, dx \right\}^{1/p} . \tag{2.3}$$

A systematic uniform approach to measuring of the distance from normality with application to discrete linear filters is presented in paper [2]. The starting point in the analysis given in [2] is the following measure of the departure from normality (in the one-dimensional case)

$$\delta(F, \Phi) = \int_{-\infty}^{+\infty} [F(x) - \Phi(x)]^2 \, dx . \tag{2.4}$$

It can be shown easily that $\delta(F,\Phi) = 0$ iff $F(x)$ is Gaussian. Using this measure Mallows defines the measure of departure for the finite- and infinite-dimensional cases (for a random process as a whole).

A simple (parametric) measure of departure from normality which turned out to be very useful in applied probability and statistics is the asymmetry and excess coefficient, defined respectively as

$$\gamma_1 = \frac{\mu_3}{\sigma^3}, \qquad \gamma_2 = \frac{\mu_4}{\sigma^4} - 3 \tag{2.5}$$

where σ is the standard deviation of the considered process and μ_3 and μ_4 are the third and fourth order central moments, respectively. For the Gaussian distribution holds $\gamma_1 = \gamma_2 = 0$. The values γ_1 and γ_2 different from zero characterize the departure of a one-dimensional distribution from normality.

Concluding these remarks it is worth noting that some extensions of Chebyshev's inequality lead to bounds of measure (2.2). Also, quantity (2.2) can be bounded in terms of (2.4). The following theorem is valid (cf.[2]):

If $\frac{dF}{dx} \le A$, $\frac{d\Phi}{dx} \le B$ for all x, then

$$|F(x) - \Phi(x)| \le \left[\frac{3AB}{A+B} \, \delta(F,\Phi)\right]^{1/3} . \tag{2.6}$$

3. Linear systems subjected to non-Gaussian excitation

3.1. Asymptotic result

Let the input to a linear system be denoted by $X(t)$. We assume that $X(t)$ is a given stationary non-Gaussian process with $t \in T \subset [0,\infty)$. The resulting response is represented as

$$Y(t) = \int_{t_0}^{t} p(t,\tau) \, X(\tau) \, d\tau \tag{3.1}$$

where $p(t,\tau)$ is a unit response function. Integral (3.1) can be approximated by a finite sum of random variables:

$$Y(t) \approx \sum_{i=1}^{n} \zeta_i , \qquad \zeta_i = \int_{(j-1)\Delta}^{j\Delta} p(t,\tau) \, X(\tau) \, d\tau \tag{3.2}$$

where Δ denotes the length of an elementary interval in the division of the interval $[0,t]$ into n parts. If the variables ζ_i were independent, then - by virtue of the central limit theorem - the distribution of $Y(t)$ could be considered approximately as Gaussian as $n \to \infty$. But the variables ζ_i are evidently dependent since they are defined by the values of the same stochastic process.

A number of central limit theorems has recently been proven for dependent random variables. This more general form of the central limit theorem imposes, however, specific conditions on the random variables in question. It is often difficult to see how these conditions are related to each other or to see their physical implications. Let us take, for illustration, the limit theorem proposed by Rosenblat [6]. He showed that, subject to appropriate conditions, weighted integrals of stationary random processes of the form (3.1) become asymptotically normal. The conditions are:

1) strong mixing: $X(t)$ is said to be strongly mixing, if there exists a non-negative, monotone decreasing function $\varphi(t)$, $\varphi(t) \to 0$ for $t \to \infty$, such that, if A is any event dependent only on values of $X(t)$ for $t \geq t_1$ and B any event dependent only on values of $X(t)$ for $t \geq t_2$, $t_2 > t_1$, then
$$|P(AB) - P(A)P(B)| < \varphi(t_2 - t_1) \;;$$
2) the correlation function and fourth order moments of $X(t)$ are absolutely integrable;
3) the spectral density is non-zero at any frequency;
4) the weighting function $P(t) = \int\limits_0^t p^2(t,\tau)\, d\tau$ increases to infinity as $t \to \infty$; the filter must be progressively narrower.

The assumptions stated above are rather involved. Especially the problem of testing the strong mixing condition 1) in real physical situations is not easy. Strong mixing is used to show that integrals over adjacent segments of $X(t)$ become statistically asymptotically independent. Only for Gaussian stationary processes this condition can be replaced by a simpler one, but we are interested in the non-Gaussian case. In spite of this it was shown in [1] that the conditions of Rosenblat's theorem concerning the properties of the filter are not satisfied for the most popular practical systems described by differential equations of first and second orders(for real parameter values). The band of these systems is not infinitely narrow and the response departs from normality significantly. In [1] it was also shown, how the departure from normality is affected by the correlation time of the excitation and the parameters of the systems.

3.2. Departure from normality

Let us consider a linear harmonic oscillator governed by the differential equation

$$\ddot{Y}(t) + 2h\,\dot{Y}(t) + \omega_0^2\, Y(t) = X(t) \tag{3.3}$$

where h and ω_o characterize, respectively, the damping and natural frequency and X(t) is a given non-Gaussian process. Let us consider two cases.

I. $X(t) = Z^2(t)$, where Z(t) is a stationary Gaussian process so that

$$\langle Z(t)\rangle = 0, \qquad K_Z(\tau) = \sigma_Z^2\, e^{-\alpha|\tau|}, \qquad \alpha > 0. \tag{3.4}$$

In this case

$$f_{X(t)}(x) = \begin{cases} \dfrac{1}{\sigma_Z\sqrt{2\pi}}\dfrac{1}{\sqrt{x}}\, e^{-\frac{x}{2\sigma_Z^2}} , & x>0, \\ 0 , & x<0 . \end{cases} \tag{3.5}$$

The coefficients of (2.5) are $\gamma_1^X = 3.9, \quad \gamma_2^X = 15.$

In [1] it was shown that

$$\gamma_1^Y = \frac{2\sqrt{2}C_3}{C_2^{3/2}}, \qquad \gamma_2^Y = \frac{12C_4}{C_2^2} \tag{3.6}$$

where

$$\begin{aligned} C_2 &= \int_0^\infty\!\!\int_0^\infty p(\tau_1)p(\tau_2)\exp\{-2\alpha|\tau_1-\tau_2|\}\, d\tau_1 d\tau_2 , \\ C_3 &= \int_0^\infty\!\!\int_0^\infty\!\!\int_0^\infty p(\tau_1)p(\tau_2)p(\tau_3)\exp\{-\alpha|\tau_1-\tau_2| - \alpha|\tau_3-\tau_1| - \alpha|\tau_3-\tau_2|\}\, d\tau_1 d\tau_2 d\tau_3 , \\ C_4 &= \int_0^\infty\!\!\int_0^\infty\!\!\int_0^\infty\!\!\int_0^\infty p(\tau_1)p(\tau_2)p(\tau_3)p(\tau_4)\exp\{-\alpha|\tau_1-\tau_3| - \alpha|\tau_1-\tau_4| - \alpha|\tau_2-\tau_3| - \alpha|\tau_2-\tau_4|\}\times \\ &\quad \times d\tau_1 d\tau_2 d\tau_3 d\tau_4 . \end{aligned} \tag{3.7}$$

The constant quantities C_2, C_3, C_4 depend on the system parameters h, ω_o and the correlation parameter α. Numerical calculations and graphical results show, for instance, that, when the damping coefficient h is finite, the departure from normality is significant.

II. $X(t) = \sum_i \delta(t-t_i)$, where the t_i are the random times of a Poisson process with intensity $\varkappa(t)$. In this case results

$$Y(t) = \sum_i p(t,t_i) \tag{3.8}$$

and (cf. [1])

$$\gamma_1^Y = \left[\frac{I_3^2(t)}{I_2^3(t)}\right]^{1/2} , \qquad \gamma_2^Y = \frac{I_4(t)}{I_2^2(t)} \tag{3.9}$$

where

$$I_n(t) = \int_{-\infty}^{+\infty} \varkappa(t)\, |p(t,\tau)|^n\, d\tau . \tag{3.10}$$

If $\varkappa(t) = \varkappa$ = const., then

$$\gamma_1^Y = \frac{8}{3}\left(\frac{4h^3\omega_o^2}{(8h^2+\omega_o^2)^2}\right)^{1/2} , \qquad \gamma_2^Y = \frac{6h\omega_o^2}{(12h^2+\omega_o^2)} . \tag{3.11}$$

When $\frac{a}{\varkappa}\to 0$ or $h\to 0$, γ_1^Y and γ_2^Y approach to zero; therefore the response can be considered as Gaussian (with respect to a one-dimensional distribution). In these cases the assumptions of Rosenblat's theorem are satisfied. However, for the values of h and ω_0 taken from experimental examination of a road vehicle suspension result $\gamma_1^Y \approx 1.76\,\varkappa^{-1/2}$, γ_2^Y $8.45\,\varkappa^{-1}$.

The results shortly presented above show that in situations when one deals with real systems the response can differ essentially from Gaussian behaviour. But it should be emphasized that in both cases linear filters 'improve normality' in the sense that $\gamma_1^Y < \gamma_1^X$, $\gamma_2^Y < \gamma_2^X$ for all values of the system parameters.

3.3. Divergence from normality

The results presented above indicate that the response of linear systems to non-Gaussian excitation is asymptotically normal (under appropriate conditions) or - if we consider real situations - that linear systems 'improve normality' in the sense that the departure from normality of the response is less than departure of the input process. It is tempting to ask whether systems do exist which deteriorate normality. To answer this question we must consider the physical mechanism of filtering random signals by linear systems more carefully.

Let us remind first that often in applications (cf. [7]) the following relations are introduced:

$$\gamma_1^Y \approx \sqrt{\frac{\tau_X}{\tau_S}}\,\gamma_1^X , \qquad \gamma_2^Y \approx \frac{\tau_X}{\tau_S}\,\gamma_2^X . \tag{3.12}$$

where τ_S denotes the characteristic time of a system and τ_X is the correlation time of the excitation. These results suggest that, if $\tau_S \gg \tau_X$ then the response is nearly Gaussian. It turns out, however, that relations (3.12) are not valid in general. In some situations the decrease of τ_X/τ_S does not imply that the one-dimensional distribution of Y(t) is close to normal.

Let the excitation process of a linear system have the form

$$X(t) = \sum_i a_i\, F_X(t-t_i) \tag{3.13}$$

where the t_i are the random times of a Poisson process with constant intensity $\varkappa$ and a_i are random variables independent of t. The excess coefficient corresponding to (3.13) is

$$\gamma_2^X = \frac{\langle a\rangle^4}{\langle a^2\rangle^2 \varkappa}\; \frac{\int_{-\infty}^{+\infty} F_X^4(\theta)\,d\theta}{\left[\int_{-\infty}^{+\infty} F_X^2(\theta)\,d\theta\right]^2} . \tag{3.14}$$

Let us introduce the effective length (or effective duration) of an impulse F(t):

$$T_{eff}^{(2)} = \int_{-\infty}^{+\infty} F^2(\theta)\, d\theta \Big/ F^2(\theta_o), \quad T_{eff}^{(4)} = \int_{-\infty}^{+\infty} F^4(\theta)\, d\theta \Big/ F^4(\theta_o) \qquad (3.15)$$

where θ_o is the argument, for which $F(\theta)$ takes its extremal value. It turns out that for a wide class of impulses the effective length (3.15) can be assumed to be equal, i.e. $T_{eff}^{(2)} \approx T_{eff}^{(4)} = T_{eff}$. For example, if $F(t) = \exp(-1/t^2)$, then $T_{eff}^{(2)}/T_{eff}^{(4)} = \sqrt{2}/2$, or, if $F(t) = \sin(bt)/bt$, then $T_{eff}^{(2)}/T_{eff}^{(4)} = 1.5$. The excess coefficient expressed in terms of (3.15) takes the form

$$\gamma_2^X \approx \frac{\langle a\rangle^4}{\langle a^2\rangle^2 \varkappa} \frac{1}{T_{eff}} . \qquad (3.16)$$

A similar expression can be derived for the asymmetry coefficient.

The response process is

$$Y(t) = \sum_i a_i F_Y(t-t_i) , \qquad F_Y(t) = \int_{-\infty}^{+\infty} p(t-\tau)\, F_X(\tau)\, d\tau . \qquad (3.17)$$

The excess coefficient of process (3.17) is given by (3.16) where T_{eff} is the effective length of $F_Y(t)$. Using this fact and (3.16) we obtain

$$\gamma_2^Y \approx (T_{eff}^X / T_{eff}^Y)\, \gamma_2^X . \qquad (3.18)$$

Hence the normalization condition is

$$T_{eff}^Y > T_{eff}^X . \qquad (3.19)$$

Therefore, the normalization takes place if the linear system increases the effective length of the input process. If, however,

$$T_{eff}^Y < T_{eff}^X \qquad (3.20)$$

holds in the system, then divergence from normality takes place.

Let the input impulse $F_X(t)$ have effective length T_{eff}^X and effective frequency band $\delta\omega_X$ and the correlation time τ_F be $\tau_F \approx 1/\delta\omega_X$. If

$$\delta\omega_X\, T_{eff}^X \approx 1 \qquad (3.21)$$

then it is seen that $\tau_X \approx T_{eff}^X$. If, however,

$$\delta\omega_X\, T_{eff}^X \gg 1 \qquad (3.22)$$

then the correlation time of the impulse is much less than T_{eff}^X, and in general it is impossible to determine the order of magnitude of the effective length of an elementary impulse on the basis of the correlation time.

Relation (3.20) can be realized by a system called compatible with the input impulse $F_X(t)$. This is a filter whose impulse response function $p(t)$ is 'specular' with respect to $F_X(t)$, i.e.

$$p(t) = F_X(t_o-t). \qquad (3.23)$$

The correlation function of the excitation process is

$$K_X(\tau) = \varkappa\langle a^2\rangle \int_{-\infty}^{+\infty} F_X(\theta)\, F_X(\theta+\tau)\, d\theta = \varkappa\langle a^2\rangle\, K_{FX}(\tau) \,. \tag{3.24}$$

The response of a compatible filter takes the form

$$\begin{aligned} Y(t) &= \int_{-\infty}^{+\infty} p(t-\theta)\, X(\theta)\, d\theta = \sum_i a_i \int_{-\infty}^{+\infty} F_X(\theta-t_i)\, F_X(t_o-t+\theta)\, d\theta \\ &= \sum_i a_i \int_{-\infty}^{+\infty} F_X(\alpha)\, F_X(\alpha-t-t_o-t_i)\, d\alpha = \sum_i a_i\, K_X(t-t_o-t_i) \end{aligned} \tag{3.25}$$

where $F_X(t)$ is an even function. Elementary response impulses (3.17) are therefore determined by the integral correlation of the input impulses, i.e.

$$F_Y(t) = K_X(t-t_o). \tag{3.26}$$

This implies that the response impulse (its main part) $F_Y(t)$ is determined on the correlation interval τ_{FX}, which for the impulse belonging to the class defined by (3.22) is much less than the effective length of the impulse $F_X(t)$. This means that the compatible filter (3.23) realizes relation (3.20) and, at the same time, performs a divergence from normality.

As an illustration let us take the excitation process (3.13) where the elementary impulse has the following form (cf. [8]) used in communication theory:

$$F_X(t) = \begin{cases} \cos(\omega_o t + \frac{\alpha}{2} t^2 \,, & |t| \le \frac{T_o}{2} \\ 0 \,, & |t| > \frac{T_o}{2} \end{cases} \,. \tag{3.27}$$

So $F_X(t)$ is an impulse with rectangular envelope of length T_o. The linear filter compatible with $F_X(t)$ is characterized by

$$p(t) = \begin{cases} A\cos\left[\omega_o(t_o-t) + \frac{\alpha}{2}(t_o-t)^2\right], & |t_o-t| \le \frac{T_p}{2} \\ 0 \,, & |t_o-t| > \frac{T_p}{2} \end{cases} \tag{3.28}$$

where T_p is the length of the envelope of p(t), and $t_o \ge \frac{T_p}{2}$. It is seen that if $T_o = T_p$, the filter (3,28) satisfies (3.23).

<u>4. Parametric noises in linear systems; departure from normality</u>. Let us consider a linear system governed by the equation

$$\ddot{Y}(t) + 2h(t)\dot{Y}(t) + \Omega^2(t)Y(t) = X(t) \tag{4.1}$$

where $h(t) = h_o+h_1(t)$, $\Omega^2(t) = \Omega_o^2+\omega_1(t)$. Let $h_1(t)$, $\omega_1(t)$, $X(t)$ be stationary stochastic processes with zero mean and with small correlation times (small compared to $1/h$, $1/\Omega_o$). The processes $h_1(t)$, $\omega_1(t)$, $X(t)$

are assumed to be statistically independent. Under these assumptions we can introduce the effective correlations

$$K_X^{eff}(\tau) = D_X\,\delta(\tau)\,, \qquad D_X = 2\int_{-\infty}^{+\infty} K_X(\tau)\,d\tau \tag{4.2}$$

and analogously for $K_{h_1}(\tau)$ and $K_{\omega_1}(\tau)$.

Equation (4.1) can be represented in the form of the system

$$\dot{Y}(t) = Z(t)\,, \qquad \dot{Z}(t) = -2h(t)Z - \Omega^2(t)Y - \Omega_o^2 X(t)\,. \tag{4.3}$$

In general the equations for the moments $\langle Y^n Z^m\rangle$ are not closed. But in the diffusion Markovian approximation (small correlation times and (4.2)) we obtain closed equations for $\langle Y^n Z^m\rangle$, especially for $\langle Y^n(t)\rangle$. For stationary moments $\langle Y^n\rangle = \lim_{t\to\infty}\langle Y^n(t)\rangle$ one obtains (when $h_1(t) \equiv 0$)

$$\langle Y^2\rangle = \frac{Pq^2}{1-\mu_{\omega_1}}\,, \qquad \langle Y^4\rangle = \frac{3Pq^2(1+\frac{3}{4q})\langle Y^2\rangle}{1+\frac{3}{4q^2}-\frac{3}{2}\mu_{\omega_1}(1+\frac{3}{2q^2})} \tag{4.4}$$

where

$$P = h_o D_X\,, \qquad q = \frac{2h_o}{\Omega_o}\,, \qquad \mu_{\omega_1} = \frac{D_{\omega_1}}{ah_o\Omega_o^2}\,. \tag{4.5}$$

If $q \ll 1$, then

$$\gamma_2^Y = \frac{\langle Y^4\rangle}{\langle Y^2\rangle^2} - 3 = \frac{3\mu_{\omega_1}}{2-3\mu_{\omega_1}} \tag{4.6}$$

and $\mu_{\omega_1} < \frac{2}{3}$ to assure the stability of a system.

When $\mu_{\omega_1} \to 0$, then $\gamma_2^Y \to 0$. Therefore, as we could expect, in the absence of parametric noises the distribution is Gaussian. However, when the effective intensity of the noise increases, the excess coefficient rapidly increases. This indicates that the normal approximation of the response of a linear system excited by parametric noises can be acceptable only for very small values of the effective noise intensity. In spite of this it turns out that the difference between $\langle Y^{2n}\rangle$ and $\langle Y^{4n}\rangle = (2n-1)!!\langle Y^2\rangle^n$ (for Gaussian processes) increases with n.

References

1 Kotulski, Z.; Sobczyk, K.: Linear systems and normality, J. Statist. Phys. 24(1981),2,359-373

2 Mallows, C.L.: Linear processes are nearly Gaussian, J. Appl. Probab. 4(1967),313-329

3 Papoulis, A.: High density shot noise and Gaussianity, J. Appl. Probab. 8(1971),118-127

4 Papoulis, A.: Narrow-band systems and Gaussianity, IEEE Trans Inform. Theory IT-18(1972),1,20-23

5 Petrov, V.V.: Sums of Independent Random Variables, Nauka, Moscow 1972

6 Rosenblat, M.: Some comments on narrow band pass filters, Quart. Appl. Math. 18(1961),387-393

7 Rytov, S.M.: Introduction to Statistical Radio Physics (in Russian), Nauka, Moscow 1966

8 Tikhonov, V.I.: Statistical Radio Technics (in Russian), Sovet. Radio, Moscow 1966

9 Volkov, V.M.: On estimation of departure of probability distribution of noise from normal distribution (in Russian), Radiotekhn. i Elektronika 15(1970),12.

APPROXIMATE ANALYSIS OF RANDOM VIBRATION PROBLEMS THROUGH STOCHASTIC AVERAGING

P-T. D. Spanos(*)

SUMMARY

Results obtained by applying the method of stochastic averaging to problems of random vibrations are discussed. This method is applicable to problems of dynamic responses of lightly damped oscillators to broad-band random excitations. Solutions pertaining to either linear or nonlinear vibrations are presented. These solutions are useful in determining oscillator response statistics and first-passage probabilities.

I. INTRODUCTION

Over a period of recent years the method of stochastic averaging has proved a quite useful tool for deriving approximate solutions of problems of lightly damped vibrations caused by broad-band random excitations. This method has been introduced by Stratonovich [1] based on a combination of physical and mathematical arguments, and in connection with nonlinear self-excited oscillations in the presence of random noise. Subsequently several efforts aiming, directly or indirectly, at establishing its validity on a mathematically rigorous foundation have been reported [2,3]. Some researchers view the method of stochastic averaging as an extension to the field of random differential equations of the well-known Bogoliubov-Mitropolski technique of deterministic vibrations [4]. The main feature of the method is that it leads to a Markovian approximation of appropriate envelopes of the oscillator response; the most common envelope is the total energy, kinetic and potential, of the oscillator. Thus, utilizing this approximation, the determination of the exact solution of the probability distribution of the <u>two-dimensional</u> Markov vector with components the displacement and the velocity of the oscillator can be circumvented. Specifically, the solution of the quite more tractable mathematically problem of evolution in time of the <u>one-dimensional</u> vector of the response envelope can be determined. This solution is quite useful in itself. Furthermore, it eventually leads to a reliable approximate solution for the probability distribution of the two-dimensional vector. An approach to interpreting the results obtained by using the method of stochastic averaging can be based on energy balance considerations [5-7]. Specifically, the method can be based on the principle that the rate of change with respect to time of the total energy of the oscillator is equal to the expected power input due to the random excitation, minus the power dissipation due to the damping mechanism. The power input and dissipation can be approximately calculated by relying on two points characterizing the problem under consideration. First, it is assumed that the damping of the oscillator is light, and the supremum of the power spectrum of the excitation is scaled accordingly. Therefore, the energy of oscillator varies slowly with respect to time and can be treated as a constant over an appropriate period of oscillation. Consequently, several oscillatory terms can be approximated by their temporal averages over a period of oscillation;

*Engineering Mechanics, WRW 201-B, University of Texas, Austin, TX., 78712, U.S.A.

this procedure is identical to that of the deterministic technique of slowly varying parameters [8]. Second, since the damping of the oscillator is small and the power spectrum of the random excitation is broad, the relaxation time of the oscillator is much greater than the correlation time of the response. Thus, it can be argued that the value of the random excitation which corresponds to a certain time, is statistically independent from the value of the energy of the oscillator which corresponds to a slightly earlier time. Detailed discussions of the aforementioned approximations and of other similar to these, can be found in references [1,5,6,9,10] and [7,11] in context with stationary and nonstationary random excitations, respectively.

In this article it is intended to present an overview of some results which can be obtained by using the method of stochastic averaging. In general, oscillators exposed to stochastic excitation possessing evolutionary power spectra [12], and exhibiting nonlinear behavior due to damping and stiffness terms, will be considered. Furthermore, special attention will be given to less general classes of problems which lend themselves more readily to solution by the method of stochastic averaging.

It is noted that, due to space limitations, the references which are used are selected based, primarily, on their immediate relevance to the stochastic averaging approach which has been followed by the author over a period of years. However, it is emphasized that other researchers have been quite active in this scientific topic.

II. MARKOVIAN MODELING

Consider the equation of motion of a randomly excited nonlinear oscillator

$$\ddot{x} + 2\zeta\omega_o\,[1 + \varepsilon g(x,\dot{x})]\dot{x} + \omega_o^2\,h(x) = w(t) \quad . \tag{1}$$

For $g(x,\dot{x}) = 0$, and $h(x) = 1$ equation (1) reduces to a randomly excited linear oscillator; in this case the symbols ω_o and ζ denote the natural frequency and the ratio of critical damping of the oscillator. The parameter ε is used to quantify the strength of the nonlinear damping term. It is assumed that $w(t)$ is a broad-band random process which possesses an evolutionary power spectrum $S(\omega,t)$. The evolutionary power spectrum can be expressed in the form

$$S(\omega,t) = |A(t,\omega)|^2\bar{S}(\omega) \quad , \tag{2}$$

where $A(t,\omega)$ and $\bar{S}(\omega)$ are appropriate functions, and the symbol $|\;|$ denotes the modulus of a complex function. Attention is restricted to problems for which

$$\zeta << 1 \tag{3}$$

and

$$\sup_{\text{over } t,\omega} \{S(\omega,t)\} = O(\zeta) \quad \text{as} \quad \zeta \to 0 \quad . \tag{4}$$

The total energy of the oscillator is equal to the sum of the potential and the kinetic energy. That is,

$$E(t) = \frac{\dot{x}^2}{2} + \omega_o^2\int_o^x h(x)dx \equiv \frac{\dot{x}^2}{2} + \Pi(x) \quad . \tag{5}$$

Furthermore, for the special case in which $h(x) = 1$ the envelope $a(t)$ and the phase

$\phi(t)$ of the response oscillator can be introduced by using the equations

$$x(t) = a(t)\cos(\omega_o t + \phi) \tag{6}$$

and

$$\dot{x}(t) = -a(t)\omega_o \sin(\omega_o t + \phi) \quad . \tag{7}$$

Clearly, in this case $E(t)$ can be determined from $a(t)$ by using the equation

$$E(t) = \frac{1}{2}\omega_o^2 a^2(t) \quad . \tag{8}$$

Based on arguments discussed in references [1,5-7,9-11] in connection with the technique of stochastic averaging, partial differential equations governing the evolution of Markovian approximations of $E(t)$, $a(t)$, and $\phi(t)$ can be derived.

Consider the case of a modulated white noise with a power spectrum provided by the equation

$$S(\omega,t) = \bar{S} \cdot f^2(t) \quad ; \tag{9}$$

the symbol $f(t)$ denotes a slowly-varying deterministic function, and $\bar{S}$ is a scaling constant. Then, it can be proved that the probability density function $q \equiv q(E,t|E_1,t_1)$, $t > t_1$, of the energy $E(t)$ satisfies the following forward Kolmogorov equation

$$\frac{\partial q}{\partial t} = \frac{\partial}{\partial E}\{[2\zeta\omega_o A(E) - \pi\bar{S}f^2(t)]q\} + \pi\bar{S}f^2(t)\frac{\partial^2}{\partial E}[B(E)q] \quad . \tag{10}$$

The functions $A(E)$ and $B(E)$ are defined by the equations

$$A(E) = \frac{1}{\sqrt{2}\,\Gamma(E)}\int_\Delta \{[1 + \varepsilon g(x,\sqrt{2}\sqrt{E-\Pi})]\}\sqrt{2(E-\Pi)}\,dx\,, \tag{11}$$

$$B(E) = \frac{1}{\Gamma(E)}\int_\Delta \sqrt{E-\Pi}\,dx \quad , \tag{12}$$

where

$$\Gamma(E) = \frac{1}{2}\int_\Delta \frac{dx}{\sqrt{E-\Pi}} \quad . \tag{13}$$

In equations (11) through (13), the symbol Δ denotes an integration interval of values of x for which

$$\omega_o^2 \int_o^x h(x)dx \le E \quad . \tag{14}$$

Consider, as well, the case for which $h(x) = 1$ and the oscillator is excited by a random process possessing a power spectrum given by equation (2). Then, it can be proved that the transition probability density functions $\bar{q} \equiv \bar{q}(a,t|a_1,t_1)$ and $\overset{*}{q} \equiv \overset{*}{q}(a,\phi,t|a_1,\phi_1,t_1)$ of the Markovian approximations of $a(t)$ and $[a(t),\phi(t)]$, respectively, satisfy appropriate forward Kolmogorov equations. Specifically, it can be shown that

$$\frac{\partial\bar{q}}{\partial t} = \frac{\partial}{\partial a}\{[\zeta\omega_o a - \frac{\pi S(\omega_o,t)}{2\omega_o^2 a}]\bar{q}\} + \varepsilon\omega_o\frac{\partial}{\partial a}(\bar{\zeta}a\bar{q}) + \frac{\pi S(\omega_o,t)}{2\omega_o^2}\frac{\partial^2\bar{q}}{\partial a^2} \tag{15}$$

and

$$\frac{\partial\overset{*}{q}}{\partial t} = \frac{\partial}{\partial a}\{[\zeta\omega_o a - \frac{\pi S(\omega_o,t)}{2\omega_o^2 a}]\overset{*}{q}\} + \varepsilon\omega_o\frac{\partial}{\partial a}(\bar{\zeta}a\overset{*}{q}) + \frac{\pi S(\omega_o,t)}{2\omega_o^2}[\frac{\partial^2\overset{*}{q}}{\partial a^2} + \frac{1}{a^2}\frac{\partial^2\overset{*}{q}}{\partial\phi^2}] \quad . \tag{16}$$

The symbol $\bar{\zeta}$ appearing in equation (15) and (16) can be viewed as an equivalent viscous damping coefficient; it is defined by the equation

$$\bar{\zeta}(a) = \frac{1}{\omega_o \pi} \int_o^{2\pi} g(a\cos\psi, -a\omega_o \sin\psi)\sin^2\psi d\psi \quad . \tag{17}$$

For the case of a linear oscillator, $\varepsilon = 0$, equations (15) and (16) reduce to the following form

$$\frac{\partial \bar{q}}{\partial t} = \frac{\partial}{\partial a}[(\zeta\omega_o a - \frac{\pi S(\omega_o,t)}{2\omega_o^2 a})\bar{q}] + \frac{\pi S(\omega_o,t)}{2\omega_o^2}\frac{\partial^2 \bar{q}}{\partial a^2} \tag{18}$$

and

$$\frac{\partial \overset{*}{q}}{\partial t} = \frac{\partial}{\partial a}[(\zeta\omega_o a - \frac{\pi S(\omega_o,t)}{2\omega_o^2 a})\overset{*}{q}] + \frac{\pi S(\omega_o,t)}{2\omega_o^2}[\frac{\partial^2 \overset{*}{q}}{\partial a^2} + \frac{1}{a^2}\frac{\partial^2 \overset{*}{q}}{\partial \phi^2}] \quad . \tag{19}$$

Appended to equations (10), (15) and (18), (16) and (19) are the initial conditions

$$\underset{t \to t_1}{q \to \delta(E-E_1)}; \quad \underset{t \to t_1}{\bar{q} \to \delta(a-a_1)}; \quad \underset{t \to t_1}{\overset{*}{q} \to \delta(a-a_1)\delta(\phi-\phi_1)} \quad . \tag{20}$$

In this equation the symbol δ denotes a delta function. The preceding forward Kolmogorov equations can be solved to determine q, $\bar{q}$ and $\overset{*}{q}$. Alternatively, the corresponding backward Kolmogorov equations can be employed for this purpose.

III SOLUTIONS FOR LINEAR OSCILLATORS

3.1 Transition-Marginal-Joint Probability Densities

It can be shown that equation (18) is satisfied by the expression

$$\bar{q}(a,t|a_1,t_1) = \frac{a}{c(1-\rho^2)}\exp[-\frac{a^2c+a_1^2\rho^2}{2cc_1(1-\rho^2)}]I_o[\frac{\rho a a_1}{\sqrt{cc_1}\,(1-\rho^2)}] \quad . \tag{21}$$

In equation (21) the symbol I_o denotes the Bessel function of zero order; the following symbols are used as well

$$\rho^2 = \frac{\int_o^{t_1} \exp(2\zeta\omega_o\tau)S(\omega_o,\tau)d\tau}{\int_o^{t} \exp(2\zeta\omega_o\tau)S(\omega_o,\tau)d\tau} \quad , \tag{22}$$

$$c = \int_o^{t} \exp(2\zeta\omega_o\tau)S(\omega_o,\tau)d\tau \quad , \tag{23}$$

and

$$c_1 = \int_o^{t_1} \exp(2\zeta\omega_o\tau)S(\omega_o,\tau)d\tau \ . \tag{24}$$

It can be proved that $\rho^2 \leq 1$ and that ρ^2 is in fact a measure of the correlation of the process $a^2(t)$. Equation (21) can be rewritten in the form

$$\bar{q}(a,t|a_1,t_1) = \sum_{n=0}^{\infty} A_n(\frac{a^2}{2c})L_n(\frac{a_1^2}{2c_1})\rho^{2n} \quad . \tag{25}$$

In equation (25) the symbol L_n denotes the Laguerre polynomial of order n defined by the equation

$$L_n(z) = \frac{1}{n!}e^z\frac{d^n}{dz^n}(e^{-z}z^n) \quad ; \tag{26}$$

the function $A_n(\frac{a^2}{2c})$ is defined by the equation

$$A_n(\frac{a^2}{2c}) = \frac{a}{c}\exp(-\frac{a^2}{2c})L_n(\frac{a^2}{2c}) \quad . \tag{27}$$

It is pointed out that the solution provided by either equation (21) or equation (25) can be derived by using the backward Kolmogorov equation which corresponds to equation (18) [13].

The transition probability density $\bar{q}$ can be used to determine joint and marginal probability densities. For example, it can be shown that for any linear oscillator with $a_1 = a_o$ at $t_1 = 0$, the marginal probability density is given by the equation

$$p(a,t) = \frac{a}{c}\exp[-\frac{a^2+a_o^2e^{-2\zeta\omega_o t}}{2c}]I_o(\frac{aa_oe^{-2\zeta\omega_o t}}{c}) \quad . \tag{28}$$

This equation yields the solutions presented in references [1,14,15] as special cases.

Expressions analogous to the preceding which have been derived by using equation (18) can be obtained, as well, by relying on equation (19). Specifically, it can be proved that the transition probability density function for the amplitude, and the phase $\Phi \equiv \omega_o t + \phi$, is

$$\overset{*}{q}(a,\Phi,t|a_1,\Phi_1,t_1) = \frac{a}{\pi d(t,t_1)}\exp\{-\frac{[a\cos\Phi-a_1\cos\Phi_1e^{-\zeta\omega_o(t-t_1)}]^2+[a\sin\Phi-a_1\sin\Phi_1e^{-\zeta\omega_o(t-t_1)}]^2}{d(t,t_1)}\}, \tag{29}$$

where

$$d(t,t_1) = \frac{2\pi}{\omega_o^2}\exp(-2\zeta\omega_o t)\int_{t_1}^{t}\exp(2\zeta\omega_o\tau)S(\omega_o,\tau)d\tau \quad . \tag{30}$$

Furthermore, using algebraic transformations, it can be shown that the transition probabity density function of $x(t)$ and the scaled velocity $y = \frac{\dot{x}(t)}{\omega_o}$ is given by the following equation

$$\hat{q}(x,y,t|x_1,y_1,t_1) = \frac{1}{\pi d(t,t_1)}\exp\{\frac{[x-x_1e^{-2\zeta\omega_o(t-t_1)}]^2+[y-y_1e^{-2\zeta\omega_o(t,t_1)}]^2}{d(t,t_1)}\} \quad . \tag{31}$$

Clearly, equations (29) and (31) can be utilized to determine joint and marginal probability densities for (a,Φ) and (x,y).

3.2 First-Passage Problems

The availability of the transition probability density function of the response envelope $a(t)$, equation (21), along with the fact that $x(t)$ is a pseudo-harmonic process with slowly varying amplitude and phase can be the basis of developing a reliable approximate solution to the so-called first-passage problem. The fundamental question of this problem pertains to the survival probability of the oscillator which is exposed to the random excitation. That is, the probability that a certain measure of the response will not reach a specified barrier during a specified time interval.

In terms of phase-plane terminology [16], single-passage type-B, double-passage type-D, and envelope-passage type-E, barriers can be considered. For example, in connection with the type-D barrier, an analysis of the associated first-passage problem can be conducted by arguing that the ith peak of $|x(t)|$ falls within the time interval $[(i-1)\hat{T}, i\hat{T}]$, where

$$\hat{T} = \frac{\pi}{\omega_o} \quad . \tag{32}$$

Furthermore, for this extreme points process $\mu(i)$ [17,18] it can be approximately assumed that the barrier is reached only at discrete times which are integer multiples of $\hat{T}$. Then, the survival probability of the oscillator at time $t = i\hat{T}$ can be expressed in the form

$$P(t) = [1-u(1)]\,[1-u(2)] \,....\, [1-u(i)] \quad ; \tag{33}$$

the symbol $u(i)$ denotes the probability that $|x(t)|$ will cross the barrier b in the time interval $[(i-1)\hat{T}, i\hat{T}]$, given that no barrier crossings have occurred prior to time $(i-1)\hat{T}$. Relying on the Markovian modeling of $a(t)$ and the preceding approximations it can be shown that

$$u(i) = \frac{\text{Prob}\ \{[\mu(i)\geq b)]\cap[\mu(i-1)<b]\}}{\text{Prob}\ \ [\mu(i-1)<b]} \equiv \frac{U_{i-1,i}}{U_{i-1}} \quad , \tag{34}$$

where the symbol $\cap$ is associated with probability of a joint event. The probabilities which appear in equation (34) and are denoted by the symbol $U_{i-1,i}$, and U_{i-1} can be calculated by relying on equation (21) for the determination of joint and marignal probabilities for the envelope $a(t)$. For example, for an initially quiescent oscillator, zero-start first-passage problem [16], it can be readily proved that

$$U_{i-1} = 1 - \exp(-\frac{b^2}{c_{i-1}}) \ ; \ c_{i-1} \equiv c[(i-1)\hat{T}] \tag{35}$$

and

$$U_{i-1,i} = \int_0^\infty d\mu(i) \int_0^b p[\mu(i-1),\mu(i)]d\mu(i-1) \quad , \tag{36}$$

where the joint probability $p(a,a_1)$ is given by the expression

$$p(a,a_1) = \frac{aa_1}{cc_1(1-\rho^2)} \exp[-\frac{a^2c+a_1^2c_1}{2cc_1(1-\rho^2)}] I_o(\frac{aa_1\rho}{\sqrt{cc_1}\ (1-\rho^2)}) \quad . \tag{37}$$

Furthermore, it can be shown by resorting to integration by parts and employing Leibnitz's rule that the calculation of the double integral which appears in equation (36) can, in fact, be accomplished by computing numerically a definite integral of a single variable [19]. It is noted that the preceding method yields step-wise solutions for the survival probability over time intervalsof width equal to $\hat{T}$. Also it is noted that for the initially quiescent oscillators the first maximum of $|x(t)|$ should be expected to occur after a time which is equal to $\pi/2\omega_o$. This fact should be taken into consideration when the preceding procedure is implemented.

Clearly, the solution method which has been outlined with regard to the type-D barrier first-passage problem, is applicable to the type-B barrier first-passage problem. However, for the latter case it should be assumed that the response reaches the barrier at discrete times which are spaced $2\pi/\omega_o$ seconds apart.

As far as the type-E barrier first-passage problem is concerned, a solution can be determined in the form of an infinite series for the case of a broad-band stationary excitation; the solution is exact for the case of white noise excitation. This solution can be determined either by solving the backward Kolmogorov equation which is associated with equation (18), classical approach, or by utilizing equation (18) and assigning appropriate boundary conditions [20]. Specifically, the survival probability at time t can be expressed in the form

$$P(a_o,t,b) = \sum_{i-1}^{\infty} T_i e^{-2\zeta\omega_o\lambda t_i}\Phi_i(a_o) \quad , \tag{38}$$

in which a_o is the value of a(t) at t = 0, and T_i are constants. The symbols λ_i and Φ_i denote the eigenvalues and the eigenfunctions, respectively, of the following Sturm-Liouville problem

$$\frac{d\Phi}{dE_o}\left(E_o e^{-E_o}\frac{d\phi}{dE_o}\right) + \lambda e^{-E_o}\Phi = 0; \; E_o = \frac{1}{2}a_o^2 \tag{39}$$

and

$$\Phi\Big|_{a_o=0} = \text{finite, and } \Phi\Big|_{a_o=\bar{b}} = 0 \quad . \tag{40}$$

In equations (39) and (40) the dimensionless amplitude

$$\bar{a}_o = \frac{2\zeta\omega_o^3}{\pi S(\omega_o)} a_o \tag{41}$$

and the dimensionless barrier

$$\bar{b} = \frac{2\zeta\omega_o^3}{\pi S(\omega_o)} \bar{b} \tag{42}$$

are used. Numerical aspects of T_i, λ_i and Φ_i have been discussed in detail in reference [21].

For the case of E-type first-passage problem of oscillators subjected to excitations with evolutionary power spectra a solution method based on numerical integration has been presented in reference [19]; another approach based on a Galerkin technique with appropriate basis functions has been developed in reference [20].

IV SOLUTIONS FOR NONLINEAR OSCILLATORS

The problem of determining exact solutions for equations (10), (15), and (16) is extremely difficult. Alternatively, methods combining analytical and numerical approaches can be used to derive reliable approximate solutions. One approach of this nature can be based on the Galerkin technique for which basis function can be selected by using the corresponding solution for the linear oscillator. For example, for the case of an initially quiescent oscillator with h(x) = 1, the marginal probability density function p(a,t) can be determined by using equation (15) and introducing the expansion [22]

$$p(\bar{a},t) = \sum_{i=1}^{N} F_i(t) A_i\left(\frac{\bar{a}^2}{2}\right) \quad , \tag{43}$$

where

$$\bar{a} = \frac{a}{\overset{*}{S}} \quad , \quad \overset{*}{S} = \text{scaling constant}, \tag{44}$$

N is an appropriate integer, and $F_i(t)$ are functions of time to be determined. Subsituting equation (43) into equation (15), and taking into consideration the orthogonality property

$$\int_0^\infty \frac{A_i A_j}{A_o} d\bar{a} = \delta_{ij} \ ; \ \delta_{ij} = \text{Kronecker delta} \quad , \tag{45}$$

ordinary differential equations can be determined for the functions $F_i(t)$. These equations can be solved numerically. Alternatively, for small values of ε they can be solved by relying on the solution for the linear oscillator, and utilizing a perturbation technique. It is noted that these functions have several interesting properties which facilitate the computation of the coefficients appearing in the ordinary differential equations governing the functions $F_i(t)$. These properties have been examined in detail in reference [23]. Two notable ones are

$$(i+1)A_{i+1} = (2i+1 - \frac{\bar{a}^2}{2})A_i - iA_{i-1} \tag{46}$$

and

$$I_{s,n} = 0 \quad \text{for} \quad n \geq \frac{s}{2} \quad , \tag{47}$$

where n is an integer, s is an even integer, and the integral $I_{s,n}$ is defined by the equation

$$I_{s,n} = \int_0^\infty \bar{a}^s A_n d\bar{a} \quad . \tag{48}$$

Equation (46) can be used as the basis of a recursive computation of A_i. Equation (47) facilitates the determination of the statistical moments of $\bar{a}$. For example, relying on this equation it can be readily proved that, independently of the value which is selected for the number $N \geq 1$ of the basis functions,

$$< \bar{a}^2 > = 2\,[F_o(t) - F_2(t)] \quad ; \tag{49}$$

the symbol < > denotes the operator of mathematical expectation. Clearly, equation (49) can be used as well for the determination of the evolution of the mean energy of the oscillator. It is noted that the scaling of the amplitude which is shown in equation (43) involves a time invariant quantity. The corresponding functions A_i are, in fact, the eigenfunctions of the boundary value problem which is formulated by following a separation-of-variables procedure in solving equation (18) for the special case of a stationary excitation. Interestingly, it has been proved recently, after some tedious algebraic and differential manipulations, that equation (18) is individually satisfied by the functions

$$\overset{*}{A}_i = \frac{a}{c(t)} \; A_i \left(\frac{a^2}{2c(t)}\right)\rho^{2i} \quad . \tag{50}$$

Conceivably, using these functions as a basis for the already discussed Galerkin approach for the case of an excitation with an evolutionary power spectrum, could lead to a simpler set of equations for $F_i(t)$. Furthermore, it could result in a smaller value of N which is required for ensuring an acceptable accuracy.

It is noted that the Galerkin approach has been outlined in connection with the solution of equation (15). A similar approach can be used for solving equation (10) upon selecting an appropriate set of basis functions. Due to the particular mathematical form of the functions A(E), B(E), and Γ(E) resorting to numerical integration will be necessary in many steps of the implementation of this approach. However, for the special case of oscillators with $g(x,\dot{x}) = 0$ and $h(x) = |x|^{\nu}\,\mathrm{sgn}(x)$ the exact solution of equation (10) can be determined analytically. For this class of nonlinear oscillators it can be readily shown that

$$A(E) = B(E) = \gamma E \quad , \tag{51}$$

where γ is a proportionality constant depending on the power ν. Thus, equation (10) becomes

$$\frac{\partial q}{\partial t} = \frac{\partial}{\partial E}\{[2\zeta\omega_o\gamma E - \pi\bar{S}f^2(t)]q\} + \pi\gamma\bar{S}f^2(t)\frac{\partial^2}{\partial E}[Eq] \quad . \tag{52}$$

The solution of this equation, subject to the initial condition specified by equation (20) is given by the following expression

$$q(E,t|E_1,t_1) = \frac{1}{c^*(t,t_1)}\left[\frac{E}{E_1e^{-2\zeta\omega_o\gamma\tau}}\right]^{\xi/2}\exp\left[-\frac{E+E_1e^{-2\zeta\omega_o\gamma\tau}}{\omega_o^2c^*(t,t_1)}\right]I_o\left(\frac{2\sqrt{EE_1}e^{-\zeta\omega_o\gamma\tau}}{\omega_o^2c^*(t,t_1)}\right) \quad , \tag{53}$$

where

$$\xi = \frac{1-\gamma}{\gamma} \quad , \quad \tau = t_1 - t_2 \tag{54}$$

and

$$c^*(t_1,t_2) = \bar{S}\int_{t_1}^{t_2} \exp(2\zeta\omega_o\gamma\tau)f^2(\tau)d\tau \quad . \tag{55}$$

Based on equation (53), joint and marginal probabilities can be found for the energy. For example, for an initially quiescent oscillator the marginal probability function is provided by the equation

$$p(E,t) = \frac{D}{c^*(t,o)^{1/\xi}}E^{\xi}\exp\left[-\frac{E}{c^*(t,o)}\right] \quad , \tag{56}$$

where D is a scaling constant. The solutions given by equations (53) and (56) cover as special cases the equivalent of equations (21) and (28), and the results of reference [24].

V. RELATIONSHIP BETWEEN ENVELOPE-DISPLACEMENT-VELOCITY STATISTICS

For the case of an initially quiescent oscillator, equation (16) can lead to a simple solution for the joint probability density $p(a,\phi,t)$, assuming that $p(a,t)$ can be determined by using equation (15). Specifically, it is found that in this case [24,25]

$$p(a,\phi,t) = \frac{1}{2\pi}p(a,t) \quad . \tag{57}$$

Based on equation (57) it can be readily proved that the joint probability density of the displacement and the velocity is

$$p(x,\dot{x},t) = \frac{\overset{*}{D}}{a} p\,(a,t) \tag{58}$$

where D^* is a scaling constant. Furthermore, based on arguments presented in references [1,24], equation (58) can be modified for the class of nonlinear oscillators with $h(x) \neq 0$. Specifically in this case the joint probability density $p(x,\dot{x},t)$ can be approximated by the expression

$$p(x,\dot{x},t) = \bar{D}\,\frac{p(E,t)}{\Gamma(E)}\,, \tag{59}$$

where $\bar{D}$ is a scaling constant. Thus, upon determining $p(E,t)$ by using equation (10), $p(x,\dot{x},t)$ can be approximated by equation (59).

ACKNOWLEDGEMENTS

The author would like to express his gratitude to the National Science Foundation and to the NASA Lewis Research Center for sustained and recent, respectively, support of his research program.

REFERENCES

1. Stratonovich, R. L., Topics in the Theory of Random Noise, Vols. I, II, Gordon and Breach, New York, 1963.
2. Khasminskii, R. Z., "A Limit Theorem for the Solutions of Differential Equations with Random Right-Hand Sides," Theory of Probability and Applications, Vol. 11, No. 3, 1966, pp. 390-405.
3. Papanicolaou, G. C., "Some Problems and Methods for the Analysis of Stochastic Equations," in Stochastic Differential Equations, SIAM-AMS Proceedings, Vol. VI, 1973, pp. 21-23.
4. Bogoliubov, N., and Mitropolski, A., Asymptotic Methods in the Theory of Nonlinear Oscillations, Gordon and Breach, New York, 1961.
5. Spanos, P-T. D., Linearization Techniques for Nonlinear Dynamical Systems, Earthquake Engineering Research Laboratory, Report EERL 76-04, September 1976, pp. 1-157.
6. Roberts, J. B., "The Energy Envelope of A Randomly Excited Non-Linear Oscillator," Journal of Sound and Vibration, Vol. 60, 1978, pp. 177-185.
7. Spanos, P-T. D., "Probabilistic Earthquake Energy Spectra Equations," Engineering Mechanics Division, American Society of Civil Engineers, Vol. 106, No. EM1, February, 1980, pp. 147-159.
8. Nayfeh, A. H., Perturbation Methods, Wiley, New York, 1973.
9. Roberts, J. B., "First-Passage Probability for Nonlinear Oscillators," Journal of the Engineering Mechanics Division, American Society of Civil Engineers, Vol. 102, EM5, October, 1976, pp. 851-866.
10. Iwan, W. D., and, Spanos, P-T. D., "Response Envelope Statistics for Nonlinear Oscillators with Random Excitation," Journal of Applied Mechanics, American Society of Mechanical Engineers, Vol. 45, March, 1978, pp. 170-174.

11. Spanos, P-T. D., "A Method for Analysis of Non-Linear Vibrations Caused by Modulated Random Excitation," Vol. 16, No. 1, 1981, pp. 1-11.

12. Priestley, M. B., "Evolutionary Spectra and Non-Stationary Processes," Journal of the Royal Statistical Society, Vol. 27, 1965, pp. 204-228.

13. Solomos, G. P., and Spanos, P-T. D., "Solution to the Backward-Kolmogorov Equation for a Non-Stationary Oscillation Problem," Journal of Applied Mechanics, American Society of Mechanical Engineers, December, 1982 (to appear).

14. Spanos, P-T. D., "Non-Stationary Response of a Linear Structure," International Journal of Solids and Structures, Vol. 14, pp. 861-867, 1978.

15. Spanos, P-T. D., and Lutes, L. D., "Probability Density of Response to Evolutionary Process, Journal of the Engineering Mechanics Division, American Society of Civil Engineers, Vol. 106, 1980, pp. 213-224.

16. Crandall, S. H., Chandiramani, K. L., and Cook, R. G., "Some First-Passage Problems in Random Vibration," Journal of Applied Mechanics, American Society of Mechanical Engineers, Vol. 33, No. 3, 1966, pp. 532-538.

17. Yang, J. N., and Shinozuka, M., "On the First-Excursion Probability in Stationary Narrow-Band Random Vibration," Journal of Applied Mechanics, American Society of Mechanical Engineers, Vol. 38, No. 4, 1971, pp. 1017-1022.

18. Roberts, J. B., "First Passage Time for the Envelope of a Randomly Excited Linear Oscillator," Journal of Sound and Vibration, Vol. 46, No. 1, 1976, pp. 1-14.

19. Solomos, G. P., and Spanos, P-T. D., "Structural Reliability Under Evolutionary Seismic Excitation," International Journal of Soil Dynamics and Earthquake Engineering, (to appear).

20. Spanos, P-T. D., and Solomos, G. P., "Circular First-Passage Problems for Oscillators Subjected to Nonstationary Stochastic Excitation, Engineering Mechanics Research Laboratory, University of Texas of Austin, Report EMRL-1163, August, 1982, pp. 1-32.

21. Spanos, P-T. D., "Numerics for Common First-Passage Problem," Journal of the Engineering Mechanics Division, American Society of Civil Engineers, Vol. 108, No. EM5, October, 1982, pp. 864-881.

22. Spanos, P-T. D., "A Method for Analysis of Non-Linear Vibrations Caused by Modulated Random Excitation," International Journal of Non-Linear Mechanics, Vol. 16, No. 1, pp. 1-11, 1981.

23. Spanos, P-T. D., and Iwan, W. D., "Computational Aspects of Random Vibration Analysis," Journal of the Engineering Mechanics Division, American Society of Civil Engineers, Vol. 104, No. EM6, 1978, pp. 1403-1415.

24. Roberts, J. B., "Transient Response of Non-Linear Systems to Random Excitation," Journal of Sound and Vibration, Vol. 74, No. 1, 1981, pp. 11-29.

25. Spanos, P-T. D., "Non-Stationary Random Response of Non-Linear Oscillators," Proceedings of the IX International Conference on Non-Linear Oscillation, Kiev, U.S.S.R., August 30 - September 6, 1981.

SOME PROBLEMS OF RANDOM VIBRATIONS AND ITS APPLICATIONS

Nguyễn Van Dao[1]; Nguyễn Dống Anh[1]

Introduction

In this review some of our main results in the field of vibrations of mechanical systems under random excitations are briefly presented. We suppose the behaviour of mechanical systems to be described by nonlinear stochastic differential or integro-differential equations. Random excitations are supposed to be either 'white noises' or responses of a linear filter to 'white noises' or Gaussian stationary processes. The results obtained by means of modified methods based on Fokker-Planck-Kolmogorov (FPK) equations, the Gaussian heuristic method, the averaging method in nonlinear mechanics are of advantage in studying random vibrations of high order stochastic mechanical systems. - Some applications to isolation of random vibrations and to random vibrations of shells are introduced.

1. The method of FPK equations

1.1. Strongly nonlinear stochastic differential equations under white noise

Let us consider the differential equation

$$\ddot{x} + \omega^2 x = F(x,\dot{x},t) + \sigma\xi(t) \tag{1.1}$$

where $\xi(t)$ stands for the white noise with uniform spectral density and $F(x,\dot{x},t)$, the strongly nonlinear function, is the following polynomial in x and $\dot{x}$:

$$F(x,\dot{x},t) = \sum_{s=0}^{m} \alpha_s \left(\sum_{i,j=0}^{i+j=s} \gamma_{ij}(t) x^i \dot{x}^j \right) . \tag{1.2}$$

The solution of equation (1.1) is a Markov process $\{x(t),\dot{x}(t)\}$ in the phase space. The density of the probability distribution $W(x,\dot{x},t)$ satisfies the FPK equation

$$\frac{\partial W}{\partial t} + \frac{\partial}{\partial x}(\dot{x}W) + \frac{\partial}{\partial \dot{x}}\left\{\left[-\omega^2 x + F(x,\dot{x},t)\right]W\right\} - \frac{\sigma^2}{2}\frac{\partial^2 W}{\partial \dot{x}^2} = 0 . \tag{1.3}$$

By means of the transformation

[1] Professor Dr. sc. Nguyễn Van Dao, Nguyễn Dống Anh, Institute of Mechanics, National Center for Scientific Researches of Vietnam
Nghia Do - Tu Liem - Hanoi

$$x = a\cos\psi\ , \quad \dot{x} = -a\sin\psi \tag{1.4}$$

the FPK equation (1.3) takes the following form:

$$\frac{\partial\pi}{\partial t} + \frac{\partial}{\partial a}B_1(a,\psi,t)\pi + \frac{\partial}{\partial\psi}B_2(a,\psi,t)\pi - \frac{1}{2}\frac{\partial^2}{\partial a^2}[B_{11}\pi] - \frac{\partial^2}{\partial a\partial\psi}[B_{12}\pi] - \frac{1}{2}\frac{\partial^2}{\partial\psi^2}[B_{22}\pi] = 0 \tag{1.5}$$

where [21]

$$B_1(a,\psi,t) = -\frac{1}{\omega}\sin\psi\ F(a\cos\psi,-a\omega\sin\psi,t) + \frac{\sigma^2\cos^2\psi}{2\,\omega^2 a}\ ,$$

$$B_2(a,\psi,t) = \omega - \frac{\cos\psi}{a\omega}F(a\cos\psi,-a\omega\sin\psi,t) - \frac{\sigma^2\cos\psi\sin\psi}{\omega^2 a^2}\ , \tag{1.6}$$

$$B_{11} = \frac{\sigma^2}{\omega^2}\sin^2\psi\ , \quad B_{12} = \frac{\sigma^2}{a\omega^2}\sin\psi\cos\psi\ , \quad B_{22} = \frac{\sigma^2\cos^2\psi}{a^2\omega^2}\ .$$

If we replace $\ln\pi(a,\psi,t)$ by $\varphi(a,\psi,t)$, equation (1.5) becomes

$$\left(\frac{\partial B_1}{\partial a} + \frac{\partial B_2}{\partial\psi} - \frac{\partial^2 B_{12}}{\partial\psi\partial a} - \frac{\partial^2 B_{22}}{\partial\psi^2}\right) + \left(B_1 - \frac{\partial B_{12}}{\partial\psi}\right)\frac{\partial\varphi}{\partial a} + \left(B_2 - \frac{\partial B_{12}}{\partial a} - \frac{\partial B_{22}}{\partial\psi}\right)\frac{\partial\varphi}{\partial\psi}$$
$$- \frac{1}{2}B_{11}\left[\left(\frac{\partial\varphi}{\partial a}\right)^2 + \frac{\partial^2\varphi}{\partial a^2}\right] - B_{12}\left[\frac{\partial\varphi}{\partial a}\frac{\partial\varphi}{\partial\psi} - \frac{\partial^2\varphi}{\partial a\partial\psi}\right] - \frac{1}{2}B_{22}\left[\left(\frac{\partial\varphi}{\partial\psi}\right)^2 + \frac{\partial^2\varphi}{\partial\psi^2}\right] = 0\ . \tag{1.7}$$

The main difficulty is to integrate equation (1.3) [or (1.5) or (1.7)]. All three versions are strongly nonlinear equations with partial derivatives. In [1,2,3,6,7] two methods of integrating these equations - the method of control parameter and the method of decomposition by the general cyclic coordinate - have been presented.

The essence of the first method is to introduce a control parameter permitting transformation of the FPK equation into a first order partial differential equation describing the first integral of some mechanical system. Then, according to every first integral (usually energetic) we can find such a mechanical system under the white noise so that this integral is a solution of the corresponding FPK equation [2,7].

In the second method we take into consideration the fact, that in (1.3), (1.5), (1.7) the coordinates $\dot{x}$ and a play the role of general cyclic coordinates. The coefficients of (1.3), (1.5), (1.7) are polynomials with integer powers of $\dot{x}$ and amplitude a. Therefore, a solution of the FPK equation can be found in the form of a polynomial with integer powers of $\dot{x}$ or a. For example, if the time t is not included in equation (1.7), its solution may be found in the form

$$\frac{\partial\varphi}{\partial a} = a^{-1} + \sum_{i=1}^{\infty} i\,\mu_i(\psi)\,a^{i-1} \tag{1.8}$$

or

$$\varphi(a,\psi) = \ln a + \sum_{i=1}^{\infty} \mu_i(\psi)\,a^i \,. \tag{1.9}$$

Substituting (1.9) into (1,7) and comparing the coefficients of a^i we obtain the equations of the separate variables μ_i, which in many cases give us the exact solutions

$$\mu_i(\psi) = h_i(\psi)\,, \qquad \mu_j(\psi) = 0\,, \quad i = \overline{1,N}\,, \quad J = \overline{N+1,\infty}\,. \tag{1.10}$$

The exact solutions of many FPK equations can be found in this way.

These methods of integrating FPK equations in combination with the asymptotic method in nonlinear mechanics give valuable results [7,13].

In [8] the influence of white noise on the self-excited system with limited energy is considered.

In [7] an exact solution of the FPK equation for a non-autonomous van der Pol equation is given. These methods are of advantage in investigating high order stochastic differential equations [9,10,11].

When random excitation is the response of a linear filter on white noise it is necessary to study the random vibration of single-degree-of-freedom mechanical systems described by high order stochastic differential equations. In fact, the differential equation of the mechanical system will be

$$\ddot{x} + \omega^2 x = f(x,\dot{x},t) + q(t) \tag{1.11}$$

where

$$L\,q(t) \equiv \frac{d^n q(t)}{dt^n} + \sum_{s=0}^{n-1} a_s \frac{d^s q(t)}{dt^s} = b\,\xi(t)\,, \tag{1.12}$$

a, b, a_s are constant, $\xi(t)$ denotes white noise.

Excluding q(t) from (1.11), (1.12), we have the (n+2)-order stochastic differential equation

$$L(\ddot{x} + \omega^2 x) = L\,f(x,\dot{x},t) + b\,\xi(t)\,. \tag{1.13}$$

1.2. Nonlinear stochastic integro-differential equations

In [14,15] we considered the dynamical system the equations of which are

$$\dot{x} = A\,x + F(x) + \varepsilon\int_0^t K(t-s)\,x(s)\,ds + f(t) \tag{1.14}$$

where x, F(x) are n-dimensional vectors, A is a constant (nxn)-matrix and f(t) a stationary Gaussian process or white noise.

The essence of the method is to approximate the linear integro-differential operator

$$\mathcal{L}(x) = \dot{x} - Ax - \varepsilon \int_0^t K(t-s)\, x(s)\, ds \tag{1.16}$$

by the differential operator

$$\tilde{\mathcal{L}}(x) = \dot{x} - (A + \varepsilon B)x , \tag{1.16}$$

and therefore, the randomly excited integro-differential equation (1.14) can be approximated by the randomly excited differential equation

$$\dot{x} = (A + \varepsilon B)x + F(x) + f(t) . \tag{1.17}$$

The error estimations for the solutions of equations (1.14) and (1.17) are considered in [14]. By using a Gaussian heuristic method the solution of integro-differential equation (1.14) is obtained [15].

2. The Gaussian heuristic method

The essence of the method is the following: If the input is a Gaussian process, then for weakly nonlinear systems the output process is supposed to be Gaussian with unknown mean value and correlation function. In this way a closed system of equations for the unknown functions can be obtained.

The Gaussian heuristic method is used in combination with the method of envelopes to solve certain problems in the theory of stochastic vibrations.

1) Let the dynamical system be described by the equation

$$\ddot{z} + 2h\,\dot{z} + \omega^2 z + \overset{\leftrightarrows}{\Phi}(A,Z) = G(t) , \qquad t \geq 0 , \tag{2.1}$$

where h, ω^2 are parameters, A is the amplitude of solution Z(t) and $\overset{\leftrightarrows}{\Phi}$ characterizes the hysteresis. In the special case, where

$$\overset{\leftrightarrows}{\Phi}(A,Z) = \varepsilon \overset{\leftrightarrows}{\alpha}(A^2 - Z^2) \tag{2.2}$$

and G(t) is a Gaussian process or white noise, the variance of the system will be determined by formula (cf. [16]):

$$\sigma^2 = \frac{\pi\sqrt{2\pi}\, S_o}{16\varepsilon\alpha^o\omega^2} , \tag{2.3}$$

with $\vec{\alpha} = \alpha^o$, $\overleftarrow{\alpha} = -\alpha^o$, $\alpha^o > 0$, S_o is the spectral density of the external force G(t).

2) The dynamical systems with hysteresis characteristics excited by a stochastic parameter and forces [17,19] may have the form

where G(t) and F(t) are stationary Gaussian processes with zero mean value and correlation functions $G_{GG}(t)$, $R_{FF}(t)$ and $R_{GF}(t)$.

A general formula for mean value, variance and correlation function of the solution process is found. The obtained results permit us to estimate the influence of hysteresis characteristics of the unknown variable.

Paper [18] is concerned with the foundation of the Gaussian heuristic method. The following theorem is proved there:

"Mean value and second order correlation functions of the solution of equation

$$\ddot{x} + 2h\,\dot{x} + \omega^2 x + \varepsilon F(x) = \xi(t) ,$$

obtained by the Gaussian heuristic method present the second order approximation with respect to a small parameter ε."

In [20] the cumulant description of random processes is used to obtain general formulas of the Gaussian heuristic method in single-degree-of-freedom and multi-degree-of-freedom systems. The Gaussian heuristic method and the statistical linearization method are compared.

Being based on the properties of the normal distribution in [23] there was found out one algorithm to choose an optimal control function which maximizes the probability of the event that the phase vector x(t) will fall into the restricted zone at the final point of time.

3. Random vibrations of shells

Random vibrations of shells have been studied in [22]. Using Hamilton's principle the stochastic differential equations of shells with consideration of geometric nonlinearities are derived. The analytic solution is obtained by means of the statistical linearization method. The values of mean, variances of deflections and normal stress and the probability of failure of the shell are determined.

In order to show possible application of the theory the dynamic response of an elastic cantilever circular cylindrical shell subjected to random wind is studied.

The results are compared with those obtained by beam theory. One can see, that the simple beam approximation gives good results only for very long shells ($\lambda = l/r \approx 60$); for shorter shells ($\lambda = l/r \approx 20$) the beam theory produces significant errors.

As a second application the random vibrations of two co-axial cylinders with finite length immersed in a fluid medium and excited by turbulent flow out of the outer cylinder is considered. The interaction between the shells and the fluid is taken into consideration, too. The variance of the pressure field in the inner cylinder is determined. Its dependence on the structure of the system is presented.

References

1 Nguyên Dông Anh: Etude des vibrations stochastiques par méthode des équations FPK, Proc. Mechanical Works, NCSR Vietnam 1978, 15-28

2 - : Etude des équations FPK par méthode des paramètres de contrôle, NCSR Vietnam, J. of Mechanics (1979), 1,2, 20-27

3 - : Etude des équations FPK par méthode d'un développement en série de MacLaurin à coordinates cycliques, ibid. 28-36

4 - ; Kiêu Thê Duc: Random vibrations of third order systems, J. Ukrain. Math. 32(1980),5,674-678

5 - : On a non-stationary solution of the FPK equation for a linear dynamical system with one degree of freedom, NCSR Vietnam, J. of Mechanics (1981),1,7-12

6 - : Some methods for integration of the FPK equations in the theory of random vibrations, J. Ukrain. Math. 33(1981),1,87-91

7 - : About the solution of the FPK equation in the mechanical non-autonomous system of one degree of freedom, NCSR Vietnam, J. of Mechanics (1982),1,1-6

8 Nguyên Van Dao: The influence of stochastic force on the self-excited system, J. of Mechanics, Vietnam 2(1966),1,43-50 (UBKHKT)

9 - : Non-linear oscillations of third order systems, Proc. VIII Intenat. Conf. on Non-Lin. Oscillations, Prague 1978

10 - : Vibrations of rigid bodies in the visco-elastic Hook-Maxwell model, NCSR Vietnam, J. of Mechanics (1979),1,2,37-44

11 - : Non-Linear Oscillations of High Order Systems, NCSR Vietnam, Hanoi 1979

12 Kiêu Thê Duc: Random vibrations of rigid bodies in the visco-elastic Hook-Maxwell model, Proc. Mechanical Works, Inst. of Communications, Vietnam (1981),1,1-7

13 - ; Nguyên Dông Anh: Non-linear non-autonomous random vibrations of third order systems, NCSR Vietnam, J. of Mechanics (1981),4,6-11

14 Nguyên Tiên Khiêm; Nguyên Dông Anh: On a method for investigation of random oscillations of the linear visco-elastic system, NCSR Vietnam, J. of Mechanics (1979),3,4,31-37

15 - : Stationary vibrations of the visco-elastic system with random excitation, NCSR Vietnam, J. of Mechanics (1981),3,17-25

16 Nguyên Cao Menh: Random vibration of the dynamical systems with hysteresis characteristics, Proc. Mechanical Works, NCSR Vietnam 1978, 63-74

17 - : The oscillations of the dynamical systems with hysteresis characteristics excited by stochastic parameters and force.I, NCSR Vietnam, J. of Mechanics (1979),3,4,46-53

18 - : On the justification of the Gaussian neuristic method in the theory of random vibrations, NCSR Vietnam, J. of Mechanics (1979),3,4,1-11

19 Nguyễn Cao Menh: The oscillations of the dynamical systems with hysteresis characteristics excited by stochastic parameters and force. II, NCSR Vietnam, J. of Mechanics (1980),1,16-22

20 - : On the comparison of the Gaussian heuristic with the statistical linearization methods in random vibrations, NCSR Vietnam, J. of Mechanics (1981),3,26-32

21 Mitropolskij, Ju.A.; Kolomiec, V.G.: The application of asymptotic methods in stochastic systems, Proc. Math. Inst. Ukrain. Acad. Sci., Kiev 1976

22 Nguyễn Xuân Hung: Random vibrations of shells, Proc. Third Nat. Conf. on Mechanics, Huế, SRV 1982

23 Nguyễn Thi Trung: Optimum isolation of a three body system with random excitation, NCSR Vietnam, J. of Mechanics (1982),3,4.

STOCHASTIC AVERAGING OF THE ENERGY ENVELOPE OF NEARLY LYAPUNOV SYSTEMS

Wei Qiu Zhu[1)]

A method based on a theorem due to Khasminskii for obtaining the stochastic differential equation and the FPK equation for the average energy envelope of a Lyapunov system with one degree of freedom subject to small random disturbances and dissipations is presented. The method is then applied to a nonlinear oscillator under random parametric and external excitations and the steady state probability densities and the moments of the average energy envelope are obtained for several cases. It is shown that the response of an oscillator which is unstable under frequency fluctuations can be limited by nonlinearity in restoring force with linear damping, nonlinear damping, and/or nonlinear inertia, while the response of a parametrically unstable oscillator due to damping fluctuations can only be limited by nonlinear damping and/or nonlinear inertia.

1. Introduction. In recent years the standard method of stochastic averaging due to Stratonovich [1], Khasminskii [2], Papanicolaou and Kohler [3] has been extensively applied to random vibration problems of quasi-linear systems, especially to systems subject to random parametric excitation [4-8]. The advantage of the method is that it simplifies the FPK equation or moment equation, and even reduces the dimension of the FPK equation or moment equation in the non-resonant case. In the latter case, however, some terms such as nonlinearities in the restoring force and inertia terms will be lost after averaging [9]. This difficulty can be overcome if the concept of energy envelope (or total energy) of a nonlinear oscillator is used.

When small random disturbances and dissipative forces are applied to a conservative (Lyapunov) system, the energy envelope varies only slightly during each period of oscillation; i.e., compared with the amplitude process the energy envelope is a slowly varying process. If the equation for the energy envelope is averaged with respect to time with the amplitude variation taken into account, then only the energy envelope is involved in the average equation and the dimension of the equation will be reduced. Under certain conditions the average energy envelope is a Markov process and its probability density can be obtained from a dimension-reduced FPK equation. The method of stochastic averaging of the energy envelope can be applied to nearly Lyapunov systems, rather than quasi-linear systems. Furthermore, for a system of one degree of freedom it yields in the first approximation the joint probability density of displacement and velocity processes, from which a wide variety of statistics

1) Visiting Scientist at Massachusetts Institute of Technology - Permanent address: Dept. of Mechanics, Zhejiang University, Hangzhou, The People's Republ. of China

such as level crossing probabilities can be calculated.

The concept of energy envelope has been used by several authors [1,10,11]. Spanos [10] applied the standard stochastic averaging method to a lightly damped linear structure under modulated random excitation to obtain a first stochastic differential equation governing approximately the total energy of the structure. Stratonovich [1] derived a one-dimensional FPK equation for the average energy envelope of the Duffing oscillator based on the argument that the conditional probability density of the oscillator displacement the given oscillator energy envelope of which is proportional to the time the oscillator spends at that displacement. Recently Roberts [11] obtained essentially the same FPK equation by simply time averaging the energy envelope equation based on the standard stochastic averaging method. Although these arguments [1,11] are physically reasonable, they still need to be justified mathematically because the limit theorems [2,3] are inapplicable in this case.

In the present paper the stochastic differential equation and corresponding FPK equation for the average energy envelope of a Lyapunov system subject to small random disturbances and dissipations is derived based on Khasminskii's theorem [12], which provides a rigorous mathematical basis for the method of stochastic averaging of the energy envelope. For the Duffing oscillator the resulting equation is the same as that obtained by Stratonovich [1] and Roberts [11], thus their arguments are justified. The method is then applied to a nonlinear oscillator subject to random parametric and external excitations. The steady state probability densities and moments (if they exist) of the average energy envelope are obtained for several cases and the effects of various nonlinearities on the boundedness of the response are discussed.

2. Development of the method. Consider a nearly Lyapunov system of one degree of freedom of the form

$$\ddot{x} + g(x) = \varepsilon F(x,\dot{x}) + \varepsilon^{1/2}\sum_{i=1}^{l} h_i(x,\dot{x})\ \xi_i(t) \tag{1}$$

where $g(x)$ represents an arbitrary nonlinear restoring force; ε is a small positive parameter; $\varepsilon F(x,\dot{x})$ includes linear damping as well as contributions from nonlinear damping and inertia terms; $\xi_i(t)$, $i = 1,\dots,l$, are assumed to be mutually independent stationary Gaussian random processes with zero mean and of wideband character; $\varepsilon^{1/2}h_i(x,\dot{x})$ are the magnitudes of the excitations, which may depend on the state of the system. Here the effects of the random disturbances and the dissipative forces over a period are assumed to be of the same order of magnitude in the parameter ε. When $\varepsilon = 0$ the system is conservative and has the energy integral

$$\frac{1}{2}\dot{x}^2 + G(x) = \text{const.} \qquad \text{where} \qquad G(x) = \int_0^x g(x)dx. \tag{2),(3}$$

With the transformation

$$x = x, \qquad E = \frac{1}{2}\dot{x}^2 + G(x) \tag{4}$$

equation (1) can be converted into the form

$$\dot{x} = \sqrt{2[E-G(x)]}, \qquad \dot{E} = \sqrt{2[E-G(x)]}\{\varepsilon F(x,\sqrt{2[E-G(x)]}) + \varepsilon^{1/2}\sum_{i=1}^{l} h_i(x,\sqrt{2[E-G(x)]})\xi_i(t)\}. \tag{5}$$

Let

$$\text{l.i.m.} \int_0^t \xi_i(\tau)d\tau = w_i(t) \tag{6}$$

where the $w_i(t)$, $i = 1,\dots,l$, are independent Wiener random processes with unit variance parameters. Then in the mean square sense system (5) is modeled by the Itô equation [13]:

$$\begin{aligned} dy &= \sqrt{2[E-G(y)]}\ dt, \\ dE &= \varepsilon\{\sqrt{2[E-G(y)]}\ F(y,\sqrt{2[E-G(y)]}) + H(y,\sqrt{2[E-G(y)]})\}dt + \\ &\quad \varepsilon^{1/2}\sqrt{2[E-G(y)]}\ \sum_{i=1}^{l} h_i(y,\sqrt{2[E-G(y)]})\ dw_i(t) \end{aligned} \tag{7}$$

where

$$H(y,\ 2[E-G(y)]) = \frac{1}{2}\sum_{i=1}^{l} \sqrt{2[E-G(y)]}\ h_i(y,\sqrt{2[E-G(y)]})\ \frac{\partial}{\partial E}\{\sqrt{2[E-G(y)]}\ h_i(y,\sqrt{2[E-G(y)]})\}. \tag{8}$$

Here y is a rapidly varying process while E is a slowly varying process. Equation (7) is exactly of the same form as that in Khasminskii's theorem [12].[1)] The conditions required by the theorem are easily satisfied by real physical systems, in which F and h_i are usually polynomials in x and $\dot{x}$. If the coefficients of the second equation of (7) are averaged with respect to time t with the first equation taken into account, the following equation governing the average energy envelope will

[1)] In Khasminskii's theorem [12] the number of independent Wiener processes is the same as the dimension of the system but this is not an essential requirement. As in the case of standard stochastic averaging method, the number of independent Wiener processes can be different from the dimension of the system [2,3].

be obtained:

$$d\bar{E} = \varepsilon U(\bar{E})\, dt + \varepsilon^{1/2} \sum_{i=1}^{l} V_i(\bar{E})\, dw_i(t) \tag{9}$$

where

$$U(\bar{E}) = \frac{1}{2\Phi(\bar{E})} \int_R \frac{1}{\sqrt{E-G(y)}} \{\sqrt{2[E-G(y)]}\; F(y,\sqrt{2[E-G(y)]}) + H(y,\sqrt{2[E-G(y)]})\} dy,$$

$$V_i(\bar{E}) = \left\{\frac{1}{\Phi(\bar{E})} \int_R [\sqrt{E-G(y)}\; h_i^2(y,\sqrt{2[E-G(y)]})] dy\right\}^{1/2}, \quad \Phi(\bar{E}) = \frac{1}{2}\int_R \frac{dy}{\sqrt{E-G(y)}} \tag{10}$$

and the region R of integration includes all values y for which $E \geq G(y)$. Thus $\bar{E}$ is a Markov process. According to Khasminskii's theorem [12] the process E in (7) will weakly converge to $\bar{E}$ in the interval $0 \leq t \leq T$, where $T \sim O(1/\varepsilon)$ when $\varepsilon \to 0$, i.e. E can be replaced in the first approximation by $\bar{E}$ for small ε.

The FPK equation corresponding to (9) is

$$\frac{\partial}{\partial t} p(\bar{E};t) = -\varepsilon \frac{\partial}{\partial \bar{E}}[U(\bar{E})p(\bar{E};t)] + \frac{\varepsilon}{2}\frac{\partial^2}{\partial \bar{E}^2} \sum_{i=1}^{l}[V_i(\bar{E})p(\bar{E};t)]. \tag{11}$$

The same FPK equation can be obtained by the classical method, i.e. without introducing Itô's stochastic differential equation. From (11) we can obtain the transition probability density $p(\bar{E};t)$ or approximately $p(E;t)$. If the restoring force $g(x)$ has power-law form and F and h_i are polynomials in x and $\dot{x}$, the integration of (10) can be completed analytically. Then the transition probability density $p(\bar{E};t)$, or at least the steady state probability density $p(\bar{E})$ can be obtained analytically.

Let h_1 = const. and the other $h_i = 0$, $i = 2,\ldots,l$, in (1), then (1) represents a Duffing oscillator with linear and/or nonlinear damping and wide-band random excitation; in this case equ. (11) is essentially the same as that obtained by Stratonovich [1] and Roberts [11]. Thus their arguments are justified.

We can also derive the equation governing the nth moment of $\bar{E}$ from (9) by using the Itô differential rule [13], or from (11) by multiplying by $\bar{E}^n$ and integration over all $0 \leq \bar{E} \leq \infty$. Usually a closure procedure must be invoked to solve the resulting infinite hierarchie of moment equations.

To get the joint probability density $p(x,\dot{x};t)$ let us at first follow Stratonovich's argument. The joint probability density $p(x,E;t)$ can be represented by

$$p(x,E;t) = p_1(x;t|E)\; p(E;t) \tag{12}$$

where $p_1(x;t|E)$ is the conditional probability density that the oscillator displacement is x at time t given that its energy envelope is E. Stratonovich argued [1] that $p_1(x;t|E)$ is proportional to the time that the oscillator whose energy envelope is E spends at x. The time spent at x is in turn inversely proportional to the velocity at x. Therefore

$$p_1(x;t|E) = \begin{cases} \dfrac{C}{\sqrt{E-G(x)}} & \text{for} \quad G(x) \le E \\ 0 & \text{for} \quad G(x) > E \end{cases} . \tag{13}$$

Normalizing this distribution we can write the joint probability density $p(x,E;t)$ as

$$p(x,E;t) = \frac{p(E;t)}{2\Phi(E)\sqrt{E-G(x)}} \tag{14}$$

where $\Phi(E)$ is defined in [11]. Finally, the joint probability density of displacement and velocity $p(x,\dot{x};t)$ can be obtained from (14) based on the transformation (4) as follows

$$p(x,\dot{x};t) = \frac{p(E;t)}{\sqrt{2}\,\Phi(E)} \approx \frac{p(\bar{E};t)}{\sqrt{2}\,\Phi(\bar{E})} \tag{15}$$

which means that a constant contour E in the phase plane is also a contour of constant probability density.

3. Example of application. Consider a nonlinear oscillator subject to random parametric and external excitation

$$\ddot{x} + g(x) = -(2\zeta+\beta x^2)\dot{x} - \alpha x(x\ddot{x}+\dot{x}^2) - \varepsilon^{1/2}[h_1\dot{x}\xi_1(t)+h_2 x\xi_2(t)-h_3\xi_3(t)] \tag{16}$$

where the first two terms on the right hand side are linear and nonlinear damping, respectively, and the third and fourth terms arise from nonlinear inertia. The parameters ζ, β and α are assumed to be of the order of ε. The excitations $\xi_1(t)$, $\xi_2(t)$ and $\xi_3(t)$ are mutually independent wide-band Gaussian processes with mean value zero; $\varepsilon^{1/2}h_1\dot{x}\xi_1(t)$, $\varepsilon^{1/2}h_2 x\xi_2(t)$ and $\varepsilon^{1/2}h_3\xi_3(t)$ are the damping fluctuation, the frequency fluctuation and the external excitation, respectively.
In the first approximation $\ddot{x}$ in the second term of the right hand side of (16) can be replaced by $-g(x)$. Then (16) becomes

$$\ddot{x} + g(x) = -(2\zeta+\beta x^2)\dot{x} - \alpha x[\dot{x}^2-xg(x)] - \varepsilon^{1/2}[h_1\dot{x}\xi_1(t)-h_2 x\xi_2(t)-h_3\xi_3(t)]. \tag{17}$$

The differential equation governing the average energy envelope $\bar{E}$ can be obtained in terms of (9), (10) and (8) as follows:

$$d\bar{E} = \left\{-2\zeta\frac{\Psi_1(\bar{E})}{\Phi(\bar{E})} - \beta\frac{\Psi_2(\bar{E})}{\Phi(\bar{E})} - \frac{\alpha}{\sqrt{2}}\frac{\Psi_3(\bar{E})}{\Phi(\bar{E})} + \varepsilon h_1^2\frac{\Psi_1(\bar{E})}{\Phi(\bar{E})} + \frac{1}{2}\varepsilon h_2^2\frac{\Psi_4(\bar{E})}{\Phi(\bar{E})} + \frac{1}{2}\varepsilon h_3^2\right\}dt +$$

$$\varepsilon^{1/2}\left\{\sqrt{2}h_1\left[\frac{\Psi_5(\bar{E})}{\Phi(\bar{E})}\right]^{1/2} dw_1(t) + h_2\left[\frac{\Psi_2(\bar{E})}{\Phi(\bar{E})}\right]^{1/2} dw_2(t) + h_3\left[\frac{\Psi_1(\bar{E})}{\Phi(\bar{E})}\right]^{1/2} dw_3(t)\right\} \tag{18}$$

where

$$\Psi_1(\bar{E}) = \int_R \sqrt{E-G(y)}\, dy, \quad \Psi_2(\bar{E}) = \int_R y^2\sqrt{E-G(y)}\, dy, \quad \Psi_3(\bar{E}) = \int_R y[2E-2G(y)-yg(y)]\,dy,$$

$$\Psi_4(\bar{E}) = \frac{1}{2}\int_R (y^2/\sqrt{E-G(y)})\, dy, \quad \Psi_5(\bar{E}) = \int_R [E-G(y)]^{3/2}dy\,. \tag{19}$$

The corresponding FPK equation is

$$\frac{\partial}{\partial t}p(\bar{E};t) = \frac{\partial}{\partial\bar{E}}\left\{\left[2\zeta\frac{\Psi_1(\bar{E})}{\Phi(\bar{E})} + \beta\frac{\Psi_2(\bar{E})}{\Phi(\bar{E})} + \frac{\alpha}{\sqrt{2}}\frac{\Psi_3(\bar{E})}{\Phi(\bar{E})} - \varepsilon h_1^2\frac{\Psi_1(\bar{E})}{\Phi(\bar{E})} - \frac{1}{2}\varepsilon h_2^2\frac{\Psi_4(\bar{E})}{\Phi(\bar{E})} - \frac{1}{2}\varepsilon h_3^2\right]p(\bar{E};t)\right\} +$$

$$\varepsilon\frac{\partial^2}{\partial\bar{E}^2}\left\{\left[h_1^2\frac{\Psi_5(\bar{E})}{\Phi(\bar{E})} + \frac{1}{2}h_2^2\frac{\Psi_2(\bar{E})}{\Phi(\bar{E})} + \frac{1}{2}h_3^2\frac{\Psi_1(\bar{E})}{\Phi(\bar{E})}\right] p(\bar{E};t)\right\}. \tag{20}$$

If the restoring force has the power-law form

$$g(x) = \gamma x^\nu \qquad \nu \text{ odd} \tag{21}$$

then the integration of (19) can be completed as follows

$$\Phi(\bar{E}) = 2\frac{\Gamma(\frac{1}{\nu+1})\Gamma(\frac{1}{2})}{\Gamma(\frac{1}{\nu+1}+\frac{1}{2})}\frac{\bar{E}^{1/2}}{\gamma}\left[\frac{\gamma}{(\nu+1)\bar{E}}\right]^{\frac{\nu}{\nu+1}}, \quad \Psi_1(\bar{E}) = 4\frac{\Gamma(\frac{1}{\nu+1})\Gamma(\frac{3}{2})}{\Gamma(\frac{1}{\nu+1}+\frac{3}{2})}\frac{\bar{E}^{3/2}}{\gamma}\left[\frac{\gamma}{(\nu+1)\bar{E}}\right]^{\frac{\nu}{\nu+1}},$$

$$\Psi_2(\bar{E}) = 4\frac{\Gamma(\frac{3}{\nu+1})\Gamma(\frac{3}{2})}{\Gamma(\frac{3}{\nu+1}+\frac{3}{2})}\frac{\bar{E}^{3/2}}{\gamma}\left[\frac{\gamma}{(\nu+1)\bar{E}}\right]^{\frac{\nu-2}{\nu+1}}, \quad \Psi_3(\bar{E}) = 8\frac{\Gamma(\frac{2}{\nu+1})\Gamma(2)}{\Gamma(\frac{2}{\nu+1}+2)}\frac{\bar{E}}{\nu+1}\left[(\frac{1}{2}+\frac{1}{\nu+1})\frac{\gamma}{\bar{E}}\right]^{-\frac{2}{\nu+1}},$$

$$\Psi_4(\bar{E}) = 2\frac{\Gamma(\frac{3}{\nu+1})\Gamma(\frac{1}{2})}{\Gamma(\frac{3}{\nu+1}+\frac{1}{2})}\frac{\bar{E}^{1/2}}{\gamma}\left[\frac{\gamma}{(\nu+1)\bar{E}}\right]^{\frac{\nu-2}{\nu+1}}, \quad \Psi_5(\bar{E}) = 4\frac{\Gamma(\frac{1}{\nu+1})\Gamma(\frac{5}{2})}{\Gamma(\frac{1}{\nu+1}+\frac{5}{2})}\frac{\bar{E}^{5/2}}{\gamma}\left[\frac{\gamma}{(\nu+1)\bar{E}}\right]^{\frac{\nu}{\nu+1}}. \tag{22}$$

The Itô equation governing the average energy envelope is

$$d\bar{E} = \left[-a_3\bar{E}^{(\frac{3}{2}+\frac{1}{\nu+1})} - a_2\bar{E}^{(1+\frac{2}{\nu+1})} - (a_1-a_4)\bar{E} + a_5\bar{E}^{\frac{2}{\nu+1}} + a_5\right]dt +$$

$$2\left[a_7\bar{E}^2\, dw_1(t) + a_8\bar{E}^{(1+\frac{2}{\nu+1})} dw_2(t) + a_9\bar{E}\, dw_3(t)\right] \tag{23}$$

and the corresponding FPK equation is

$$\frac{\partial}{\partial t}p(\bar{E};t) = \frac{\partial}{\partial \bar{E}}\left\{\left[a_3\bar{E}^{\left(\frac{3}{2}+\frac{1}{\nu+1}\right)} + a_2\bar{E}^{\left(1+\frac{2}{\nu+1}\right)} + (a_1-a_4)\bar{E} - a_5\bar{E}^{\frac{2}{\nu+1}} - a_6\; p(\bar{E};t)\right]\right\} + \frac{\partial^2}{\partial \bar{E}^2}\left\{\left[a_7\bar{E}^2 + a_8\bar{E}^{\left(1+\frac{2}{\nu+1}\right)} + a_9\bar{E}\right]p(\bar{E};t)\right\} \tag{24}$$

where

$$a_1 = 4\zeta\frac{\nu+1}{\nu+3}, \quad a_2 = \beta\left(\frac{\nu+1}{\gamma}\right)^{\frac{2}{\nu+1}}\frac{\Gamma(\frac{3}{\nu+1})\Gamma(\frac{1}{\nu+1}+\frac{1}{2})}{\Gamma(\frac{1}{\nu+1})\Gamma(\frac{3}{\nu+1}+\frac{3}{2})},$$

$$a_3 = 2\sqrt{2}\left[\gamma(\nu+1)\right]^{-\frac{1}{\nu+1}}\left(\frac{1}{2}+\frac{1}{\nu+1}\right)^{-\frac{2}{\nu+1}}\frac{\Gamma(\frac{2}{\nu+1})\Gamma(2)\Gamma(\frac{1}{\nu+1}+\frac{1}{2})}{\Gamma(\frac{1}{\nu+1})\Gamma(\frac{1}{2})\Gamma(\frac{2}{\nu+1}+2)}, \quad a_4 = 2\varepsilon h_1^2\frac{\nu+1}{\nu+3},$$

$$a_5 = \frac{1}{2}\varepsilon h_2^2\left(\frac{\nu+1}{\gamma}\right)^{\frac{2}{\nu+1}}\frac{\Gamma(\frac{3}{\nu+1})\Gamma(\frac{1}{\nu+1}+\frac{1}{2})}{\Gamma(\frac{1}{\nu+1})\Gamma(\frac{3}{\nu+1}+\frac{1}{2})}, \quad a_6 = \frac{1}{2}\varepsilon h_3^2, \quad a_7 = \varepsilon h_1^2\frac{3(\nu+1)}{(3\nu+5)(\nu+3)}, \tag{25}$$

$$a_8 = \frac{1}{2}\varepsilon h_2^2\left(\frac{\nu+1}{\gamma}\right)^{\frac{2}{\nu+1}}\frac{\Gamma(\frac{3}{\nu+1})\Gamma(\frac{1}{\nu+1}+\frac{1}{2})}{\Gamma(\frac{1}{\nu+1})\Gamma(\frac{3}{\nu+1}+\frac{3}{2})}, \quad a_9 = \varepsilon h_3^2\frac{\nu+1}{\nu+3}.$$

The transition probability density function and a closed set of moment equations can be obtained analytically from (24) or (23) only in certain special cases. The steady state probability density and moments (if they exist), however, can always be determined by direct integration of (24). In view of the complexity of the solution in the general case, in the following we discuss only the steady state behavior of the system for some particular cases.

3.1. $\nu = 1$ and $\alpha = \beta = 0$. In this case (16) represents a linear oscillator subject to random parametric and external excitation. The steady state probability density of the average energy envelope is

$$p(E) = K\left[\left(\frac{3}{16}\varepsilon h_1^2 + \frac{1}{4\gamma}\varepsilon h_2^2\right)E + \frac{1}{2}\varepsilon h_3^2\right]^{-\mu} \tag{26}$$

where

$$\mu = \left(2\zeta - \frac{5}{8}\varepsilon h_1^2\right) \Big/ \left(\frac{3}{16}\varepsilon h_1^2 + \frac{1}{4\gamma}\varepsilon h_2^2\right). \tag{27}$$

The requirement for normalizability is

$$2\zeta > \frac{13}{16}\varepsilon h_1^2 + \frac{1}{4\gamma}\varepsilon h_2^2 \tag{28}$$

and the normalization constant is

$$K = (2\zeta - \frac{13}{16}\varepsilon h_1^2 - \frac{1}{4\gamma}\varepsilon h_2^2)(\frac{1}{2}\varepsilon h_3^2)^{\mu-1} \quad . \tag{29}$$

Here and hereafter we omit the bar over 'E' for simplicity.

The differential equation governing the mean value of the average energy envelope can be obtained from (23) by taking the ensemble average or from (24) by multiplying by E and integrating over all $0 \le E \le \infty$. The result is

$$\frac{d}{dt}\langle E\rangle = (-2\zeta + \varepsilon h_1^2 + \frac{1}{2\gamma}\varepsilon h_2^2)\langle E\rangle + \frac{1}{2}\varepsilon h_3^2 \tag{30}$$

from which the condition for "energy stability" or "second moment stability" is easily deduced to be

$$2\zeta \ge \varepsilon h_1^2 + \frac{1}{2\gamma}\varepsilon h_2^2 \, . \tag{31}$$

It is seen that in this case the effects of the damping fluctuation and the frequency fluctuation on the stability of the system are of the same significance and the external force has no effect on the stability of the system.

The steady state nth moment can be obtained from (26) by multiplying E^n and integrating over all $0 \le E \le \infty$. The condition that the finite nth moment exists is

$$2\zeta \ge (n+1)(\frac{3}{16}\varepsilon h_1^2 + \frac{1}{4\gamma}\varepsilon h_2^2) + \frac{5}{8}\varepsilon h_1^2 \tag{32}$$

which reduces to the stability condition (31) when $n = 1$. Because the random disturbances and the dissipative forces are assumed to be of the same order, there is always one value of m such that all the nth moments converge for $n \le m$ and diverge for $n > m$. The nth moment when it exists has a magnitude that is proportional to $(\varepsilon h_3^2)^n$.

3.2. $\nu = 1$ and $\alpha, \beta \neq 0$. The oscillator in this case is one with linear restoring force, linear and nonlinear damping, and nonlinear inertia, subject to random parametric and external excitation. The steady state probability density of the average energy envelope is

$$p(E) = K\left[(\frac{3}{16}\varepsilon h_1^2 + \frac{1}{4\gamma}\varepsilon h_2^2)E + \frac{1}{2}\varepsilon h_3^2\right]^{-\mu'} e^{-\eta E} \tag{33}$$

where K is a normalization constant and

$$\mu' = \frac{2\zeta - \frac{5}{8}\varepsilon h_1^2}{\frac{3}{16}\varepsilon h_1^2 + \frac{1}{4\gamma}\varepsilon h_2^2} - \frac{1}{2}\varepsilon h_3^2 \frac{\frac{\alpha}{\pi\sqrt{\gamma}} + \frac{\beta}{2\gamma}}{(\frac{3}{16}\varepsilon h_1^2 + \frac{1}{4\gamma}\varepsilon h_2^2)^2} \, , \quad \eta = \frac{\frac{\alpha}{\pi\sqrt{\gamma}} + \frac{\beta}{2\gamma}}{(\frac{3}{16}\varepsilon h_1^2 + \frac{1}{4\gamma}\varepsilon h_2^2)^2} \, . \tag{34}$$

When $\varepsilon h_3^2 = 0$, $\mu' = \mu$ and the normalization constant is

$$K = \frac{\eta^{(1-\mu)}}{\Gamma(1-\mu)} \left(\frac{3}{16}\varepsilon h_1^2 + \frac{1}{4\gamma}\varepsilon h_2^2\right)^{\mu}, \quad \mu < 1 . \tag{35}$$

Then the nth moment in this case is

$$\langle E^n \rangle = \frac{(n-\mu)(n-1-\mu)\dots(1-\mu)}{\eta^n}, \quad \mu < 1, \tag{36}$$

which is always finite for finite n when $\alpha, \beta > 0$[1]. This means that no matter how large the damping fluctuation and/or frequency fluctuation, and how small (even zero) the linear damping, the nonlinear damping and/or nonlinear inertia can always limit the response of the system and make all response moments finite. This conclusion remains correct when $\varepsilon h_3^2 \neq 0$.

3.3. $\nu = 3$ and $\alpha = \beta = 0$. In this case the steady state behavior of the oscillator due to damping fluctuations is quite different from that due to frequency fluctuations. The steady state probability densities of the average energy envelope of the oscillator under frequency fluctuation and external excitation are

$$p(E) = \frac{1}{\sqrt{\pi}}\left[\frac{5}{3}\frac{\zeta\gamma^{1/2}}{\varepsilon h_2^2}\left(\frac{\Gamma(\frac{1}{4})}{\Gamma(\frac{3}{4})}\right)^2\right]^{3/2} E^{-1/4} \exp\left[-\frac{5\zeta\sqrt{\gamma}}{3\varepsilon h_2^2}\left(\frac{\Gamma(\frac{1}{4})}{\Gamma(\frac{3}{4})}\right)^2 E^{1/2}\right] \tag{37}$$

and

$$p(E) = \frac{1}{\Gamma(\frac{3}{4})}\left(\frac{4\zeta}{\varepsilon h_3^2}\right)^{3/4} E^{-1/4} \exp\left(-\frac{4\zeta}{\varepsilon h_3^2} E\right), \tag{38}$$

respectively. The probability distribution due to frequency fluctuation is broader than that due to external excitation as evidenced by the $E^{1/2}$ factor in the exponent of (37) in contrast to the E in (38). The nth moments in these two cases are

$$\langle E^n \rangle = \frac{2\Gamma(2n+\frac{3}{2})}{\pi^{1/2}}\left[\frac{3\varepsilon h_2^2}{4\zeta\gamma^{1/2}}\left(\frac{\Gamma(\frac{3}{4})}{\Gamma(\frac{1}{4})}\right)^2\right]^{2n} \quad \text{and} \quad \langle E^n \rangle = \frac{\Gamma(n+\frac{3}{4})}{\Gamma(\frac{3}{4})}\left(\frac{\varepsilon h_3^2}{4\zeta}\right)^n, \tag{39),(40}$$

respectively, which are finite for any finite n. This means that the nonlinearity in restoring force with linear damping can make any nth response moment finite when the system is unstable due to frequency fluctuation.

The same conclusion can be drawn when the oscillator is subject to both frequency fluctuation and external excitation. The steady state probability density of the average energy envelope in this case is

---

[1] It can be seen from (33) that for the case $\mu > 1$ the probability density approaches to zero more rapidly when $E \to \infty$ than for the case $\mu < 1$. Thus for $\mu > 1$ and for finite n, $\langle E^n \rangle$ should be also finite. But in this case there is some difficulty with integrating in the evaluating of $\langle E^n \rangle$.

$$p(E) = K\,E^{-1/4}\left[\frac{16}{5}\,\frac{\varepsilon h_2^2}{\gamma^{1/2}}\left(\frac{\Gamma(\frac{3}{4})}{\Gamma(\frac{1}{4})}\right)^2 E^{1/2} + \frac{2}{3}\varepsilon h_3^2\right]^{\frac{25\zeta\gamma\varepsilon h_3^2}{72\varepsilon^2 h_2^4}\left(\frac{\Gamma(\frac{1}{4})}{\Gamma(\frac{3}{4})}\right)^4} \exp\left[-\frac{5}{3}\,\frac{\zeta\gamma^{1/2}}{\varepsilon h_2^2}\left(\frac{\Gamma(\frac{1}{4})}{\Gamma(\frac{3}{4})}\right)^2 E^{1/2}\right] \tag{41}$$

which approaches (37) when $\varepsilon h_3^2 \longrightarrow 0$. It is seen from comparison between (37) and (41) that the probability distribution is broadened due to external excitation. When the oscillator is excited by both damping fluctuations and external excitation the steady state probability density of the average energy envelope is

$$p(E) = K\,E^{-1/4}\left(\frac{1}{7}\varepsilon h_1^2 E + \frac{2}{3}\varepsilon h_3^2\right)^{7\left(\frac{13}{12} - \frac{8\zeta}{3\varepsilon h_1^2}\right)} . \tag{42}$$

The condition that a finite nth moment exists is

$$2\zeta > \frac{3n + 25}{28}\,\varepsilon h_1^2 \tag{43}$$

and, in particular, the condition for energy stability is

$$2\zeta > \varepsilon h_1^2 \tag{44}$$

which is the same as in linear case (see (31) when $\varepsilon h_2^2 = 0$). Furthermore, as in the linear case there always exists an m such that all the nth moments converge for $n \leq m$ and diverge for $n > m$. It can be inferred based on (24) that this conclusion remains unchanged for any ν; i.e. , the nonlinearity in restoring force can not limit the response of a system which is unstable due to damping fluctuations. This conclusion remains correct when the system is unstable due to both damping and frequency fluctuations.

3.4. $\nu = 3$ and $\alpha, \beta \neq 0$. While a nonlinear restoring force can not limit the response of an unstable system due to damping fluctuations, it can be shown as in subsection 3.2 that in this case nonlinear damping and/or nonlinear inertia can make any nth moment of the average energy envelope finite.

Finally, when $g(x) = \gamma x + \delta x^3$ it is difficult to obtain analytic expressions for the steady state probability density of the average energy envelope because of the complexity of the integration in (19). However, from the above discussion we can infer that the response of the oscillator under frequency fluctuations and external excitation is always limited and has finite nth response moment for any finite n only if the linear damping is not zero; the response of the oscillator excited by damping fluctuations (with or without frequency fluctuations and external excitation) can only be limited by nonlinear damping and/or nonlinear inertia if the linear damping is not large enough to stabilize the system. The steady state probability distribution will be intermediate between the case $\nu = 1$ (subsections 3.1 or 3.2) and the case $\nu = 3$ (subsections 3.3 or 3.4); for small energy average envelope the distribution will be more like the one for $\nu = 1$, whereas it will be more like the one for $\nu = 3$ for large average energy envelope.

Acknowledgement. The author would like to express his gratitude to Professor S.H. Crandall of the Massachusetts Institute of Technology for his encouraging comments and helpful advice.

References

1 Stratonovich, R.L., Topics in the Theory of Random Noise. Vol.1, Gordon and Breach, New York 1963

2 Khasminskii, R.Z., A limit theorem for solution of differential equations with random right-hand side, Theor. Probab. Appl. 11(1966),390

3 Papanicolaou, G.C.; Kohler, W., Asymptotic theory of mixing stochastic ordinary differential equations, Commun. pure appl. Math. 27(1974),641

4 Ariaratnam, S.T.; Tam, D.S.F., Moment stability of coupled linear systems under combined harmonic and stochastic excitation, in 'Stochastic Problems in Dynamics', IUTAM Symposium, ed. by B.L. Clarkson, Pitman, London 1977, pp. 90-105

5 -,-; -,-, Parametric random excitation of damped Mathieu oscillators, ZAMM 56 (1976),449-452

6 -,-; -,-, Random vibration and stability of a linear parametrically excited oscillator, ZAMM 59(1979),79-84

7 Lin, Y.K.; Fujimori, Y.; Ariaratnam, S.T., Rotor blade stability in turbulence flows. Part I, AIAA Journal 17(1979),545-554

8 -,-, Motion of suspension bridges in turbulent winds, J Engg. Mech. Division, ASCE 105, N. EM6, proceedings paper 15044, Dec. 1979, pp.921-932

9 Mickens, R.E., Generalization of the method of slowly varying amplitude and phase to non-linear oscillatory systems with two degrees of freedom, J. Sound Vibrations 74(1982),3,455-458

10 Spanos, P.-T.D., Energy analysis of structural vibrations under modulated random excitation, J, Struc. Mech. 6(1978),3,289-302

11 Roberts, J.B., The energy envelope of a randomly excited non-linear oscillator, J. Sound Vibrations 60(1978),2,177-185

12 Khasminskii, R.Z., On the averaging principle for stochastic differential Itô equations, Kibernetika 4(1968),3,260-279

13 Jazwinski, A., Stochastic Processes and Filtering Theory, Academic Press, New York 1970.

RANDOM VIBRATIONS OF LIQUID-FILLED CONTAINERS (EXCITED BY EARTHQUAKES)

FRANZ ZIEGLER *)

Summary

Stationary and instationary responses of a fluid-filled free standing cylindrical container are determined by spectral methods, when excited by the horizontal component of an earthquake. Instationary forcing functions are modeled by stationary random processes with deterministic shape functions. Results are given for maximum responses, like base shearing force, wall- and overturning moment in case of stationary random vibrations and the probability of excess sloshing(first passage probability within a fixed quake duration) is calculated for instationary excitation. A simple mechanical model (recently discussed by M.A.Haroun and G.W.Housner) replaces the fluid (to first order) also in case of combined translational and rotational container motion.

INTRODUCTION

Recently the size of free standing cylindrical storage tanks has been dramatically enlarged. Therefore, wind- and earthquake loadings become more pronounced design criteria and the solution of the dynamic interaction problem of fluid and shell structure determines the required safety against buckling and excess sloshing. In this paper earthquake loadings are considered by stationary and instationary random processes. For engineering applications, a simple mechanical model of the liquid filled tank is derived from the "substructure technique". Considering at first the vibrating ideal fluid body with a proper series solution for the velocity potential, the hydrodynamic pressure frequency response at the tank wall is calculated. The important modes of the tank vibrations consist of horizontal and rotational motion, in contrast to several papers (see e.g. the references in D. Fischer[1]). A first order mechanical model adds an impulsive mass to the tank which, due to the rotation, has also a nonvanishing moment of inertia, and a convective mass with a spring support, respectively. For sake of brevity the deformation modes of the tank-shell which add additional mass-spring systems to the model are neglected. Also the simplest possible model is chosen for the soil-tank interaction: An anisotropic 2-dimensional effective spring replaces the elastic foundation half-space. For modal analysis of the linearized 3-degree-of-freedom vibrational system still another simplifying assumption is necessary to change the effective damping (by geometric spreading of waves in the half-space) to a modal damping. After all, structural safety is conveniently described via earthquake response spectra for individual stationary vibrating oscillators, and superposition by RSS (root of the sum of squares) renders the desired result. Additionally, a stochastic analysis by the method of instationary spectra is presented, and applied to the calculation of the probability of excess sloshing.

*) Professor, Department of Civil Engineering, Technical University Vienna, Karlsplatz 13. A-1040 WIEN, AUSTRIA.

VIBRATIONS OF THE FLUID BODY.

The theory of linear vibrations of ideal fluids in moving containers has been summarized by Moiseyev and Rumyantsev [2] and independently by Abramson [3]. Recently, Parkus [4], studied the free vibrations in a cylindrical shell tank. A non-linear numerical analysis of the problem by FEM (Finite Element Method) was given in [5], a linear semi-numerical solution by boundary-integral-equation was presented by Höllinger in [6]. Using Stokes' linearized boundary condition at the free surface $z = H$, see Fig.1, the hydrodynamic excess pressure p becomes (quasi-static), in case of time-harmonic vibrations, the factor $\exp(i\omega t)$ is understood,

$$p(r,\varphi,z=H,\omega) = \rho g[z_S(r,\varphi,\omega) - H] \quad , \tag{1}$$

ρ fluid density, $z_S - H$ amplitude of surface wave.

Introducing the velocity potential Φ with $\vec{v}$ = grad Φ (absolute velocity), and considering $p = -\rho\frac{\partial\Phi}{\partial t} = -\rho\dot{\Phi}$, the linearized boundary condition becomes after differentiation with respect to time (keeping same notation for frequency response functions):

$$-\omega^2\Phi + g\Phi,_z\,]_{z=H} = 0 \quad , \qquad i\omega z_S = \frac{\partial\Phi}{\partial z} = \Phi,_z\,]_{z=H} \quad . \tag{2}$$

An incompressible fluid without sources requires div $\vec{v} = 0$, and Φ is a proper solution of Laplace's equation $\Delta\Phi = 0$. At the container walls coincidence of the normal velocity renders inhomogeneous Neumann boundary conditions:

$$r=R: \quad \Phi,_r = i\omega(x_g + z\theta_g)\cos\varphi + i\omega w(\varphi,\omega) \quad , \tag{3}$$

$$z=0: \quad \Phi,_z = -i\omega\theta_g r\cos\varphi - i\omega \underline{w}(r,\varphi,\omega) \quad , \tag{4}$$

x_g and θ_g are the rigid body coordinates associated with the horizontal translational and rotational motion, respectively, w denotes deflection of the container walls (it is neglected in the random analysis). Solution is greatly simplified by superposition of potentials

$$\Phi = \Phi_1 + \Phi_2 + \Phi_3 \quad , \tag{5}$$

where

$$\Phi_1(r,\varphi,z,\omega) = i\omega(x_g + z\theta_g)\, r\cos\varphi \tag{6}$$

and a proper choice of Φ_3 renders a homogeneous Neumann problem for Φ_2. Separation of the potential equation gives for the first circumferential wave

$$\Phi_2 = \sum_{j=1,2} [A_j(\omega)\cosh\lambda_j z/R + B_j(\omega)\sinh\lambda_j(z-H)/R] J_1(\lambda_j r/R)\cos\varphi \quad , \tag{7}$$

where λ_j, $j=1,2,\ldots$, are the roots of $J_1'(\lambda)=0$, and J_1 is Besselfunction of first kind and order, $\lambda_1 = 1,8412$. Neglecting w and representing the linear function r by a Dini-series, cf.[7],

$$r = \sum_j C_j J_1(\lambda_j r/R), \quad C_j = 2R\lambda_j J_2(\lambda_j)/(\lambda_j^2 - 1)J_1^2(\lambda_j) = 2R/(\lambda_j^2 - 1)J_1(\lambda_j) \quad , \tag{8}$$

$$J_2(\lambda_j) = J_1(\lambda_j)/\lambda_j \quad ,$$

boundary condition (4) renders explicitly the coefficients B_j due to rotation,

$$B_j(\omega) = -4i\omega\theta_g R^2/\lambda_j(\lambda_j^2-1)J_1(\lambda_j)\cosh\lambda_j H/R \quad , \quad j = 1,2,... \tag{9}$$

Inserting Eqs.(6), (7), (9) into Eq.(2) and applying Eq.(8) to

$$\frac{\partial\Phi_1}{\partial z} = i\omega\theta_g r\cos\varphi \tag{10}$$

the complex solution is

$$A_j(\omega) = i\frac{2R\omega^3}{\nu_j^2-\omega^2}\ \{x_g + H\theta_g[1-\frac{g}{H\omega^2}(1-2/\cosh\lambda_j H/R)]\}/(\lambda_j^2-1)J_1(\lambda_j)\cosh\lambda_j H/R \quad ,$$

$$j = 1,2,..., \tag{11}$$

showing resonances at the sloshing frequencies

$$\nu_j^2 = \frac{g\lambda_j}{R}\tanh\lambda_j H/R \quad . \tag{12}$$

Hence, the real hydrodynamic pressure frequency response function in case of $w=0$ becomes, $p= -i\omega\rho\Phi$,

$$\frac{p(\omega,r,\varphi,z)}{\rho R\omega^2} = (x_g + z\theta_g)\frac{r}{R}\cos\varphi + \sum_j[\frac{2\omega^2}{\nu_j^2-\omega^2}\{x_g+H\theta_g[1-\frac{g}{H\omega^2}(1-2/\cosh\lambda_j H/R)]\}\cosh\lambda_j z/R-$$

$$-\frac{4R\theta_g}{\lambda_j}\sinh\lambda_j(z-H)/R]J_1(\lambda_j r/R)\cos\varphi/(\lambda_j^2-1)J_1(\lambda_j)\cosh\lambda_j H/R \quad , \tag{13}$$

and the sloshing amplitude at r=R is given by

$$\frac{g}{R\omega^2}(z_S-H) = (x_g+H\theta_g)\cos\varphi + \sum_j\frac{2\omega^2}{\nu_j^2-\omega^2}\{x_g+H\theta_g[1-\frac{g}{H\omega^2}(1-2/\cosh\lambda_j H/R)]\}\cos\varphi/(\lambda_j^2-1) \tag{14}$$

Exact frequency responses of resultant hydrodynamic forces and moments are found by integration and superposition of a hydrostatic moment due to a rigid rotation of the free surface through θ_g.

Shearing force at z=0 :

$$\frac{Q}{m_F\omega^2} = \frac{1}{m_F\omega^2}\int_0^H\int_0^{2\pi} p]_{r=R}\cos\varphi R d\varphi dz = x_g + \frac{H}{2}\theta_g(1+\sum_j\alpha_j) + \sum_j\frac{\beta_j\omega^2}{\nu_j^2-\omega^2}\{x_g+H\theta_g[1-\frac{g}{H\omega^2}(1-$$

$$-2/\cosh\lambda_j H/R)]\} \quad , \tag{15}$$

$$\alpha_j = \frac{8R^2}{H^2}\frac{\cosh\lambda_j H/R - 1}{\lambda_j^2(\lambda_j^2-1)\cosh\lambda_j H/R} \quad , \qquad \beta_j = \frac{2R}{H}\frac{\tanh\lambda_j H/R}{\lambda_j(\lambda_j^2-1)} \quad .$$

Buckling(wall)moment at z=0 :

$$\frac{M_b}{m_F \frac{H}{2}\omega^2} = 2\frac{Q}{m_F\omega^2} - x_g - \frac{H}{3}\theta_g\left[1-\sum_j \frac{12R^2}{\lambda_j^2H^2}\left(\beta_j+\frac{2}{\lambda_j^2-1}\right)\right] - \sum_j \frac{\alpha_j\omega^2}{2(\nu_j^2-\omega^2)}\{x_g+H\theta_g[1-\frac{g}{H\omega^2}(1-$$

$$-2/\cosh\lambda_j H/R)]\} . \tag{16}$$

Overturning moment at z=0 :

$$\frac{M_o}{m_F\frac{H}{2}\omega^2} = \frac{M_b}{m_F\frac{H}{2}\omega^2} + \left(\frac{1}{m_F\frac{H}{2}\omega^2}\int_0^R\int_0^{2\pi} r^2 p]_{z=0}\cos\varphi\, d\varphi dr\right) + \frac{g\theta_g}{\omega^2}(1+R^2/2H^2) \quad . \tag{17}$$

The above included bottom pressure moment (.) can be expressed in closed form by noting the integral, cf.[7],

$$\int_0^R r^2 J_1(\lambda_j r/R)\, dr = \frac{R^3}{\lambda_j^3}\int_0^{\lambda_j} t^2 J_1(t)dt = \frac{R^3}{\lambda_j^2} J_1(\lambda_j)$$

to become

$$(\ .\ ) = \frac{R^2}{2H^2}x_g + R\theta_g\sum_j\frac{4R\beta_j}{H\lambda_j^2} + \sum_j\frac{\gamma_j\omega^2}{\nu_j^2-\omega^2}\{x_g + H\theta_g[1-\frac{g}{H\omega^2}(1-2/\cosh\lambda_j H/R)]\} \quad ,$$

$$\gamma_j = \frac{4R^2/H^2}{\lambda_j^2(\lambda_j^2-1)\cosh\lambda_j H/R} \quad . \tag{18}$$

For engineering calculations a simple mechanical model of the fluid action Q, M_o (in a one-term approximation j=1) can be determined, see Fig.2, with impulsive (rigid) mass $m_o = m_F(1-\beta_1)$, radius of rotational inertia

$$i_o^2 = \frac{H^2}{3}\{1+\frac{3}{8}[\alpha_1-4\beta_1(1-\frac{8R^2}{\lambda_1^2H^2}) - 10\gamma_1]\}/(1-\beta_1) - H_o^2 \quad , \tag{19}$$

placed at effective height

$$H_o = \frac{H}{2}\{1+\frac{R^2}{2H^2}+\frac{1}{4}[\alpha_1 - 4\beta_1 - 2\gamma_1]\}/(1-\beta_1) \quad , \tag{20}$$

and convective oscillating mass $m_1 = m_F\beta_1$ placed at height

$$H_1 = H\{[-\alpha_1 + 4\beta_1 + 2\gamma_1]/8\beta_1 - g/H\nu_1^2\} \tag{21}$$

with spring support, - effective stiffness is

$$k = m_1\nu_1^2 \quad . \tag{22}$$

Note from Fig. 2 for the above used correspondence of coefficients

$$\omega^2 m_o(x_g + H_o\theta_g) + \omega^2 m_1(x_g + H_1\theta_g + \xi) = Q ,$$

$$\omega^2 m_o[H_o x_g + (H_o^2 + i_o^2)\theta_g] + \omega^2 m_1[H_1(x_g+\xi) + H_1^2\theta_g] = M - (m_o H_o + m_1 H_1)g\theta_g + m_1 g\xi$$

and

$$\xi = \frac{\omega^2}{k/m_1 - \omega^2}[x_g + \theta_g(H_1 + g/\omega^2)] .$$

APPROXIMATED INTERACTIVE MODEL AND STATIONARY RESPONSE.

Replacing the fluid by the rigidly attached mass m_o, i_o and the convective oscillating mass m_1 and adding the rigid masses of the container, $m_C = m_2 + m_3$, m_2, i_2 at $H_2 = L/2$ of walls, and m_3, i_3 at $z=0$ of base and foundation, respectively, and considering horizontal and rotational stiffness of the supporting elastic half-space, a 3-degree-of-freedom vibratory system is established, see Fig.2. Using the absolute coordinate θ_g and the relative coordinates x_g and ξ, the equations of motion with horizontal earthquake excitation $x_G(t)$ become (damping-force excitation is neglected),

$$\underset{\sim}{m}\ddot{\vec{X}} + \underset{\sim}{c}_e\dot{\vec{X}} + \underset{\sim}{k}_e\vec{X} = -\underset{\sim}{m}\vec{r}\ddot{x}_G(t) \quad , \tag{23}$$

$$\vec{X}^T = \{x_g, \xi, R\theta_g\} \quad , \vec{r}^T = \{1,0,0\} \quad ,$$

where

$$\underset{\sim}{m} = \begin{pmatrix} m_F + m_C & m_1 & (m_o H_o + m_1 H_1 + m_2 H_2)/R \\ (m_o H_o + m_1 H_1 + m_2 H_2)/H_o & m_1 H_1/H_o & [m_o(H_o^2 + i_o^2) + m_1 H_1^2 + m_2(H_2^2 + i_2^2) + m_3 i_3^2]/RH_o \\ m_1 & m_1 & m_1 H_1/R \end{pmatrix} ,$$

$$\underset{\sim}{k}_e = \begin{pmatrix} k_x & 0 & 0 \\ 0 & -m_1 g/H_o & [k_\theta - g(m_o H_o + m_1 H_1 + m_2 H_2)]/RH_o \\ 0 & k & -m_1 g/R \end{pmatrix} ,$$

$$\underset{\sim}{c}_e = \begin{pmatrix} c_x & 0 & 0 \\ 0 & 0 & c_\theta/RH_o \\ 0 & 0 & 0 \end{pmatrix} .$$

See e.g.[8], - for low frequencies a static effective half-space stiffness is chosen to be $k_x = 32\, GR \frac{1-\nu}{7-8\nu}$, $k_\theta = 8\, GR^3/3(1-\nu)$. Instead of $\underset{\sim}{c}_e$ proper model

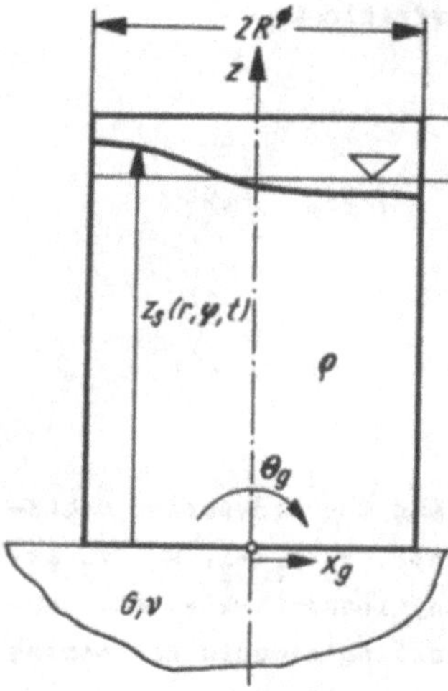

Fig.1

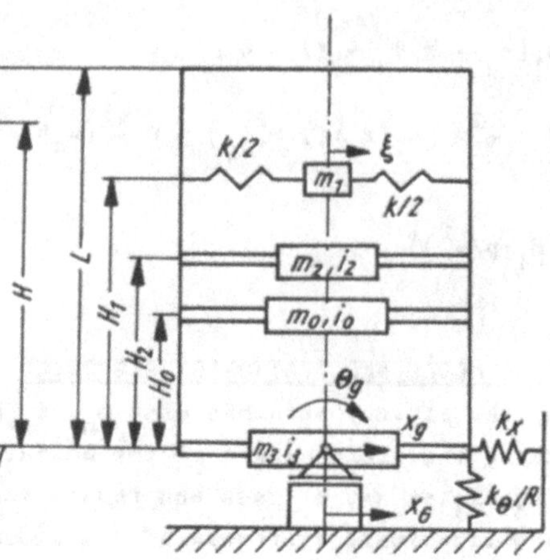

Fig.2

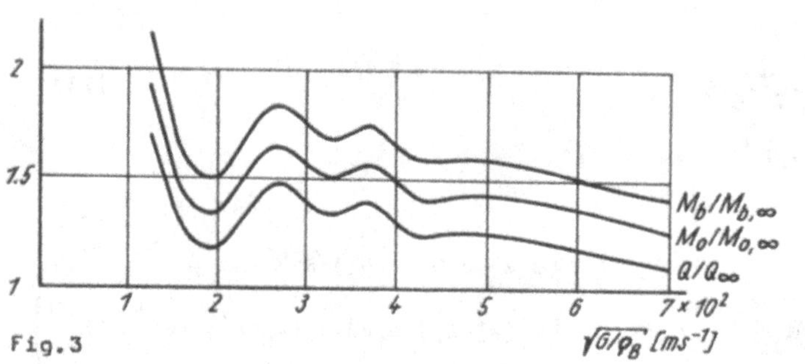

Fig.3

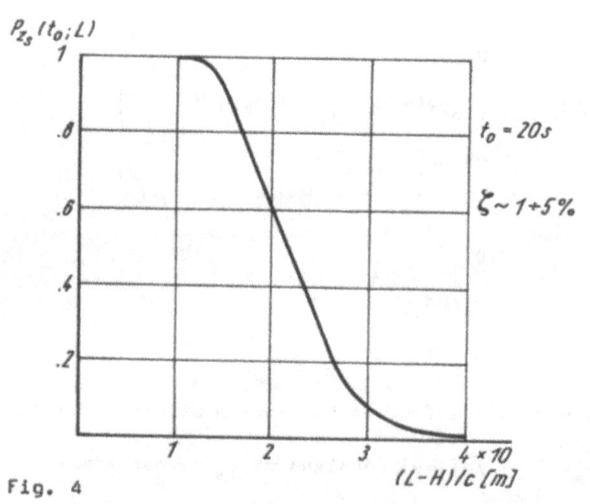

Fig. 4

damping is preferable. Following a report of M.A.Haroun and G.W.Housner, modal analysis is applied and the maximum resultant forces and moments are calculated representing $x_G(t)$ through the response spectrum of the 1940 N-S El Centro earthquake, see e.g.[9]. Superposition of the modal maxima is made by RSS-rule (sum of variances of uncorrelated random modal responses). For rigid half-space the reference values are $Q_\infty = 0{,}42 m_F g$, $M_{b,\infty} = 0{,}16 m_F gH$, $M_{o,\infty} = 0{,}21 m_F gH$ and the relative maximum response for elastic half-space is shown in Fig.3, (H/R=2, R=9,14m). These analysis corresponds to a stationary stochastic treatment of the earthquake problem which, from a practical point of view, is best done by spectral methods. Given the power spectral density

$$S_{xx}(\omega) = \frac{1}{2\pi}\int_{-\infty}^{\infty} R_{xx}(t)e^{-i\omega t}dt \tag{24}$$

of the excitation process $x_G(t)$, $\langle x_G \rangle = 0$, the spectral density of any output of the vibratory system y(t) is represented by

$$S_{yy}(\omega) = H(i\omega)H(-i\omega)S_{xx}(\omega) \quad , \tag{25}$$

when the deterministic frequency response of y, $H(i\omega)$, is known. Hence,

$$\sigma_{yy}^2 = \int_{-\infty}^{\infty} S_{yy}(\omega)\, d\omega \quad , \quad \sigma_{\dot y\dot y}^2 = \int_{-\infty}^{\infty} \omega^2 S_{yy}(\omega)\, d\omega \quad \text{etc.} \tag{26}$$

For narrow band processes y(t) the probability density for crossing a barrier λ is

$$\theta(t,\lambda) = n_\lambda e^{-n_\lambda t} \quad , \tag{27}$$

where, see e.g.[10] ,

$$n_\lambda^+ = \frac{\sigma_{\dot y\dot y}}{2\pi\sigma_{yy}} e^{-\lambda^2/2\sigma_{yy}^2} \quad . \tag{28}$$

INSTATIONARY RESPONSE.

Representing the earthquake excitation by an instationary random process

$$x_G(t) = \Psi(t)X_G(t) \tag{29}$$

where $X_G(t)$ is stationary with $\langle X_G \rangle = 0$ and spectral density

$$S_{XX}(\omega) = \frac{1}{2\pi}\int_{-\infty}^{\infty} R_{XX}(t)\, e^{-i\omega t}dt \quad , \tag{30}$$

and $\Psi(t)$ is a "simple" deterministic shape function, the technique of instationary spectra (using evolutive power spectral density) proves to be economic, cf. [11]. Also the output y(t) is instationary with time dependent power spectral density

$$S_{yy}(t,\omega) = G(t,\omega)\bar{G}(t,\omega)S_{XX}(\omega) \quad , \tag{31}$$

where the deterministic function

$$G(t,\omega) = \int_0^t \Psi(t-\tau)h(\tau)e^{-i\omega\tau}d\tau \quad (32)$$

is determined via the impulse response

$$h(\tau) = \frac{1}{2\pi}\int_{-\infty}^{\infty} H(i\nu)e^{i\nu\tau}d\nu \quad . \quad (33)$$

Interchanging integration we find

$$G(t,\omega) = \frac{1}{2\pi}\int_{-\infty}^{\infty} H(i\nu)F(t,\omega,\nu)\, d\nu \quad , \quad (34)$$

where

$$F(t,\omega,\nu) = \int_0^t e^{i(\nu-\omega)\tau}\Psi(t-\tau)\, d\tau \quad , \quad (35)$$

in many cases of simple shape functions $\Psi(t)$, is analytically integrabel. E.g. for a simple switch at t=0, $\Psi(t)=H(t)$, we find

$$F(t,\omega,\nu) = \frac{i}{\omega-\nu}\,[e^{-i(\omega-\nu)t}-1] \quad . \quad (36)$$

Hence, $G(t,\omega)$ is determined by a single integration of the weighted frequency response of $y(t)$.

Further, all moments $s_n(t)$ are given by

$$s_n(t) = \int_{-\infty}^{\infty} \omega^n S_{yy}(t,\omega)\, d\omega \quad . \quad (37)$$

Vanmarcke [12] gave the first-passage-probability of a barrier λ by

$$V(t,\lambda) = 1 - e^{-t_o\alpha(t)} \quad , \quad (38)$$

where

$$t_o = t \exp\left[-2 \left(\frac{s_o(t)}{s_o(t/2)} - 1\right)\right] \quad ,$$

$$\alpha(t) = \frac{\Omega}{\pi}\{1 - \exp[-\delta\sqrt{\pi\lambda^2/2s_o(t)}\;]\}/\exp[\lambda^2/2s_o(t)] - 1 \quad ,$$

$$\Omega(t) = \sqrt{s_2/s_o} \quad , \quad \delta(t) = \sqrt{1-s_1^2/s_os_2} \quad .$$

Parkus [13] applied those formulas the first time to the instationary excess sloshing $(z_S-H)\,]_{r=R,\varphi=0}$ with barrier λ=L-H. Assuming $\sigma^2_{\ddot{X}_G} = (cg)^2$ and $\Psi(t)=H(t)$ the probability of first passage within 20 seconds for R/H=1 and R=20m is graphically presented in Fig. 4.

REFERENCES

[1] D.Fischer, Dynamic fluid effects in liquid-filled flexible cylindrical tanks. Earthquake Engng. & Struct.Dynamics 7, 587-601 (1979).

[2] N.N.Moiseyev and V.V.Rumyantsev, Dynamic stability of bodies containing fluid. Springer New York 1968 . Cf. in Advances in Applied Mechanics 8 (1964).

[3] H.N.Abramson, The dynamic behaviour of liquids in moving containers. NASA-Rep SP-106, 1966 .

[4] H.Parkus, Modes and frequencies of vibrating liquid-filled cylindrical tanks. Int.J.Engng.Sci.20(2), 319-326 (1982).

[5] T.Nakayama and K.Washizu, Nonlinear analysis of liquid motion in a container subjected to forced pitching oscillations. Int.J.Num.Meth.Engng 15, 1207-1220(1980).

[6] F.Höllinger, Earthquake excited vibrations of elastic dam-fluid systems (in German), Dissertation, Techn.Univ.Vienna, Austria, 1982 .

[7] M.Abramowitz and Irene A.Stegun (eds.), Handbook of Mathematical Functions. Dover, New York 1965 .

[8] F.E.Richart, Jr., et al., Vibrations of Soils and Foundations. Civil Engng. and Engng.Mechanics Series (W.J.Hall & N.M.Newmark,eds.), Prentice Hall, Englewood Cliffs 1970 .

[9] N.M.Newmark and E.Rosenblueth, Fundamentals of Earthquake Engineering. Prentice Hall, Englewood Cliffs 1971 .

[10] Y.K.Lin, Probabilistic Theory of Structural Dynamics. McGraw-Hill, New York 1967 .

[11] M.B.Priestley, Power spectral analysis of non-stationary random processes. J.Sound & Vibrations, 6(1), 86-97 (1967).

[12] E.H.Vanmarcke, Structural response to earthquakes. In: Seismic Risk and Engng Decisions, C.Lomnitz and E.Rosenblueth(eds.). Amsterdam 1976

[13] H.Parkus, Excess-sloshing probability for a liquid-filled container under earthquake excitation, (in German) Ingenieur-Archiv 49, 179-185 (1980).

[14] M.A.Haroun and G.W.Housner, Dynamic interaction of liquid storage tanks and foundation soil. Lecture Notes.

These Proceedings contain all the papers presented at this international Symposium read by top ranking researchers in this field. The lectures dealt with three main topics:

Reliability of mechanical systems (Results on reliability in connection with random, especially earthquake and wind excitations, first-passage failure of randomly excited structures, reliability prediction applications to the construction of structures in mechanical and civil engineering);

Identification of random systems (Parameter estimation, structural identification, identifications to the dynamics of machines, vehicles and structures);

Random vibrations (Forced, parametrically excited and self-excited random vibrations with various nonlinear influences, stochastic stability, narrow-band and wide-band excitations, applications of Markov's theory, comparison of different analytical and simulation methods, applications to mechanical and civil engineering)

The lectures covered theoretical as well as applied problems of practical interest.

MIX
Papier | Fördert gute Waldnutzung
FSC® C083411

Zeitfracht Medien GmbH
Ferdinand-Jühlke-Straße 7
99095 Erfurt, Deutschland
produktsicherheit@kolibri360.de